Progress in Water
Footprint Assessment

Progress in Water Footprint Assessment

Special Issue Editors

Arjen Y. Hoekstra
Ashok K. Chapagain
Pieter R. Van Oel

MDPI • Basel • Beijing • Wuhan • Barcelona • Belgrade

MDPI

Special Issue Editors
Arjen Y. Hoekstra
University of Twente
The Netherlands

Ashok K. Chapagain
University of the Free State
South Africa

Pieter R. Van Oel
Wageningen University & Research
The Netherlands

Editorial Office
MDPI
St. Alban-Anlage 66
4052 Basel, Switzerland

This is a reprint of articles from the Special Issue published online in the open access journal *Water* (ISSN 2073-4441) from 2018 to 2019 (available at: https://www.mdpi.com/si/water/progress_water_footprint_assessment)

For citation purposes, cite each article independently as indicated on the article page online and as indicated below:

LastName, A.A.; LastName, B.B.; LastName, C.C. Article Title. *Journal Name* **Year**, *Article Number*, Page Range.

ISBN 978-3-03921-038-1 (Pbk)
ISBN 978-3-03921-039-8 (PDF)

Cover image courtesy of Arjen Y. Hoekstra.

Contents

About the Special Issue Editors

Arjen Y. Hoekstra is Professor in Water Management at the University of Twente in the Netherlands, and co-founder of the Water Footprint Network. As creator of the water footprint concept, he introduced supply chain thinking in water management and was the first to highlight the global dimension of wise water governance.

Ashok K. Chapagain is Senior Professor at the University of the Free State South Africa. Originally an irrigation engineer in Nepal, Chapagain received his Master and doctoral degrees from the UNESCO-IHE Institute for Water Education, thereafter working for WWF-UK as Senior Water Advisor and Water Footprint Network as Science Director.

Pieter R. Van Oel is Assistant Professor at Wageningen University in the Netherlands. He is an expert in sociohydrology, water resources management, and water footprint assessment, and specializes in spatial and temporal water scarcity patterns.

·**water**

MDPI

Editorial

Progress in Water Footprint Assessment: Towards Collective Action in Water Governance

Arjen Y. Hoekstra [1,2,*], **Ashok K. Chapagain** [3] **and Pieter R. van Oel** [4]

1. Twente Water Centre, University of Twente, 7500AE Enschede, The Netherlands
2. Institute of Water Policy, Lee Kuan Yew School of Public Policy, National University of Singapore, Singapore 259770, Singapore
3. Agricultural Economics, University of the Free State, Bloemfontein 9301, South Africa; chapagainak@ufs.ac.za
4. Water Resources Management Group, Wageningen University, P.O. Box 47, 6700AA Wageningen, The Netherlands; pieter.vanoel@wur.nl
* Correspondence: a.y.hoekstra@utwente.nl

Received: 9 April 2019; Accepted: 21 May 2019; Published: 23 May 2019

Abstract: We introduce ten studies in the field of water footprint assessment (WFA) that are representative of the type of papers currently being published in this broad interdisciplinary field. WFA is the study of freshwater use, scarcity, and pollution in relation to consumption, production, and trade patterns. The reliable availability of sufficient and clean water is critical in sustaining the supply of food, energy, and various manufactured goods. Collective and coordinated action at different levels and along all stages of commodity supply chains is necessary to bring about more sustainable, efficient, and equitable water use. In order to position the papers of this volume, we introduce a spectrum for collective action that can give insight in the various ways different actors can contribute to the reduction of the water footprint of human activities. The papers cover different niches in this large spectrum, focusing on different scales of governance and different stages in the supply chain of products. As for future research, we conclude that more research is needed on how actions at different spatial levels and how the different players along supply chains can create the best synergies to make the water footprint of our production and consumption patterns more sustainable.

Keywords: water footprint assessment; multi-level governance; value chain; consumption; international trade; river basin management; sustainability; water accounting; water productivity; water footprint benchmarks

---

1. Introduction

We present here the fifth special collection of papers in the field of water footprint assessment (WFA). The first collection was a special issue published in *Water* over the years 2010–2011 [1]. A second volume followed in the journal *Water Resources and Industry* in 2013 [2], a third volume in *Sustainability* in 2015 [3], and a fourth volume in *Water* in the years 2016–2017 [4]. Each of the volumes contains a snapshot of what was being researched in the field at the time of publication. This is also true for the current collection of papers. The red line over the years is the interdisciplinarity of the studies and diversity of the subjects researched. The progress lies in the gradual shift in focus from accounting to what we can learn from the accounts for better water governance at different levels and better supply chain management, as illustrated by this latest collection of papers. Water footprint assessment (WFA) is the study of freshwater use, scarcity, and pollution in relation to consumption, production, and trade [5]. By nature, the field is integrative, bringing together different disciplines and perspectives, for instance, natural sciences, policy studies, and geographical and supply-chain perspectives. It links water issues to food, energy, and climate and addresses issues of sustainability, efficiency, and equitability of resource use. All these themes come back in the various papers in the current volume. What makes this new field of research so exciting is that it opens up ways to analyze linkages between previously disconnected fields of study and that it offers a much broader perspective on how we can approach the solution of the water scarcity and pollution problems that people are facing in so many places today, in either direct or indirect ways.

Historically, interventions in response to water shortages have mostly aimed at increasing either water supply or water-use efficiency, interpreting efficiency narrowly as the ratio of output to input [6]. Unfortunately, the scope for finding sustainable, equitable, and resilient solutions through these types of interventions is limited. Moreover, because of the complexities and feedbacks in human-environmental interactions, it is less than straightforward to understand the redistributive effects of building reservoirs [7] and promoting micro-irrigation technologies [8]. Water demand is projected to grow because of continued population and economic growth while water availability in critical periods is expected to decrease in many places because of climate change [9] so that the need to act and mitigate water scarcity only becomes more pressing. Apart from actors in water resource management (e.g., irrigation boards, water boards, river basin committees, water ministries) and agricultural water management (e.g., farmers, farmer associations, agricultural ministries), there are multiple others that have an effect on the way we mobilize the world's water resources for producing the goods and services we wish to consume. Patterns of water use, scarcity, and pollution are intricately related to the way we have organized our economies. As a result, wise water governance inevitably means that we have to look beyond managing water resources use itself. We need to consider also indirect drivers of water problems, like incentives to produce water-intensive products in water-scarce regions for export, governmental subsidy programs to shift from fossil fuels to biofuels, and lack of mechanisms to reduce wastage of food along all stages of the supply chain. Next, we need to look at ways in which actors outside the water field can contribute to the indirect solution of the water problems. Figure 1 shows the spectrum of collective action that we need to consider to understand how we can effectively reduce water footprints of human activities to sustainable levels, by interventions through different actors along supply chains and at different scale levels. We can distinguish different types of interventions:

(1) Interventions at different scale levels: from the field or production-line level to the farm or factory level, the river basin level, the country level, and the international level.
(2) Interventions in different stages of the supply chain: from production, trade, processing, international markets and auctions to distribution, sale and household management;

Furthermore, we can distinguish between different types of actors:

(1) Actors at different governance levels: from the individual water user (e.g., farmer or factory manager) to irrigation and water boards, governmental policy makers, and international agreements;
(2) Actors in the different stages of the supply chain: from stockholders, investors, producers, processors, and traders to retailer and consumers.

The collection of papers that are presented here offers ten studies that form a reflection of the type of papers currently being published in the field of WFA. They illustrate the range of spatial intervention levels, each with different players, and show the relevance of considering different supply-chain stages, each of which can be identified with different actors again. In the following two sections, the papers will be positioned in the spectrum for collective action introduced here. Each paper falls in a niche in this large spectrum. In Section 2, we present seven papers at different levels of governance; in Section 3, we present three papers that address supply chain management. In Section 4, we conclude by reflecting on major challenges in future research.

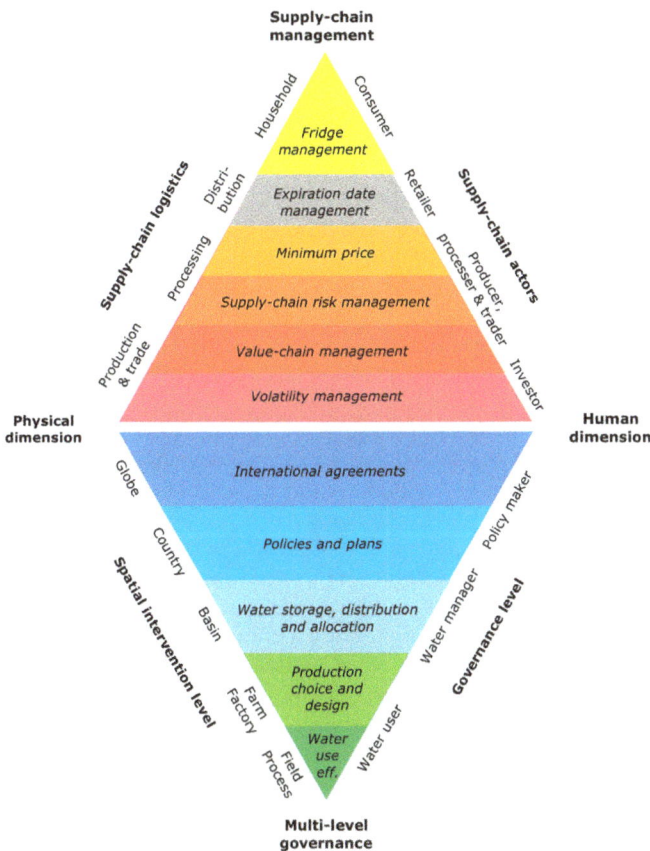

Figure 1. The spectrum of collective action to reduce water footprints of human activities to sustainable levels, along supply chains, and at different scale levels.

2. Water Footprint Reduction through Multi-Level Governance

The collection of papers in this volume is illustrative of the fact that water footprint assessments are carried out at different levels. One paper focusses at the farm level with an outlook to the catchment level, considering the water footprint of silk production, which relies on the cultivation of mulberry shrubs that provide leaves to feed the silkworms [10]. Another paper at the farm level analyzes the water footprint of palm oil [11]. The next paper focusses on water footprint assessment at the urban level [12]. Three papers take a river basin perspective on sustainable water use; one of them considers a basin in China [13], another one a basin in Brazil [14], while the third basin-level paper is more theoretical in nature [15]. The last paper presented in this section is a water footprint application at the country level, with an international perspective by including the domestic water implications of international trade [16]. In this bundle of articles, we have no paper that focusses on the global level, but there are plenty of examples outside this volume (e.g., [17,18]).

In a farm-focused study, Hogeboom and Hoekstra [10] estimate water and land footprints and economic productivity as factors in local crop choice, for a case in Malawi where farmers consider to shift from traditional rainfed crops to irrigated sericulture (silk production). For farmers, it is interesting to look at how they can best use the water and land resources that they have access to. The authors explore how information on water and land footprints, and on economic water and land productivity can inform micro-level decision making of crop choice, in the macro-level context of

3

sustainable resource use at the catchment level. For a proposed sericulture project in Malawi, they calculate water and land footprints and economic water and land productivities of silk production. They compare the growing of mulberry trees to current crops and address the implications of water consumption at the catchment scale. The study finds that farmers may prefer irrigated mulberry cultivation for silk production over currently grown rain-fed staple crops because its economic water and land productivity is higher than that for the crops currently cultivated. However, because the water footprint will be higher, sericulture will increase the pressure on local water resources. The authors point out that optimizing water and land use at the farm level may result in total water and land footprints at the catchment level that are in conflict with sustainable resource use. In the case studied, however, water consumption in the catchment does not exceed the maximum sustainable footprint (i.e., water consumption remains below the amount that can sustainably be consumed without affecting the environmental flow requirements in the catchment), so that sericulture seems a viable alternative crop for farmers, as long as the production remains small-scale.

In a second farm-based study, Safitri et al. [11] analyze the variation of water use by oil palm plants for different crop ages and different soil types, in a village in Central Kalimantan, Indonesia. They conclude that water use depends on crop age. As root density increases with crop age, root water uptake also increases. Furthermore, they find that the water footprint of oil palm fresh fruits for spodosol soils is considerably smaller than for inceptisol and ultisol soils. The results of this study suggest that there are relevant differences in water footprints, both in space and time, which could be relevant for farmers and other actors aiming at more sustainable agricultural water management.

In their urban-focused paper, Fialkiewicz et al. [12] propose a simplified direct water footprint model to support urban water management. The paper explores how WFA can help to formulate strategies for urban water management with case studies from three different cities in the European Union, namely Wroclaw (Poland), Innsbruck (Austria), and Vicenza (Italy). The study focuses on the internal availability and use of water resources in the urban area to support decision-making in managing local resources. It excludes water used in urban agriculture. It simplifies the accounting phase by schematizing the urban area into different zones based on land cover, including impermeable surfaces, permeable surfaces, and water areas. Green, blue, and grey water footprints are estimated following Hoekstra et al. [19]. The authors compare the results with a more detailed modular approach. The study finds that the simplified water footprint accounting results are within ±3% to 28% of more detailed studies. It is shown how the results obtained for the three cities could be the base for drawing up urban water management plans or strategies. The study is a good example of how even without a complex and detailed data-rich assessment, a simplified water footprint assessment can also help decision makers to take effective measures in local water management.

In a basin-focused study, Han et al. [13] assess the green, blue, and grey water footprint of wheat and maize production, in the Haihe River Basin in Northern China. They analyze the temporal trends and spatial variations of the water footprint during the period 1956–2015. It is shown that the water footprint per unit of wheat and maize have decreased, but that increased production has led to a growing pressure on the environment, mainly due to increasing loads of nitrogen into the environment. Given that the grey water footprint is larger than the blue water footprint in recent decades it seems evident that reducing fertilizer use and increasing fertilizer use efficiency could substantially contribute to the reduction of the overall water footprint in the basin. The authors show that high spatial and temporal detail can be helpful to inform basin-level water management decisions aiming to (prioritize locations for taking measures that can) increase water use efficiency and reduce water pollution.

Lathuillière et al. [14] take a basin focus as well and evaluate water use for agricultural intensification in Xingu Basin of Mato Grosso (XBMT), which is part of the Amazon basin in Brazil, using the water footprint sustainability assessment method. They analyze the sustainability of water use based on the resultant green and blue water scarcity in the years 2000 and 2014, and under deforestation and climate change scenarios for 2030 and 2050. The study finds that although the blue water footprint in the basin is currently within the limit of what is still sustainable, under the future expansion of irrigation and cattle confinement, blue water scarcity will move from low to moderate, making the production system vulnerable in dry years. Both options for future production changes (either the expansion of agricultural land use or the intensification of productions) have consequences for future water availability, e.g., continued reduction in natural vegetation cover, which is accompanied by reduced water vapor supply to the atmosphere affecting terrestrial ecosystems that rely on precipitation for ecosystem functioning, while dry season water consumed in intensified livestock and irrigation systems will impact aquatic ecosystems downstream. This study provides an important case for estimating blue and green water scarcities in the context of land use change, climate change, and agricultural production scenarios applied for a river basin in Brazil.

Ruddell [15] proposes the use of a theoretical model with a threshold that determines the point beyond which the blue water footprint in a river basin starts to have adverse environmental impacts. The impacts grow exponentially with increasing blue water footprint beyond the threshold. He introduces a volumetric threshold-based water footprint (TWF), defined as the part of the blue water footprint in a basin that exceeds the adverse-impact threshold. The part of the blue water footprint below the threshold is called "free footprint" because it is not associated with adverse environmental impacts. The TWF indicator is compared with the volumetric blue water footprint (BWF) and the water-scarcity weighted BWF indicator that has been used in the life-cycle assessment community. The paper is in line with earlier publications that propose to set a cap on the blue water footprint in a river basin as a policy instrument to prevent adverse environmental impacts from water consumption [20]. An important question remains on how to define the threshold or cap practically.

Karandish and Hoekstra [16] take a national perspective and explore how national food and water security policy can be informed through water footprint assessment, for the case of Iran. They argue that Iran's focus on food self-sufficiency has resulted in investments directed towards increasing water supplies to farmers while neglecting the role of consumption and trade. The authors quantify the green and blue water footprint of crop production in the country, per province, for 26 crops over the period 1980–2010, as well as the water footprint related to crop consumption per province. Furthermore, they quantify provincial virtual water imports and exports in relation to international and inter-provincial crop trade and subsequently estimate the water saving per province associated with this trade. They find that, over the period considered, the water footprint per unit of crop increased rather than decreased for many crops in various regions, with the blue share in the total water footprint increasing nearly everywhere (because of increased irrigation). Combined with the increased production, this led to an increase in the total water footprint of national crop production with a factor 2.2. By 2010, about a quarter of the total water consumption in the semi-arid parts of Iran served the production of crops for export to other regions within the country (mainly cereals) or abroad (mainly fruits and nuts). The authors argue that Iran's food and water policy could be enriched by reducing the water footprints of crop production to certain benchmark levels per crop and climatic region and aligning cropping patterns to spatial differences in water availability and productivities, and by paying due attention to the increasing food consumption per capita in Iran.

3. Water Footprint Reduction through Supply-Chain Management

Three papers in this volume approach the water footprint from a supply-chain rather than from a geographic perspective. One paper considers water consumption along the supply chain of wheat bread [21], while another paper considers the water footprint of food waste along the chain [22]. A third paper studies the relation between the demand for biofuels and green water resources use and scarcity [23]. To start with, Mohlotsane et al. [21] quantify and assess green, blue, and grey water footprints along the wheat-bread value chain in South Africa. Water footprints are analyzed in the context of economic water productivity. The authors find that the wheat-bread value chain is becoming more blue-water intensive, which is a critical factor in water resources management as South Africa is experiencing higher frequency and degrees of blue water scarcity recently. As the local context defines the sustainability of water footprint in the value chain, the study is a fine example highlighting the need for catchment- or region-specific water footprint benchmarks.

Roux et al. [22] assess the blue water footprint of vegetable crop wastage along the supply chain in a case for South Africa. They focus on food waste in the value chain rather than the food finally consumed at the end of the value chain. This study aimed to quantify indirect blue water losses through the wastage of vegetable crops produced in a major production region on the Steenkoppies Aquifer located west of Tarlton in Gauteng, South Africa. The total water withdrawal from the aquifer is 25 million m^3 per year while the sustainability threshold of this aquifer is only 17 million m^3 per year. In this context, the estimated blue water footprint of 4 million m^3 per year resulting from the wastage of carrots, cabbage, beetroot, broccoli, and lettuce, is significant in managing the water scarcity in the region. The paper concludes that reducing such wastage not only contributes to the sustainability of water use in the region but also reduces other negative environmental impacts resulting from the use of fertilizers, pesticides, and energy.

Xu and Wu [23] make a first estimation of county-based green water availability and its implications for agriculture and bioenergy production in the United States. Water resources assessments still often focus on blue water consumption versus blue water availability, while comparing green water consumption versus green water availability is as relevant [24]. Xu and Wu [23] define a green water availability index as the fraction of green water resources that, after the green water demand of specified sectors (e.g., agriculture) has been met, is available to the remaining green water users (e.g., timber, pasture, ecosystem services). In the paper, they quantify, for each county in the US, the fraction of green water resources needed if the water demands of three major crops (corn, soybeans, and wheat) in the county are met by green water, and the fraction of green water resources in the county that is available to remaining green water users (other crops, grassland, forest, and ecosystem services). They also estimate how much green water resources are available for non-bioenergy purposes after fulfillment of the crop water demand of all corn and soybeans grown specifically for biofuel feedstock.

4. Conclusions

Water footprint studies are available at all levels, as indicated in Figure 1, from studies that consider how to reduce the water footprint in a crop field or industrial production unit to global-level studies. The current volume presents several illustrative examples at different levels, notably the farm, urban, basin, and country level. Still, a weak point in water footprint literature is how these different levels are connected. One relevant question is how local-scale actions (e.g., crop choice or water use efficiency increase) together can contribute to the solution of problems at higher levels (e.g., water scarcity at basin level or international burden shifting through trade) and how global actions (e.g., international agreement on sustainable trade) can contribute to the solution of local water problems. Another relevant question is how actions at different levels can create synergies to make the water footprint of our production and consumption patterns more sustainable, with due attention to local carrying capacities.

The water footprint concept is integrative by nature by its applicability at different levels (local to global) and along supply chains (from investment and production to processing, sales, and consumption). The full potential of water footprint analysis along supply chains needs to be realized still though. As the papers in the current collection show, methods to localize and quantify water footprints in the various steps of the supply chain of a product are improving; for many products, particularly agricultural products, we understand how water footprints vary with climate, soil and production practices, and how water footprints can be reduced by changing policies and applying better technologies and practices. Obtaining good local data, however, remains difficult. Perhaps even more challenging is to improve our understanding of the interactions between actors along the supply chain, for instance: how can consumers motivate retailers and producers to reduce the water footprint of products in the hotspots along the product supply chain; how can companies influence suppliers through sustainable procurement strategies that include water criteria; how will water pricing affect final commodity prices and thus potentially influence consumer decisions; and how could the inclusion of water criteria in environmental product labels bring about sustainable water use along the supply chain.

Despite the considerable progress in the water footprint assessment research field, the research field is still largely focusing on challenges in modeling and quantification, with major challenges remaining in the translation of data and insights to coherent mechanisms of governance and ways of intervention, along supply chains and at different levels.

Author Contributions: The authors contributed equally to the writing of this editorial.

Conflicts of Interest: The authors declare no conflict of interest.

References

1. Feng, K.; Hubacek, K.; Minx, J.; Siu, Y.L.; Chapagain, A.; Yu, Y.; Guan, D.; Barrett, J. Spatially explicit analysis of water footprints in the UK. *Water* **2011**, *3*, 47–63. [CrossRef]
2. Zhang, G.P.; Hoekstra, A.Y.; Mathews, R.E. Water Footprint Assessment (WFA) for better water governance and sustainable development. *Water Resour. Ind.* **2013**, *1–2*, 1–6. [CrossRef]
3. Hoekstra, A.Y.; Chapagain, A.K.; Zhang, G.P. Water footprints and sustainable water allocation. *Sustainability* **2016**, *8*, 20. [CrossRef]
4. Hoekstra, A.Y.; Chapagain, A.K.; Van Oel, P.R. Advancing water footprint assessment research: Challenges in monitoring progress towards Sustainable Development Goal 6. *Water* **2017**, *9*, 438. [CrossRef]
5. Hoekstra, A.Y. Water footprint assessment: Evolvement of a new research field. *Water Resour. Manag.* **2017**, *31*, 3061–3081. [CrossRef]
6. Savenije, H.H.G.; Hoekstra, A.Y.; Van der Zaag, P. Evolving water science in the Anthropocene. *Hydrol. Earth Sys. Sci.* **2014**, *18*, 319–332. [CrossRef]
7. Di Baldassarre, G.; Wanders, N.; AghaKouchak, A.; Kuil, L.; Rangecroft, S.; Veldkamp, T.I.E.; Garcia, M.; van Oel, P.R.; Breinl, K.; Van Loon, A.F. Water shortages worsened by reservoir effects. *Nat. Sustain.* **2018**, *1*, 617–622. [CrossRef]
8. Grafton, R.Q.; Williams, J.; Perry, C.J.; Molle, F.; Ringler, C.; Steduto, P.; Udall, B.; Wheeler, S.A.; Wang, Y.; Garrick, D.; et al. The paradox of irrigation efficiency. *Science* **2018**, *361*, 748. [CrossRef]
9. Distefano, T.; Kelly, S. Are we in deep water? Water scarcity and its limits to economic growth. *Ecol. Econ.* **2017**, *142*, 130–147. [CrossRef]
10. Hogeboom, R.J.; Hoekstra, A.Y. Water and land footprints and economic productivity as factors in local crop choice: The case of silk in Malawi. *Water* **2017**, *9*, 802. [CrossRef]
11. Safitri, L.; Hermantoro, H.; Purboseno, S.; Kautsar, V.; Saptomo, S.K.; Kurniawan, A. Water footprint and crop water usage of oil palm (Eleasis guineensis) in Central Kalimantan: Environmental sustainability indicators for different crop age and soil conditions. *Water* **2019**, *11*, 35. [CrossRef]
12. Fialkiewicz, W.; Burszta-Adamiak, E.; Kolonko-Wiercik, A.; Manzardo, A.; Loss, A.; Mikovits, C.; Scipioni, A. Simplified direct water footprint model to support urban water management. *Water* **2018**, *10*, 630. [CrossRef]
13. Han, Y.; Jia, D.; Zhuo, L.; Sauvage, S.; Sánchez-Pérez, J.-M.; Huang, H.; Wang, C. Assessing the water footprint of wheat and maize in Haihe River Basin, Northern China (1956–2015). *Water* **2018**, *10*, 867. [CrossRef]
14. Lathuillière, M.J.; Coe, M.T.; Castanho, A.; Graesser, J.; Johnson, M.S. Evaluating water use for agricultural intensification in Southern Amazonia using the Water Footprint Sustainability Assessment. *Water* **2018**, *10*, 349. [CrossRef]
15. Ruddell, B.L. Threshold based footprints (for water). *Water* **2018**, *10*, 1029. [CrossRef]
16. Karandish, F.; Hoekstra, A.Y. Informing national food and water security policy through water footprint assessment: The case of Iran. *Water* **2017**, *9*, 831. [CrossRef]
17. Wang, R.; Zimmerman, J. Hybrid analysis of blue water consumption and water scarcity implications at the global, national, and basin levels in an increasingly globalized world. *Environ. Sci. Technol.* **2016**, *50*, 5143–5153. [CrossRef]
18. Yang, C.; Cui, X. Global changes and drivers of the water footprint of food consumption: A historical analysis. *Water* **2014**, *6*, 1435–1452. [CrossRef]

19. Hoekstra, A.Y.; Chapagain, A.K.; Aldaya, M.M.; Mekonnen, M.M. *The Water Footprint Assessment Manual: Setting the Global Standard*; Earthscan: London, UK, 2011.
20. Hoekstra, A.Y. Sustainable, efficient and equitable water use: The three pillars under wise freshwater allocation. *WIREs Water* **2014**, *1*, 31–40. [CrossRef]
21. Mohlotsane, P.M.; Owusu-Sekyere, E.; Jordaan, H.; Barnard, J.H.; Van Rensburg, L.D. Water footprint accounting along the wheat-bread value chain: Implications for sustainable and productive water use benchmarks. *Water* **2018**, *10*, 1167. [CrossRef]
22. Roux, B.L.; Van der Laan, M.; Vahrmeijer, T.; Annandale, J.G.; Bristow, K.L. Water footprints of vegetable crop wastage along the supply chain in Gauteng, South Africa. *Water* **2018**, *10*, 539. [CrossRef]
23. Xu, H.; Wu, M. A first estimation of county-based green water availability and its implications for agriculture and bioenergy production in the United States. *Water* **2018**, *10*, 148. [CrossRef]
24. Schyns, J.F.; Hoekstra, A.Y.; Booij, M.J.; Hogeboom, H.J.; Mekonnen, M.M. Limits to the world's green water resources for food, feed, fibre, timber and bio-energy. *Proc. Natl. Acad. Sci. USA* **2019**, *116*, 4893–4898. [CrossRef]

water

MDPI

Article

Water and Land Footprints and Economic Productivity as Factors in Local Crop Choice: The Case of Silk in Malawi

Rick J. Hogeboom [1,*] and Arjen Y. Hoekstra [1,2]

[1]   Twente Water Centre, University of Twente, P.O. Box 217, 7500 AE Enschede, The Netherlands;
      a.y.hoekstra@utwente.nl
[2]   Institute of Water Policy, Lee Kuan Yew School of Public Policy, National University of Singapore,
      Singapore 259770, Singapore
*   Correspondence: h.j.hogeboom@utwente.nl; Tel.: +31-053-489-3911

Received: 31 August 2017; Accepted: 10 October 2017; Published: 18 October 2017

Abstract: In deciding what crops to grow, farmers will look at, among other things, the economically most productive use of the water and land resources that they have access to. However, optimizing water and land use at the farm level may result in total water and land footprints at the catchment level that are in conflict with sustainable resource use. This study explores how data on water and land footprints, and on economic water and land productivity can inform micro-level decision making of crop choice, in the macro-level context of sustainable resource use. For a proposed sericulture project in Malawi, we calculated water and land footprints of silk along its production chain, and economic water and land productivities. We compared these to current cropping practices, and addressed the implications of water consumption at the catchment scale. We found that farmers may prefer irrigated silk production over currently grown rain-fed staple crops, because its economic water and land productivity is higher than that for currently grown crops. However, because the water footprint of irrigated silk is higher, sericulture will increase the pressure on local water resources. Since water consumption in the catchment generally does not exceed the maximum sustainable footprint, sericulture is a viable alternative crop for farmers in the case study area, as long as silk production remains small-scale (~3% of the area at most) and does not depress local food markets.

Keywords: water footprint; land footprint; economic water productivity; economic land productivity; crop choice; CSR; sericulture; silk; Malawi

---

1. Introduction

Suppose you are a farmer in Malawi. What crops would you grow, and on what factors would you base that decision? You would probably consider the availability, quality and cost of seeds, labour, land, water, fertilizers and technology, the access to markets, available capital to invest, insurance, and what alternative options you have to feed your family if crops fail. Now, you are aware that pressures on water and land resources are increasing—due to climate change, growing populations and more demanding lifestyles—and you want to find out how your operations affect overall questions of sustainability, efficient resource use, and equity. How can you make sure you maximize your farming operations' profitability, while at the same time minimizing harmful impacts on both others in your area and on the next generation? After all, they will also need the natural resources to support their livelihoods.

This stream-of-thought sketches the tension between micro-level decision making in agriculture and its macro-level effects. Much research has been done to identify factors that influence local crop choice [1–7]. In the current study, we focus on water and land availability and consider indicators such as water and land footprints and economic water and land productivity [8–11]. Water footprints (WF) and land footprints (LF) of crop production represent the volume of water (m^3) and area of land (m^2) that are appropriated to produce a crop (kg) [12]. Footprints inform the farmer how much water and land the intended crop requires in absolute terms, or, if compared to a benchmark footprint for that crop, in relative terms [13,14]. Economic water productivity (EWP, in € m^{-3}) and economic land productivity (ELP, in € m^{-2}) address economic considerations, by showing how much money each cubic meter of water or square meter of land generates.

Whereas micro-level questions focus on efficiency and productivity, macro-level questions are concerned with the sustainability and equity of resource use at the higher system level, such as the catchment, biome or even global level [15]. Total footprints at the system level result from the pressures placed on the system by all individual water and land using activities combined. Studies concerned with macro-level questions typically try to quantify total pressure limits of the system, also termed assimilation capacity, operation space or boundaries [15–17]. Exceeding these lead to undesirable consequences. Defining maximum sustainable footprints is one way to quantify such macro-level limits to resource use [13,18]. If farmers are only guided by micro-level factors—such as local water and land footprints, or economic and land productivities of their intended crops—then maximum sustainable system footprints may eventually be violated at the macro-level. On the other hand, total footprint limits at the system level only become practical if they can be translated to implications at the local level.

The aim of this study is therefore to explore how data on water and land footprints and economic water and land productivity can inform micro-level decision making on crop choice, in the context of macro-level sustainability of resource use, for a case study of proposed silk production in Mzimba District in Malawi. Malawi is economically poor, but relatively rich in arable land and water resources. It has a large untapped potential for irrigation expansion [19]. Nevertheless, agricultural output is low and about a quarter of the population is unable to secure its minimum daily recommended food intake, despite enough food being produced at the national level [20]. The Malawian government therefore wants to diversify the current low-value, staple-crop-only agricultural portfolio, in order to boost overall productivity and possibly increase exports. Introducing sericulture can help achieve the desired diversification, while holding the promise of providing better livelihoods to rural families. Cultivating silk is labour intensive, requires low skill levels, and silk has had and is expected to have a steady global market for years to come [21]. However, sericulture has implications for land and water resource use, both locally for the farmers' operations and for the wider catchment. In this study, we explore the local implications of silk production based on water and land productivity, and we place water footprints in the context of catchment-level water availability. We conclude with a discussion of whether farmers should appropriate local water and land resources to sericulture based on these factors.

2. Method and Data

2.1. The Production Chain of Raw Silk

The production chain of raw silk has several steps, each of which may have a water or land footprint associated with it. The total water or land footprint of raw silk is the sum of the respective footprints in each step [12]. The first step of silk production is the cultivation of mulberry shrubs for their leaves and the rearing of silkworms (*Bombyx mori*). The leaves serve as feed for the silkworms, which are raised on rearing beds in special nurseries. When the worms reach maturity, they form cocoons, which, once pupation is about to complete, are harvested. After each harvest (4–7 per year), the nurseries have to be thoroughly cleaned to prevent the spread of diseases and promote general hygiene before a new batch of worms is reared [2]. The harvested cocoons are stifled to kill the pupae inside without disturbing the structure of the silk shell. This is usually done by means of hot air-conditioning, which is why the process is referred to as drying. After drying, the cocoons are heated in boiling water in order to soften the gummy protein sericin to a point where unravelling of the silk filament is possible. The dry raw silk is then reeled onto bobbins and is ready for further processing, dyeing or direct sale. The processes that require water and land are shown in Figure 1. In the case of water use, we distinguish between the green WF, representing the consumptive use of rainwater, and the blue WF, referring to the consumptive use of surface or groundwater [12].

Cultivating	Cleaning	Drying cocoons	Cooking cocoons	Reeling silk
mulberry shrubs & silk worms				
Blue WF related to irrigation	Blue WF related to water used for cleaning	Blue WF related to electricity	Blue WF related to water used for cooking	Blue WF related to water used for reeling
Green WF related to evapotranspiration of rainwater			Blue WF related to energy to operate the multi-end reeling machine	
LF related to crop land	LF related to area required for nursery and equipment storage			

Figure 1. Water and land footprints along the production chain of raw silk.

2.2. Study Area

The choice for the case study in Malawi is borne out of an intended sericulture project by a non-governmental organisation (NGO) based in The Netherlands. This project is to be implemented around three estates and roughly 200 surrounding smallholder farms in the Mzimba District in the Northern Region of Malawi (Figure 2). The study area is within the Nyika Plateau catchment, with an elevation of about 1200 m above mean sea level and temperatures ranging between 9 °C and 30 °C. With an average annual precipitation of 644 mm and an average annual potential evapotranspiration of 1350 mm, the climate can be classified as subtropical highland variety [22]. The wet season starts in November and ends in April, and the dry season is from May to October. The main soil types are sandy loam and silty clay loam. These climate and soil conditions are favourable for mulberry cultivation [23]. The perennial Runyina River close to the study location is the preferred source of irrigation water.

Smallholder farmers currently grow crops such as tobacco, groundnuts and maize, while the estates mainly grow chillies and paprika. The project intends to replace currently grown crops with mulberry shrubs for silk production on about 20 hectares of the estates, and on half a hectare of each of the smallholder farms.

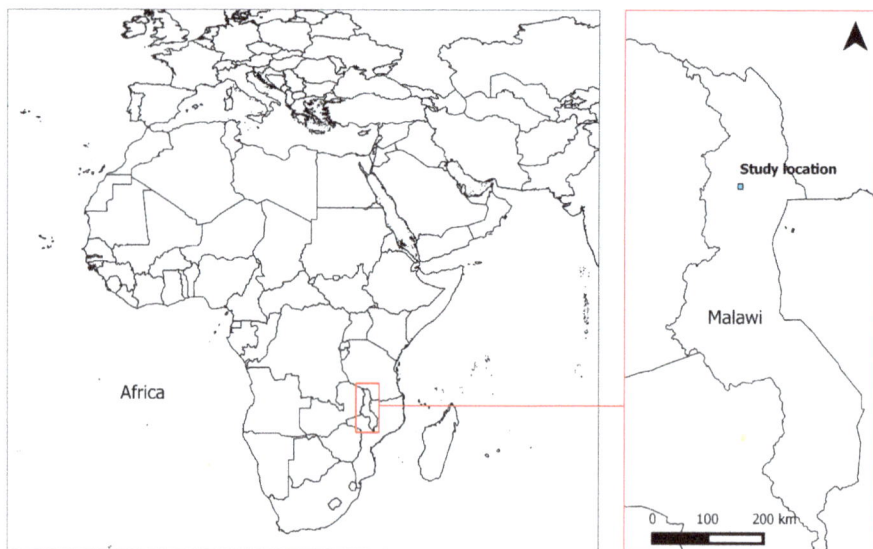

Figure 2. Location of the study area where switching from currently grown crops (maize, chillies, paprika, groundnuts, and tobacco) to sericulture is being considered.

2.3. Calculation of Water and Land Footprints and Economic Productivities

Water and land footprints were assessed along each step of the production chain of raw silk (Figure 1), following the global water footprint standard [12]. To estimate the WF of mulberry cultivation and the currently grown crops (maize, chillies, paprika, groundnut and tobacco), we used the method as in Mekonnen and Hoekstra [24], but replaced the CropWat model with the more advanced AquaCrop model developed by the Food and Agriculture Organisation of the United Nations (FAO) [25]. AquaCrop simulates the daily soil water balance and biomass growth, in order to estimate crop water use and yield. Because mulberry is a perennial crop—and AquaCrop is developed for annuals—we set crop parameters such that AquaCrop mainly simulates canopy development and reflects local (projected) cropping practice. For mulberry shrubs, yield refers to the tonnes of leaves that can be harvested per year per hectare (note: not to the yield in terms of mulberries). For currently grown crops, simulated yields are scaled based on average local yields in the study area (Figure 2). We calculated land footprints (m^2 kg^{-1}) by taking the inverse of the yield, and we distinguished between green and blue WF based on the method described in Chukalla et al. [26]. To account for inter-annual variation in WFs, we simulate crop production for each year in the period 1986–2016. We ignored the blue WF related to energy for pumping water to the fields in case mulberry shrubs are irrigated, because the exact location, setting and types of pumps are not yet decided. We also ignored the grey WF, because of a lack of sensible data and its high dependency on local, actual practices.

We assumed that the leaves represent the full value gained from the mulberry plantation, so no value or WF is attributed to by-products such as berries. Based on estimates from the International Centre of Insect Physiology and Ecology (ICIPE, pers. comm. via email), we assumed that 187.5 kg of fresh mulberry leaves are needed to harvest 9.1 kg of dry cocoons, which after processing yield 1 kg of dry raw silk.

Data on soil properties are taken from De Lannoy et al. [27] and local data. We assumed that soil fertility is good and does not hamper crop production. Crop calendars were taken from Chapagain and Hoekstra [28] and Portmann et al. [29]. Climate data have been taken from global high-resolution datasets by Harris et al. [30] and Dee et al. [31]. These daily fields—evaluated at the location of the estates—have been scaled such that the monthly averages match monthly fields that were observed locally, at the nearby Bolero climate station.

We evaluated five mulberry cultivation scenarios, in which we compare various irrigation strategies and techniques for growing mulberry shrubs (Table 1), to assess the effect of farming practice on WFs and LFs.

Table 1. Different scenarios of cultivating mulberry shrubs evaluated in this study.

Scenario	Irrigation Strategy	Irrigation Technique	Expected Effect
Rain-fed	No irrigation	None	Sensitive to climate variability; a dry year leads to lower leaf yields.
Full-furrow	Full irrigation	Furrow	No water stress; optimum yields. High evaporation because large part of soil is wetted.
Full-drip	Full irrigation	Drip	No water stress; optimum yields. Lower evaporation because small part of soil is wetted.
Deficit-drip	Deficit irrigation	Drip	Some water stress, leading to lower yields. Lower evaporation because small part of soil is wetted. Smaller water footprint per tonne of leaves.
Deficit-drip-organic mulching	Deficit irrigation	Drip	Some water stress, leading to lower yields. Very low evaporation because of protective organic mulching layer covering the soil. Minimum water footprint per tonne of leaves.

The blue WF associated with cleaning, drying, cooking and reeling is highly dependent on local factors and practices. Due to the lack of a credible source, we assumed a water footprint of 100 L per harvest for cleaning the premises and five harvests per year, based on a one-hectare operation and a consumptive fraction of 10%. Generating electricity requires water, which needs to be accounted for [12]. Singh [32] estimates that electricity consumption of cocoon drying is 1.0 kWh per kg cocoons. We assumed a conservative blue WF of the energy mix for Malawi at 400 m^3 TJ^{-1} (or 0.00144 m^3 kWh^{-1}) based on a study by Mekonnen et al. [33]. Kathari et al. [34] report that—using a multi-end reeling machine—cocoon cooking consumes 57 L of water per kg of raw silk and reeling 100 L per kg of raw silk. We adopted these estimates here as well, since a similar centrally operated multi-end reeling machine is anticipated to be used in the Malawi project. This machine—if wood-powered—requires 2.6 kg of wood per kg of cocoon for the cooking and reeling processes [35]. We calculated the WF related to wood using the average (green) WF of wood in Malawi of 74 m^3 per m^3 of wet round-wood (or 137 L kg^{-1} dry firewood) as determined by Schyns et al. [36]. However, solar power is the project's preferred source of energy to power the machine. We therefore estimated the blue WF of cooking with solar energy as well, by converting the caloric value of wood into an equivalent amount of solar energy, and multiplying solar energy demand with the blue WF of solar energy of 150 m^3 TJ^{-1} as estimated by Mekonnen et al. [33]. For the lack of a better estimate, the LF of silk processing (for the rearing facilities and equipment storage) is assumed at 100 m^2 per hectare of mulberry shrubs.

We calculated the economic water productivity (EWP, in € m^{-3}) and economic land productivity (ELP, in € m^{-2}) of silk and of the currently grown crops, by dividing the local market price (€ kg^{-1}) by the WF (m^3 kg^{-1}) or LF (m^2 kg^{-1}), respectively.

Finally, we placed the WF in the context of water availability at the catchment level. Due to the lack of local hydrological assessments for the Nyika Plateau catchment, we took data on local water scarcity levels from the high-resolution global study by Mekonnen and Hoekstra [37] to see if sustainability levels are currently being exceeded. In addition, we drew up a hypothetical case based on local precipitation figures to obtain a rough estimate of water availability levels in the catchment.

3. Results

3.1. The Water and Land Footprint of Silk Production

The total WF and LF of silk production is a summation of all WFs and LFs along the production chain of silk, as shown in Figure 1. We summarized all steps into two major components: (1) the WF and LF of silk related to cultivation of mulberry leaves; and (2) the WF and LF of silk related to the silk processing steps of cleaning, drying, cooking and reeling.

3.1.1. The Water and Land Footprint of Mulberry Cultivation

The WF of rain-fed mulberry leaves is 423 m^3 t^{-1} and the LF 820 m^2 t^{-1}—on average over the period 1986–2016 (Table 2). The WF is 100% green, because only rainwater stored in the soil is consumed. Since there is no irrigation in this scenario to keep plants from suffering water stress, footprints strongly depend on the prevailing weather conditions in a given year. Temporal variability of both water and land footprints is high, as shown by their respective standard deviations of 169 m^3 t^{-1} and 537 m^2 t^{-1}.

If the mulberry fields are irrigated, the LF of leaf production goes down considerably, to 236 m^2 t^{-1} on average, and the total WF shrinks by at least 25%. The WF associated with full irrigation using the furrow technique is 314 m^3 t^{-1}, and becomes smaller with each improvement in irrigation practice. In the best-practice scenario in terms of water consumption per metric ton of leaves—i.e., deficit irrigation using drip systems while applying a layer of organic mulching—the WF is 254 m^3 t^{-1}. Temporal variability of footprints is much lower than under rain-fed conditions, because the shrubs do not suffer water stress as they do under rain-fed conditions. For example, under full drip irrigation, standard deviations are 19 m^3 t^{-1} and 10 m^2 t^{-1} for WF and LF, respectively. However, the WF does have a blue component in these scenarios.

Footprints expressed per tonne of mulberry leaves are converted to footprints per kg of raw silk based on the assumed feed requirement of 187.5 kg of mulberry leaves per kg of final raw silk. Water and land footprints of silk related to mulberry leaf production are listed in Table 3. It shows that rain-fed silk has a green water consumption of 79,300 L kg^{-1} and irrigated silk has a total water consumption between 47,500 and 58,900 L kg^{-1}. Land footprints range from 154 m^2 kg^{-1} under rain-fed condition to 44 or 45 m^2 kg^{-1} under irrigation scenarios.

Table 2. Green and blue water footprint (WF) and average, minimum and maximum total WF and land footprint (LF) of mulberry leaf production per metric ton of leaf for five different scenarios. Average WF and LF are production weighted over the period 1986–2016.

Scenario	$WF_{avggreen}$ (m³ t⁻¹)	$WF_{avgblue}$ (m³ t⁻¹)	$WF_{avgtotal}$ (m³ t⁻¹)	WF_{min} (m³ t⁻¹)	WF_{max} (m³ t⁻¹)	LF_{avg} (m² t⁻¹)	LF_{min} (m² t⁻¹)	LF_{max} (m² t⁻¹)
Rain-fed	423	0	423	340	1336	820	532	3704
Full-furrow	117	197	314	278	356	236	217	254
Full-drip	117	180	297	265	339	236	217	254
Deficit-drip	129	142	271	239	308	243	216	278
Deficit-drip-organic mulching	122	132	254	223	288	242	212	279

Table 3. Green and blue WF and average, minimum and maximum total WF and LF of raw silk related to mulberry leaf production per kg of raw silk for five different scenarios. Average WF and LF are production weighted over the period 1986–2016.

Scenario	$WF_{avggreen}$ (L kg⁻¹)	$WF_{avgblue}$ (L kg⁻¹)	$WF_{avgtotal}$ (L kg⁻¹)	WF_{min} (L kg⁻¹)	WF_{max} (L kg⁻¹)	LF_{avg} (m² kg⁻¹)	LF_{min} (m² kg⁻¹)	LF_{max} (m² kg⁻¹)
Rain-fed	79,300	0	79,300	63,800	250,500	154	100	694
Full-furrow	22,000	37,000	58,900	52,100	66,800	44.2	40.7	47.7
Full-drip	22,000	33,700	55,700	49,600	63,500	44.2	40.7	47.7
Deficit-drip	24,100	26,600	50,800	44,800	57,800	45.6	40.5	52.1
Deficit-drip-organic mulching	22,800	24,800	47,500	41,900	54,000	45.4	39.8	52.2

3.1.2. The Water and Land Footprint of Cleaning, Drying, Cooking and Reeling

Table 4 shows the WF of cleaning, drying, cooking and reeling, which in each process step is fully blue. The reeling process is the major water consuming step, but this is only so if we assume that the multi-end machine runs on solar power. Alternatively, the reeling machines may run on firewood, or small-scale sericulture farmers—who cannot afford a multi-end reeling machine at all—may simply heat water in pots on firewood stoves. The use of firewood profoundly alters the water footprint. While a solar-energy powered silk processing has a total blue WF of 180 L kg^{-1}, using firewood results in a much larger green WF of firewood of over 3200 L kg^{-1}. The choice of energy source to heat water for cooking therefore has a substantial influence on the total WF of the processing of silk.

Table 4. Green, blue and total water footprint (WF) related to cleaning, drying, cooking and reeling per kg of raw silk, assuming water for cooking is heated using solar energy.

Process Step	WF$_{green}$ (L kg^{-1})	WF$_{blue}$ (L kg^{-1})	WF$_{total}$ (L kg^{-1})
Cleaning	0	2	2
Drying electricity	0	13	13
Cooking cocoons	0	57	57
Reeling silk	0	100	100
Multi-end machine energy when solar powered	0	8	8
Alternative: multi-end machine energy when wood powered	*3200*	*0*	*3200*
Total	0	180	180

The land footprint of the rearing facilities and equipment storage was estimated at 100 m^2 per hectare of mulberry plantation.

3.1.3. The Total Water and Land Footprint of Silk Production

The total footprint of raw silk is the sum of the footprint of mulberry leaf production and the footprint of silk processing (Table 5). The total WF of silk decreases with each mulberry cultivation scenario, while the blue portion of 62.8% in the full-furrow irrigation scenario decreases to 52.3% in the best-practice scenario of deficit drip irrigation with organic mulching. For each scenario, a full WF split per colour and stage of the production chain is shown in Figure 3. We find that the largest parts of both the total LF and WF are the result of the mulberry cultivation component. The LF related to processing is around 1% of the total, while the WF related to processing is 0.2–0.4% of the total.

Table 5. Green, blue and total water footprint (WF) and land footprint (LF) of silk under five mulberry cultivation scenarios per kg of raw silk.

Scenario	WF$_{green}$ (L kg^{-1})	WF$_{green}$ (%)	WF$_{blue}$ (L kg^{-1})	WF$_{blue}$ (%)	WF$_{total}$ (L kg^{-1})	LF$_{total}$ (m^2 kg^{-1})
Rain-fed	79,300	99.7	180	0.3	79,500	155
Full-furrow	22,000	37.2	37,200	62.8	59,200	44.7
Full-drip	22,000	39.4	33,900	60.6	55,900	44.7
Deficit-drip	24,100	47.3	26,800	52.7	50,900	46.1
Deficit-drip-organic mulching	22,800	47.7	25,000	52.3	47,800	45.9

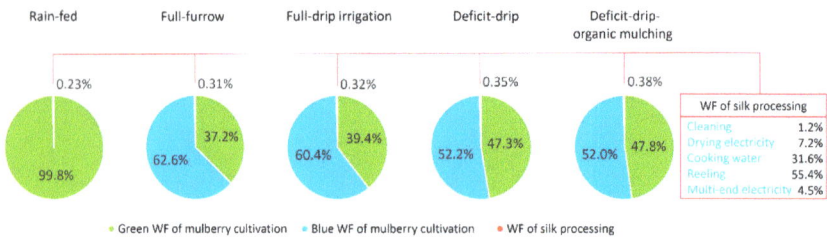

Figure 3. The composition of the water footprint (WF) of raw silk, by colour and by production stage, for five mulberry cultivation scenarios.

3.2. Economic Water and Land Productivity

Producing one kg of silk requires far more water and land than to produce one kg of the crops currently grown by farmers (Table 6). The market price of silk, on the other hand, is much higher than for the other crops. Comparing economic water and land productivities of silk with those of currently grown crops confirms that silk generates more economic value per unit of natural resource used. The average ELP of silk—0.37 € m^{-2} for the rain-fed scenario and 1.24–1.28 € m^{-2} for the drip irrigation scenarios—is considerably higher than the ELP of currently grown crops, which ranges from 0.04 € m^{-2} for maize to 0.19 € m^{-2} for chillies. The average EWP of silk for the rain-fed scenario, 0.72 € m^{-3}, is much larger than the EWP of maize, groundnuts and tobacco, slightly larger than the EWP of paprika and similar as the EWP of chillies. Under drip irrigation, the EWP of silk is estimated at 1.02 to 1.20 € m^3, which is much higher than for all currently grown rain-fed crops. The large range for the EWP of rain-fed silk (0.23–0.89 € m^{-3}) compared with, for example, silk production under full drip irrigation (0.90–1.15 € m^{-3}), demonstrates the higher variability of rain-fed versus irrigated production.

Table 6. Economic water productivity (EWP) and land productivity (ELP) for silk under three scenarios, and for five currently grown crops. Minimum and maximum EWP are based on highest and lowest WF over the period 1986–2016, respectively. Silk yields are simulated; yields of current crops and market prices are based on local data.

Crop	WF_{total} (L kg^{-1})	$Yield_{avg}$ (kg ha^{-1})	Market Price (€ kg^{-1})	EWP_{min} (€ m^{-3})	EWP_{avg} (€ m^{-3})	EWP_{max} (€ m^{-3})	ELP_{avg} (€ m^{-2})
Silk, rain-fed	79,500	65	57.00	0.23	0.72	0.89	0.37
Silk, full drip irrigation	55,900	226	57.00	0.90	1.02	1.15	1.28
Silk, def. drip irr., organic mulch	47,800	220	57.00	1.05	1.20	1.35	1.24
Maize	2500	1500	0.26	0.01	0.10	0.16	0.04
Chilly	3400	750	2.50	0.42	0.74	0.84	0.19
Paprika	1900	1350	1.20	0.36	0.64	0.73	0.16
Groundnuts	3300	1250	0.48	0.03	0.15	0.23	0.06
Tobacco	3300	1250	1.05	0.00	0.32	0.40	0.13

EWP and ELP vary with WF and LF, respectively, as well as with changing market prices. With a local estimate of a bottom market price for raw silk of 54 € kg^{-1}, average EWP and ELP of rain-fed silk (the least productive form of silk production) reduce to 0.68 € m^{-3} and 0.35 € m^{-2}, respectively. When we assume a low market price of raw silk of 42 € kg^{-1}, as has been reported in India [38], EWP and ELP of rain-fed silk would be 0.53 € m^{-3} and 0.27 € m^{-2}, respectively. Under such low silk prices, average water productivities of chillies and paprika—if unchanged themselves—become higher than for rain-fed silk; land productivity of silk remains higher than for currently grown crops regardless such low silk prices. Both average EWP and average ELP of irrigated silk remain higher than those for currently grown crops even under low silk price estimates.

3.3. Macro-Level Sustainability

Current consumption of blue water resources for agricultural and domestic purposes in the Nyika Plateau watershed is low and remains within sustainable limits for most of the year according to Mekonnen and Hoekstra [37]. Only toward the end of the dry season, in October and November, total blue WFs in the watershed are slightly higher than the volume of water that is sustainably available, potentially causing moderate water scarcity in that part of the year. This estimate is based on the assumption that 80% of runoff is to be reserved to maintain environmental flows. Due to the lack of a reliable catchment-level assessment, no exact sustainability limit could be given. However, small-scale mulberry cultivation in the order of magnitude proposed in the project is not expected to cause water scarcity in the catchment.

To sketch out what would happen if silk production in the area takes off on a larger scale, we considered the following hypothetical case. Based on local data, average rainfall over the period

1986–2016 is 644 mm per year. The Malawi Government estimates the local runoff coefficient at 20% [39]; Ghosh and Desai report a runoff coefficient of 25% for the nearby Rukuru River and 34% for the also nearby Luweya River [40]. We conservatively assume here that 20% of annual precipitation around the study location becomes runoff, and thus becomes a blue water resource. In addition, from a precautionary principle, we assume that 80% of this runoff is to remain in rivers and streams to protect riparian ecosystems [41]. Given these assumptions, local total blue WFs are sustainable as long as they do not exceed about 25 mm per year on average (the macro-level sustainability limit). The blue WF of mulberry shrubs under full drip irrigation is about 750 mm per year. This implies that up to 3.3% of the local watershed area could be used for irrigated mulberry cultivation, before water consumption exceeds 20% of annual runoff potentially and environmental flow requirements are violated. Coverage of the area with irrigated mulberry shrubs beyond this share could lead to moderate water scarcity. In this scenario, we did not consider the blue WF of other activities, such as the presence of other irrigated agriculture. However, we know that the agricultural area equipped for irrigation (in the whole of Malawi) is low, at only 2.3% of the total [19]. Unfortunately, we could not evaluate locally what flow is sustainably available throughout the year in the Runyina River.

4. Discussion

We calculated WFs and LFs of silk and currently grown crops using FAO's AquaCrop model, which yielded several uncertainties. Firstly, AquaCrop is not calibrated for mulberry shrubs or for local Malawian circumstances. Secondly, although we accounted for variations in time by performing multi-year analyses, the sensitivities of yield and biomass build-up to specific weather conditions in a given year may not be fully captured by the model. Leaf yield will also depend on crop genetic make-up, since different mulberry varieties respond differently to different conditions. Nonetheless, simulated yields were about the same as anticipated yields of mulberry shrubs (International Centre of Insect Physiology and Ecology, ICIPE, pers. comm.).

Another source of uncertainty is the conversion factor of mulberry leaves to raw silk. The estimate of 187.5 kg of leaves to produce 9.1 kg of cocoons and 1 kg of raw silk (as expressed by ICIPE, pers. comm.) is slightly lower than the estimate by Astudillo et al. [35] of 238 kg leaves per kg raw silk and slightly higher than the 8.6 kg of cocoons per kg of silk by Patil et al. [42]. Any changes in this conversion factor directly translate into changes in the footprints of silk. Literature estimates of water consumption in silk processing also show a spread. For example Kathari et al. [34] estimate that 100 L of water is needed per kg of raw silk in the reeling process versus 1000 L by FAO [43] for the same process. However, since processing hardly contributes to overall footprints, the associated uncertainty is negligible.

There are no other studies to our knowledge quantify the total WF of silk. Astudillo et al. [35] estimated the blue WF component of silk in an Indian setting at 54.0 m^3 kg^{-1} and 26.7 m^3 kg^{-1}, for conditions following recommended guidelines and under actual farm practices, respectively. These numbers match our estimates (25.0–37.2 m^3 kg^{-1} for irrigation scenarios), but it has to be noted that climatic conditions are not necessarily comparable among the studies. Karthik and Rathinamoorthy [44] and Central Silk Board [38] estimate the LF of silk at 256 m^2 kg^{-1} and 103 m^2 kg^{-1}, respectively. Especially for irrigated scenarios, our estimate is significantly lower (around 45 m^2 kg^{-1}), which can probably be explained by the previously mentioned leaves-to-cocoons-to-silk conversion factors. This provides one more argument to assess thoroughly these conversion factors before embarking on sericulture.

We only considered the green and blue WF of silk production, and not the grey WF related to pollution. Sericulture has more than once been associated with pollution [2,43]. Depending on farming practices, such as fertilizer and pesticides application, this component may therefore add to the total WF. In addition, chemicals and disinfectants used in the silk processing stages may increase the WF if wastewater is not treated properly before disposal.

Like cotton, silk is a fibre harnessed by the apparel sector, so we thought it relevant to compare the water and land implications of silk versus cotton fibre. The global average WF of cotton of 9100 L kg^{-1} and LF of 4.2 m^2 kg^{-1} [45] are much lower than those for silk. Silk therefore is not the preferred source of fibre to replace cotton on a large scale. The cotton market price in Malawi estimated by Bisani [46] is 0.46 € kg^{-1}. Therefore, the economic value of cotton is much lower than that of silk. EWP and ELP of cotton (0.05 € m^{-3} and 0.11 € m^{-2}, respectively) are lower still than their silk equivalents (see Table 6). Considering only water and land, this implies that farmers would prefer sericulture to cotton production if they act as rational economic agents.

The same argument goes for the currently grown crops. Land and water requirements of silk—which is a luxury item—are higher than for low-value staple food crops, but the monetary added value per unit of resource is higher still for sericulture. Silk's advantages hold as long as: (1) market prices for silk remain high; (2) sericulture does not depress local food markets; and (3) total (blue) water consumption does not exceed sustainability limits at the catchment level. The implication is that silk has to remain a marginally produced product, in the case of our study area at no more than 3% of available land in the catchment area.

Clearly, water and land are not the sole factors a farmer considers in choosing what crop to grow [8,18]. However, footprints and economic productivities—calculated at the local level and placed in the wider environmental context of catchment-level sustainability—proved useful factors in our Malawi case study. It helps farmers to link implications of their crop choice to natural resources use and catchment-level sustainability limits [47]. Especially the estate owners could thereby—however partially and by no means exhaustively—give substance to their Corporate Social Responsibility (CSR) programs.

5. Conclusions

This study set out to explore how data on water and land footprints and economic productivity can inform micro-level decision making on crop choice—in the context of macro-level sustainability of resource use—with a study of proposed silk production in Malawi.

The total WF and LF of silk depend on the farming practices under which mulberry shrubs are cultivated. We found the total WF and LF of silk at the study location ranges from 79,500 L kg^{-1} and 155 m^2 kg^{-1}, respectively, under rain-fed conditions, to 47,800 L kg^{-1} and 45 m^2 kg^{-1} under the best farming practices. Here, best practice entails the use of deficit drip irrigation with organic mulch application. Over 99% of both the WF and LF relates to mulberry leaf production. The rest relates to silk processing, that is cleaning the nurseries, drying and cooking of the cocoons and reeling the silk. The WF of mulberry cultivation is all green in rain-fed agriculture and a mix of green and blue under irrigated conditions. The blue WF makes up 52 to 63% of the total WF, depending on the irrigation strategy and technique. Variability in time is considerably lower in irrigated than in rain-fed agriculture. A more constant silk production is therefore expected under irrigated farming conditions.

The WF and LF of silk are higher than those of currently grown rain-fed crops (maize, groundnuts, chilly, paprika and tobacco) and cotton, but the economic water and land productivities are also higher. Average EWP of silk ranges from 0.72 € m^{-3} (rain-fed conditions) to 1.20 € m^{-3} (deficit drip irrigation with mulching). EWP of cotton is much lower at 0.05 € m^{-3}, and EWPs of currently grown crops range from 0.10 € m^{-3} (maize) to 0.74 € m^{-3} (chilly). Average ELP of silk ranges from 0.37 € m^{-2} (rain-fed conditions) to 1.24 € m^{-2} (deficit drip irrigation with mulching) and is considerably higher than ELP of the currently grown crops (0.04–0.19 € m^{-2}).

The blue WF resulting from the introduction of irrigated mulberry plantations will increase the pressure on blue water resources compared with current rain-fed cropping practices. Current total water footprints in the Nyika Plateau catchment remain below the maximum sustainable footprint during most months of the year; only toward the end of the dry period is a moderate scarcity reported. Therefore, as long as irrigated mulberry cultivation takes place on a relatively small scale—not exceeding ~3% of the catchment area—no harmful environmental effects are expected.

Sericulture holds the promise of creating agricultural diversity, income and employment for the rural Malawian setting of our study case. Based on our assessment of water and land productivity, we conclude that sericulture is a viable alternative for farmers to currently grown crops—especially if they can irrigate their fields. This conclusion holds as long as prices of silk stay high, production remains marginal, and local food markets are not repressed. We recommend, however, to more closely evaluate both catchment hydrology and mulberry leaves-to-cocoons-to-raw silk conversion factors before a decision to grow silk is made.

With the case study of proposed silk production in Malawi, we have shown how water and land footprints and economic productivity data can be useful to farmers in choosing their crops. Moreover, these indicators provide a means for the farmers to give substance to their Corporate Social Responsibility (CSR) programs. However, final decision making should include considerations of other relevant factors (about seeds, labour, technology, access to markets, capital and so on) for a fully comprehensive assessment.

Acknowledgments: We would like to thank Marianne Löwik from The Netherlands-based NGO Sympany, Duncan McDavid from the estates at the study location, and Bonaface Ngoka and Everlyn Nguku from ICIPE for their valuable insights into sericulture and the provision of local data. This study greatly benefited from their contributions. The research was funded by Sympany; the writing of the article was funded by NWO Earth and Life Sciences (ALW), project 869.15.007. The work was partially developed within the framework of the Panta Rhei Research Initiative of the International Association of Hydrological Sciences (IAHS).

Author Contributions: Rick J. Hogeboom and Arjen Y. Hoekstra conceived and designed the study; Rick J. Hogeboom performed the modelling and analysed the data; and Rick J. Hogeboom and Arjen Y. Hoekstra wrote the paper.

Conflicts of Interest: The authors declare no conflict of interest and the founding sponsors had no role in the design of the study; in the collection, analyses, or interpretation of data; in the writing of the manuscript, and in the decision to publish the results.

References

1. Dercon, S. Risk, crop choice, and savings: Evidence from Tanzania. *Econ. Dev. Cult. Chang.* **1996**, *44*, 485–513. [CrossRef]
2. Raina, S.K. *The Economics of Apiculture and Sericulture Modules for Income Generation*; International Centre of Insect Physiology and Ecology: Nairobi, Kenya, 2000.
3. Qiu, H.G.; Wang, X.B.; Zhang, C.P.; Xu, Z.G. Farmers' seed choice behaviors under asymmetrical information: Evidence from maize farming in China. *J. Integr. Agric.* **2016**, *15*, 1915–1923. [CrossRef]
4. Sherrick, B.J.; Barry, P.J.; Ellinger, P.N.; Schnitkey, G.D. Factors influencing farmers' crop insurance decisions. *Am. J. Agric. Econ.* **2004**, *86*, 103–114. [CrossRef]
5. Wineman, A.; Crawford, E.W. Climate change and crop choice in Zambia: A mathematical programming approach. *NJAS Wagening. J. Life Sci.* **2017**, *81*, 19–31. [CrossRef]
6. Schmautz, Z.; Loeu, F.; Liebisch, F.; Graber, A.; Mathis, A.; Bulc, T.G.; Junge, R. Tomato productivity and quality in aquaponics: Comparison of three hydroponic methods. *Water* **2016**, *8*, 533. [CrossRef]
7. Tan, Q.; Zhang, S.; Li, R. Optimal use of agricultural water and land resources through reconfiguring crop planting structure under socioeconomic and ecological objectives. *Water* **2017**, *9*, 488. [CrossRef]
8. Hoekstra, A.Y. Water footprint assessment (WFA): Evolvement of a new research field. *Water Resour. Manag.* **2017**, *31*, 3061–3081. [CrossRef]
9. Bruckner, M.; Fischer, G.; Tramberend, S.; Giljum, S. Measuring telecouplings in the global land system: A review and comparative evaluation of land footprint accounting methods. *Ecol. Econ.* **2015**, *114*, 11–21. [CrossRef]
10. Aldaya, M.M.; Garrido, A.; Llamas, M.R.; Varela-Ortega, C.; Novo, P.; Casado, R.R. Water footprint and virtual water trade in Spain. In *Water Policy in Spain*; Garrido, A., Llamas, R., Eds.; CRC Press: Leiden, The Netherlands, 2010; pp. 49–59.
11. Gutierrez-Martin, C.; Borrego-Marin, M.M.; Berbel, J. The economic analysis of water use in the water framework directive based on the system of environmental-economic accounting for water: A case study of the Guadalquivir river basin. *Water* **2017**, *9*, 180. [CrossRef]

12. Hoekstra, A.Y.; Chapagain, A.K.; Aldaya, M.M.; Mekonnen, M.M. *The Water Footprint Assessment Manual: Setting the Global Standard*; Earthscan: London, UK, 2011.

13. Hoekstra, A.Y. The sustainability of a single activity, production process or product. *Ecol. Indic.* **2015**, *57*, 82–84. [CrossRef]

14. Chukalla, A.D.; Krol, M.S.; Hoekstra, A.Y. Marginal cost curves for water footprint reduction in irrigated agriculture: Guiding a cost-effective reduction of crop water consumption to a permit or benchmark level. *Hydrol. Earth Syst. Sci.* **2017**, *21*, 3507–3524. [CrossRef]

15. Hoekstra, A.Y.; Wiedmann, T.O. Humanity's unsustainable environmental footprint. *Science* **2014**, *344*, 1114–1117. [CrossRef] [PubMed]

16. Steffen, W.; Richardson, K.; Rockstrom, J.; Cornell, S.E.; Fetzer, I.; Bennett, E.M.; Biggs, R.; Carpenter, S.R.; de Vries, W.; de Wit, C.A.; et al. Sustainability. Planetary boundaries: Guiding human development on a changing planet. *Science* **2015**, *347*, 1259855. [CrossRef] [PubMed]

17. Vörösmarty, C.J.; McIntyre, P.B.; Gessner, M.O.; Dudgeon, D.; Prusevich, A.; Green, P.; Glidden, S.; Bunn, S.E.; Sullivan, C.A.; Liermann, C.R.; et al. Global threats to human water security and river biodiversity. *Nature* **2010**, *467*, 555–561. [CrossRef] [PubMed]

18. Chenoweth, J.; Hadjikakou, M.; Zoumides, C. Quantifying the human impact on water resources: A critical review of the water footprint concept. *Hydrol. Earth Syst. Sci.* **2014**, *18*, 2325–2342. [CrossRef]

19. International Water Management Institue (IWMI). *Trends and Outlook—Agricultural Water Management in Southern Africa—Country Report of Malawi*; International Water Management Institue: Colombo, Sri Lanka, 2010.

20. Food and Agricultural Organization (FAO). *Review of Food and Agricultural Policies in Malawi*; Food and Agricultural Organization: Rome, Italy, 2015.

21. International Trade Centre (ITC). *Trade Map for Product: 5002 Raw Silk. Trade Statistics for International Business Development*; International Trade Centre: Geneva, Switzerland, 2017.

22. Kottek, M.; Grieser, J.; Beck, C.; Rudolf, B.; Rubel, F. World map of the Köppen-Geiger climate classification updated. *Meteorol. Z.* **2006**, *15*, 259–263. [CrossRef]

23. Jian, Q.; Ningjia, H.; Yong, W.; Zhonghuai, X. Ecological issues of mulberry and sustainable development. *J. Resour. Ecol.* **2012**, *3*, 330–339. [CrossRef]

24. Mekonnen, M.M.; Hoekstra, A.Y. The green, blue and grey water footprint of crops and derived crop products. *Hydrol. Earth Syst. Sci.* **2011**, *15*, 1577–1600. [CrossRef]

25. Steduto, P.; Hsiao, T.C.; Raes, D.; Fereres, E. Aquacrop-the fao crop model to simulate yield response to water: I. Concepts and underlying principles. *Agron. J.* **2009**, *101*, 426–437. [CrossRef]

26. Chukalla, A.D.; Krol, M.S.; Hoekstra, A.Y. Green and blue water footprint reduction in irrigated agriculture: Effect of irrigation techniques, irrigation strategies and mulching. *Hydrol. Earth Syst. Sci.* **2015**, *19*, 4877–4891. [CrossRef]

27. De Lannoy, G.J.M.; Koster, R.D.; Reichle, R.H.; Mahanama, S.P.P.; Liu, Q. An updated treatment of soil texture and associated hydraulic properties in a global land modeling system. *J. Adv. Model. Earth Syst.* **2014**, *6*, 957–979. [CrossRef]

28. Chapagain, A.K.; Hoekstra, A.Y. *Virtual Water Flows between Nations in Relation to Trade in Livestock and Livestock Products*; UNESCO-IHE: Delft, The Netherlands, 2003.

29. Portmann, F.T.; Siebert, S.; Bauer, C.; Döll, P. *Global Dataset of Monthly Growing Areas of 26 Irrigated Crops: Version 1.0.*; University Institute of Physical Geography: Frankfurt, Germany, 2008.

30. Harris, I.; Jones, P.D.; Osborn, T.J.; Lister, D.H. Updated high-resolution grids of monthly climatic observations—The cru ts3.10 dataset. *Int. J. Climatol.* **2014**, *34*, 623–642. [CrossRef]

31. Dee, D.P.; Uppala, S.M.; Simmons, A.J.; Berrisford, P.; Poli, P.; Kobayashi, S.; Andrae, U.; Balmaseda, M.A.; Balsamo, G.; Bauer, P.; et al. The ERA-Interim reanalysis: Configuration and performance of the data assimilation system. *Q. J. R. Meteorol. Soc.* **2011**, *137*, 553–597. [CrossRef]

32. Singh, P.L. Silk cocoon drying in forced convection type solar dryer. *Appl. Energy* **2011**, *88*, 1720–1726. [CrossRef]

33. Mekonnen, M.M.; Gerbens-Leenes, P.W.; Hoekstra, A.Y. The consumptive water footprint of electricity and heat: A global assessment. *Environ. Sci. Water Res.* **2015**, *1*, 285–297. [CrossRef]

34. Kathari, V.P.; Mahesh, K.N.; Basu, A. Energy efficient silk reeling process using solar water heating system and ushma shoshak unit in multiend silk reeling unit. *Sericologia* **2013**, *53*, 219–224.

35. Astudillo, M.F.; Thalwitz, G.; Vollrath, F. Life cycle assessment of indian silk. *J. Clean. Prod.* **2014**, *81*, 158–167. [CrossRef]

36. Schyns, J.F.; Booij, M.J.; Hoekstra, A.Y. The water footprint of wood for lumber, pulp, paper, fuel and firewood. *Adv. Water Resour.* **2017**, *107*, 490–501. [CrossRef]

37. Mekonnen, M.M.; Hoekstra, A.Y. Four billion people facing severe water scarcity. *Sci. Adv.* **2016**, *2*, e1500323. [CrossRef] [PubMed]

38. Central Silk Board. Indian Silk April–June 2015: Impressive Achievements. Available online: http://www.csb.gov.in/publications/indian-silk/contents/current-issue-editorial/ (accessed on 13 October 2017).

39. Malawi Government. *Malawi State of Environment and Outlook Report: Environment for Sustainable Economic Growth*; Ministry of Natural Resources, Energy and Environment, Malawi Government: Lilongwe, Malawi, 2010.

40. Ghosh, S.N.; Desai, V.R. *Environmental Hydrology and Hydraulics—Eco-Technological Practices for Sustainably Development*; Taylor and Francis: Boca Raton, FL, USA, 2006.

41. Richter, B.D.; Davis, M.M.; Apse, C.; Konrad, C. A presumptive standard for environmental flow protection. *River Res. Appl.* **2012**, *28*, 1312–1321. [CrossRef]

42. Patil, B.R.; Singh, K.K.; Pawar, S.E.; Maarse, L.; Otte, J. *Sericulture—An Alternative Source of Income to Enhance the Livelihoods of Small-Scale Farmers and Tribal Communities*; Food and Agriculture Organization: Rome, Italy, 2009.

43. Food and Agriculture Organization (FAO). *Conservation Status of Sericulture Germplasm Resources in the World—II. Conservation Status of Silkworm (Bombyx mori) Genetic Resources in the World*; Food and Agriculture Organization: Rome, Italy, 2003.

44. Karthik, T.; Rathinamoorthy, R. 6-Sustainable silk production a2-Muthu, Subramanian Senthilkannan. In *Sustainable Fibres and Textiles*; Woodhead Publishing: Cambridge, UK, 2017; pp. 135–170.

45. Hoekstra, A.Y. *The Water Footprint of Modern Consumer Society*; Routledge: Abingdon, UK, 2013.

46. Bisani, L. Malawi Cotton Farmers Bemoan Poor Prices. Available online: https://malawi24.com/2016/06/06/malawi-cotton-farmers-bemoan-poor-prices/ (accessed on 13 October 2017).

47. Herva, M.; Franco, A.; Carrasco, E.F.; Roca, E. Review of corporate environmental indicators. *J. Clean. Prod.* **2011**, *19*, 1687–1699. [CrossRef]

water

MDPI

Article

Water Footprint and Crop Water Usage of Oil Palm (*Eleasis guineensis*) in Central Kalimantan: Environmental Sustainability Indicators for Different Crop Age and Soil Conditions

Lisma Safitri [1,*]**, Hermantoro Hermantoro** [1]**, Sentot Purboseno** [1]**, Valensi Kautsar** [1]**, Satyanto Krido Saptomo** [2] **and Agung Kurniawan** [3]

[1] Department of Agricultural Engineering, Stiper Agricultural University, Daerah Istimewa Yogyakarta 55281, Indonesia; sastro2mantoro@gmail.com (H.H.); sentot.purboseno@gmail.com (S.P.); valkauts@instiperjogja.ac.id (V.K.)

[2] Department of Civil and Environmental Engineering, Bogor Agricultural University, Jawa Barat 16680, Indonesia; saptomo.sk@gmail.com

[3] Research Centre PT Bumitama Gunajaya Agro, Pundu, Kotawaringin Timur, Kalimantan Tengah 74359, Indonesia; agung.kurniawan@bumitama.com

* Correspondence: lismasafitri86@gmail.com or lisma@instiperjogja.ac.id; Tel.: +62-852-2781-2729

Received: 28 September 2018; Accepted: 20 December 2018; Published: 25 December 2018

Abstract: Various issues related to oil palm production, such as biodiversity, drought, water scarcity, and water and soil resource exploitation, have become major challenges for environmental sustainability. The water footprint method indicates that the quantity of water used by plants to produce one biomass product could become a parameter to assess the environmental sustainability for a plantation. The objective of this study is to calculate the water footprint of oil palm on a temporal scale based on root water uptake with a specific climate condition under different crop age and soil type conditions, as a means to assess environmental sustainability. The research was conducted in Pundu village, Central Kalimantan, Indonesia. The methodology adopted in carrying out this study consisted of monitoring soil moisture, rainfall, and the water table, and estimating reference evapotranspiration (ETo), root water uptake, and the oil palm water footprint. Based on the study, it was shown that the oil palm water usage in the observation area varies with different crop ages and soil types from 3.07–3.73 mm/day, with the highest contribution of oil palm water usage was in the first root zone which correlates to the root density distribution. The total water footprint values obtained were between 0.56 and 1.14 m^3/kg for various plant ages and soil types. This study also found that the source of green water from rainfall on the upper oil palm root zone delivers the highest contribution to oil palm root water uptake than the blue water from groundwater on the bottom layer root zone.

Keywords: water footprint; root water uptake; oil palm (*Eleasis guineensis*); crop ages; soil type; environmental sustainability

---

1. Introduction

Oil palm plantations in Indonesia are well developed. Based on analysis, it was found that in 2015, the total area of oil palm in Indonesia was 5,980,982 ha, and it increased by about 13.7% by 2017 to 6,798,820 ha [1]. Various environmental issues related to oil palm commodity production, such as biodiversity, drought, water scarcity, and water and soil resource exploitation, have become major challenges for environmental sustainability. One of the persistent and recurring developing issues is that of water usage. One plant water usage efficiency parameter is the water footprint (WF), which indicates the quantity of water used by plants to produce one mass unit of biomass product. The water footprint of plants consists of a green water footprint (from rainfall), a blue water footprint (from aquifers, rivers, irrigation, etc.), and a gray water footprint (certain quantity of water used to dissolve chemical substances in order to make it appropriate with the environmental threshold) [2]. The water footprint is generally affected by the water usage of plants and its generated production. Two techniques are used to determine the rate of plant water usage or water utilization, which use a crop water requirement (ETP, potential evapotranspiration) and crop water usage (ETA, actual evapotranspiration) [2,3].

Plant water usage using actual evapotranspiration assumes more representative value to the real condition than to potential evapotranspiration. There are various values of the oil palm water footprint, which are usually based on geographical location and the climate condition. It should also be noted that the value of the water footprint tends to vary based on soil type, plant age, etc.

Nowadays, the water footprint has become an indication of environmental sustainability. There is an urgent need to develop oil palm plantations as a way to sustain plantations and encourage efforts to analyze the water footprint condition in each plantation location. Water footprint analysis could be conducted by various methods, such as the eco-scarcity method, the Milai Canals approach, the Pfister approach, etc. [4]. The water footprint represents the total sum of water used in a supply chain, which comprises blue, green, and gray water [3]. Lower water usage input without having a significant impact on yield will decrease the water footprint in milk production [5].

However, the limitation of climate data, which is the main factor used to analyze the water footprint in oil palm plantations, has become a major challenge. Moreover, the temporary cultivation of crops, and the various impacts associated with it, have been neglected in analyzing the water footprint in oil palm plantations globally [4]. Consequently, an annual assessment might be misleading regarding crop choices within and among different regions. A temporal resolution is therefore essential for proper life cycle assessment (LCA) or assessing the water footprint of crop production. For this purpose, a water stress index (WSI) was developed on a monthly basis for more than 11,000 watersheds with global coverage [6].

On the other hand, the water footprint has been calculated using an evapotranspiration and productivity approach which gave different ranges of variation between each region [7]. This analysis was based on geographical location, climate condition, plant condition factors, soil types, etc. As a result of this, and in order to develop the water footprint as a factor of environmental sustainability, a description of the water footprint value in the specific location with various soil types and plant ages is needed. The limitations of climate data for analyzing the water footprint in a specific time frame could be solved by developing a method of water footprint analysis using primary data.

Another factor that affects water usage as the main factor of the water footprint is root density. The oil palm root architecture consists of primary vertical and horizontal roots, secondary horizontal roots, vertical upward and downward growing secondary roots, superficial and deep tertiary roots, and quaternary roots [8–10]. The distribution of root density on the structure of oil palm root architecture varies between different soil types and crop ages. Therefore, in this research study, the water footprint of oil palms growing in various soil types and with different plant ages were analyzed. Based on the above-stated problems, it could be stated that variations of oil palm water footprint values are developed based on actual climate and production data in specified locations. It is of interest to find how oil palm water usage varies with different crop ages and soil types, how it distributes between the upper, middle, or lower part of the oil palm root zone, and which one is the highest. Furthermore, it is also important to determine how the water footprint varies temporally and which is greater between consumption from green water from rainwater and blue water from the subsurface layer. The current study is the first to propose the water footprint estimation under varying soil types and crop ages at a specific developmental stage of the oil palm plant in order to provide detailed information about the water footprint of oil palm and as an indicator of environmental sustainability in oil palm plantation.

Site Specific Features of Biophysics and Production

The research was conducted in Pundu village, Central Kalimantan, Indonesia, located at 11°58′01″ S and 113°04′32″ E at an altitude of 27 m above sea level. The site experiences a tropical climate and is represented by: (1) average annual rainfall of 3002 mm/year; (2) average annual temperatures between 21.4 and 33.8 °C; and (3) yearly average daily sunshine hours of around 5.9. The various observed plant ages and soil types were obtained from an oil palm plantation of 22,457.7 hectares in area.

Generally, the soil types used for oil palm plantations in Indonesia are spodosols and inceptisols. These soils are spread across Kalimantan and Sumatra island [11], where most oil palm plantations in Indonesia are located, making it feasible to analyze the effect of the utilization of spodosols and inceptisols and their relation to the water footprint.

There are limiting factors associated with the use of spodosols, such as the depth of the spodic layer, its sandy soil texture, and its acidic texture associated with the tropical area. The depth of the spodic layer is the main factor contributing to poor root growth. This is because it depends on the roots to penetrate the soil, whereas the sandy soil texture will reduce the soil's ability to retain water and produce a greater chance for the soil to leach its nutrients. Other limiting factors that could possibly hinder plant growth include poor drainage and soil acidity. The depth of the spodic layer in spodosols ranges from 30 to 70 cm below the soil surface [12]. Oil palm requires solum depths greater than or equal to 80 cm without layers of rock for optimal growth and development [13]. In some marginal area, the oil palm needs a minimum depth of 75 cm to grow well without additional land improvement [14].

Inceptisols are acid mineral soils with low nutrient availability. The productivity of oil palm planted in this soil is low, and there are symptoms of decreased productivity in certain months of the year. The use of inceptisols for agricultural purposes has resulted in many physical, biological, and chemical inconsistency properties of the soil. The problem associated with the physical properties of inceptisols is related to the coarse texture of the topsoil, which happens to be less coarse in the lower layer. Therefore, the permeability value is bigger on the top surface and smaller in the lower layer. The topsoil structure is granular or crumby with a lower unstructured layer. Its density is lower on the surface and increases with depth. The cation exchange capacity is relatively moderate at about 14.1–17.3 me/100 g, while base saturation is low, between the range of 24% to 29% [15].

2. Materials and Methods

The methodology used in this study was accomplished through the following stages:

2.1. Observing Soil Moisture, Rainfall, and the Water Table for Varying Crop Age and Soil Type

In order to better understand the water balance system in the oil palm, a set of computerized instruments were installed to observe water balance parameters such as rainfall, water table depth, and soil moisture in the oil palm root zone. The data were used to predict the crop water usage of oil palm as a main variable to determine the water footprint. The lateral water flux in and out of the oil palm system was neglected.

The observation was undertaken for various soil types and crop ages as shown below:

1. Soil type: inceptisol; crop age: 8 years old
2. Soil type: inceptisol; crop age: 13 years old
3. Soil type: spodosol; crop age: 8 years old
4. Soil type: spodosol; crop age: 13 years old
5. Soil type: spodosol; crop age: 7 years old
6. Soil type: ultisol; crop age: 9 years old

Rainfall was measured using an automatic double-tipping bucket rain gauge while the water was measured using an automatic water level. Soil moisture was observed using a soil moisture sensor which was spread horizontally and vertically in the root zone based under varying crop age and soil types by referring to oil palm root architecture [16].

2.2. Reference Evapotranspiration (ETo) Analysis by Penman–Monteith

The reference evapotranspiration (ETo) is the main parameter of crop water usage. In this study, the ETo was calculated using the Penman–Monteith equation according to study analysis by References [17–21], using inputs of hourly climate data, including solar radiation, sunshine, wind speed, temperature, and relative humidity, that were observed using an automatic weather station (AWS). Climate data were collected from April–May 2017 and June–August 2017. The ETo was predicted using Equation (1) by the standardization for grass crops [17–21].

$$ETo = \frac{0.408 \times \Delta(Rn - G) + \gamma \frac{Cn}{T+273} u_2 (e_s - e_a)}{\Delta + \gamma(1 + Cd)} \tag{1}$$

where:

ETo = reference evapotranspiration (mm day^{-1});
R_n = net radiation at the crop surface (MJ m^{-2} day^{-1});
G = soil heat flux density (MJ m^{-2} day^{-1});
T = mean daily air temperature at 2 m height (°C);
u_2 = wind speed at 2 m height (m s^{-1});
e_s = saturation vapor pressure (kPa);
e_a = actual vapor pressure (kPa);
$e_s - e_a$ = saturation vapor pressure deficit (kPa);
D = slope vapor pressure curve (kPa °C^{-1});
g = psychrometric constant (kPa °C^{-1});
Cn = numerator constant for reference type and calculation time step, aerodynamic resistance where the constant was 900 for daily, and 37 for hourly daytime and night-time;
Cd = denominator constant for reference type and calculation time step. Bulk surface resistance and aerodynamic resistance where the constant was 0.34 for daily, 0.24 for hourly daytime, and 0.96 for hourly night-time.

2.3. Root Water Uptake Analysis under Varying Crop Age and Soil Type

The crop water usage was the major formula used to calculate the water footprint in this study. The root water uptake in the oil palm root zone could be represented by the actual evapotranspiration in each root zone layer. The distribution of this root water uptake could also be determined using the water used by the oil palm which led to the emission of green or blue water. There are several methods used for measuring actual evapotranspiration (crop water usage), such as change in soil water, lysimetry, Bowen ratio-energy balance (BREB), eddy covariance, water balance, and remote sensing energy balance [21]. We have already observed the oil palm in fields, as well as the water balance at small scales (plant and root zone), using the Penman–Monteith equation. Following the procedure in [17] and R [21], we installed several soil moisture sensors in the root zone layer, following the result of oil palm root architecture [16], which we used to measure the change in soil water in the root zone. The recorded change in soil moisture was then used to adjust the coefficient of oil palm compared to the change in soil moisture using Richards' equation [22–25].

2.3.1. Calculation of Soil Moisture Change Based on the Richards' Equation Model under Varying Soil Type Which Depends on Soil Properties

The soil properties of the oil palm field study are denoted in Table 1.

Table 1. Soil properties and van Genuchten parameter of soil type variation.

Soil Type	Ultisol	Spodosol	Inceptisol
Sand (%)	33.3	89.29	52.38
Silt (%)	30.32	3.44	16.24
Loam (%)	36.39	7.28	31.38
Bulk Density (g/cm^3)	1.33	1.42	1.38
Porosity (%)	49.91	46.59	47.86
Ks (cm/hour)	10.31	36.49	8.24
Vg Parameters			
θs	0.439	0.404	0.418
θr	0.142	0.147	0.169
alpha	0.011	0.009	0.011
n	1.356	1.821	1.605

The distribution of soil moisture in the oil palm root zone was analyzed using Richards' equation [22–28] as shown below:

- Water retention calculated by van Genuchten [24]:

$$\theta(h) = \theta_r + \frac{\theta_s - \theta_h}{(1+|\alpha h|^n)^m}$$
$$m = 1 - \frac{1}{n}$$
(2)

- Water capacity (Darcy's law and Richards' equation):

$$C(h) = \frac{d\theta}{dh} = \frac{\alpha^n (\theta_s - \theta_r)(n-1)(|h|)^{n-1}}{[1+(\alpha|h|)^n]^{2-1/n}}$$
(3)

$$K(S_e) = K_s \times S_e^\lambda \times (1 - [1 - S_e^{1/m}]^m)^2$$
(4)

- S degree of saturation [24]:

$$S_e = \frac{\theta(h) - \theta_r}{\theta_s - \theta_r}$$
(5)

- Water flux by vertical flow of Richards' equation:

$$J_w = -\left(K\frac{\partial h}{\partial z} - K\right)$$
(6)

- Richards' equation (positive downward): 1D vertical flow

$$\frac{\partial \theta}{\partial t} = -\frac{\partial J_w}{\partial z} - S$$
$$\frac{\partial \theta}{\partial t} = -\frac{\partial}{\partial z}\left(-\left(K\frac{\partial h}{\partial z} - K\right)\right) - S\frac{\partial \theta}{\partial h} = C_w$$
$$C_w\frac{\partial h}{\partial t} = -\frac{\partial}{\partial z}\left(-\left(K\frac{\partial h}{\partial z} - K\right)\right) - S$$
(7)

where:

K = hydraulic conductivity (cm/hour);
h = water pressure head (Pa);
θ_s = saturated water content (cm^3/cm^3);
θ_r = residual water content (cm^3/cm^3);
α = air entry value (ha = α−1);

n = curve gradient;
λ = pore-size distribution index;
$C(h)$ = water capacity;
S_e = effective saturation/degree saturation;
m = empirical parameters;
J_w = total flux (cm/hour);
S = sink factor, root water uptake/accumulative actual evapotranspiration (cm/hour).

2.3.2. Determination of Root Water Uptake Distribution in Root Zone

The value of root water uptake, which is considered as the actual evapotranspiration, was analyzed by the varied crop coefficient (Kc) value and ETo as shown in Equation (8) below:

$$ET = K_c \times ET_o \tag{8}$$

where:

ET = evapotranspiration (root water uptake) (mm/hour);
K_c = crop coefficient;
ETo = reference evapotranspiration (mm/hour).

The crop coefficient based on a grass crop is the ratio of ET to ETo, which depends on nonlinear interactions of soil, crop, atmospheric conditions, and irrigation management practices [17,21]. A major uncertainty associated with using this approach is that a significant number of Kc values used in the literature were empirical and often not adapted to local conditions. Therefore, in this study, Kc was determined through the calibration between the soil moisture change model based on Richards' equation and the soil moisture change observation based on the soil moisture sensor placed in the root zone. Based on the results, it can be deduced that the values of Kc vary between 0.68 and 0.7 for different crop ages of oil palm (7–13 years). Furthermore, the total root water uptake was partitioned along the oil palm root zone [16], which was referred to as the root density distribution.

2.4. The Monthly Oil Palm Water Footprint Analysis under Varying Crop Age and Soil Type

The water footprint concept includes green, blue, and gray water footprints (Equations (9)–(12)). Due to the absence of fertilization during the observation period, the gray water footprint was disregarded. The monthly water footprint of oil palm was determined based on Equations (10). This involved the root water uptake as evapotranspiration (mm/month) and oil palm yields (kg/month).

$$WF\ green = \frac{area \times ET green}{Y} \left(m^3/kg\right) \tag{9}$$

$$WF\ blue = \frac{area \times ET blue}{Y} \left(m^3/kg\right) \tag{10}$$

$$WF\ grey = \frac{\alpha \times AR/(Cmax - Cnat)}{Y} \left(m^3/kg\right) \tag{11}$$

$$WF_{Total} = WF_{green} + WF_{blue} + WF_{grey} \tag{12}$$

The *ET green* was considered as root water uptake from rainfall and *ET blue* from groundwater. The contribution of groundwater to oil palm was neglected because the water level depth was below 10 m, which is where the root zone only reaches a maximum of 2 m. The oil palm absorbed the water from capillary only in shallow ground water in this case < 2 m from the top soil. As well as the value of crop water used, it is necessary to determine the yield data production of oil palm in order to determine the oil palm water footprint.

3. Results

3.1. Reference Evapotranspiration (ETo) as the Main Parameter of Crop Water Usage

The ETo is the main factor used to determine the crop water requirements based on the rate of transpiration in the area. Figure 1 demonstrates the result of the ETo (mm/hour) in the study area for two consecutive observations. Figure 1a shows that the average ETo (mm/hour) between 3 April and 24 May 2017 was 0.17 mm/hour, with the minimum recorded value being 0.0068 mm/hour during the night and the maximum value of 1.099 mm/hour during the day. For the daily rate, the average ETo value was 4.18 mm/day. Figure 1b shows that the average ETo (mm/hour) during 22 June–31 August 2017 was 0.16 mm/hour, with the minimum value obtained being 0 mm/hour during the night and the maximum value being 1.114 mm/hour during the day. For the daily rate, the average ETo was 3.87 mm/day. This shows that in the field study, which is categorized in the tropical rainforest zone, there is insignificantly different rates of reference evapotranspiration.

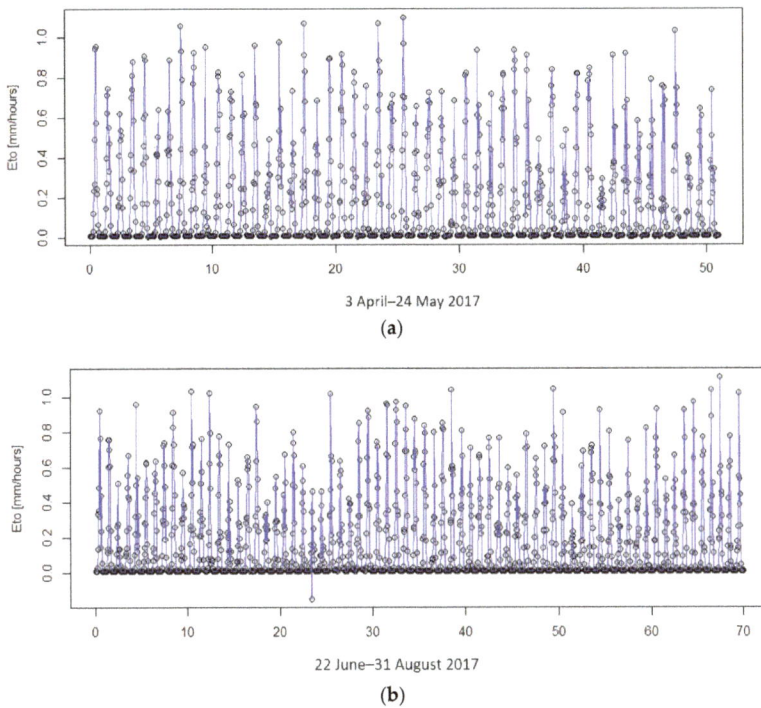

3 April–24 May 2017

(**a**)

22 June–31 August 2017

(**b**)

Figure 1. The reference evapotranspiration (ETo) (mm/hour) of the oil palm plantation area in Pundu, Central Kalimantan, during (**a**) April–May 2017, (**b**) June–August 2017.

3.2. Oil Palm Root Water Uptake (mm/day) Analysis and Its Distribution in the Root Zone

The term evapotranspiration has become more common compared to the term consumptive use. ET is the same as consumptive use, with the only difference between the two being that the latter includes minor water retained in the plant tissue that is relative to the total ET [21].

In this study, the term consumptive use was represented by the root water uptake of oil palm, which is obtained from the water absorption of the root spread along the root zone. Many studies such as References have shown that the highest contribution of root extraction comes from the smaller/finer root [29–31]. Among the fourth level oil palm root architecture, the sizes (primary, secondary, tertiary, and quarterly root) of the tertiary and quarterly absorb more water than others [8–10]. A classified oil palm root zone in several types of soil and oil palm crop ages in oil palm fields in Pundu, Central Kalimantan [16], shown in Table 2.

Table 2. The oil palm root zone for varying soil type and crop age.

	Soil Type and Crop Age					
	Inceptisol 8 Years	Inceptisol 13 Years	Spodosol 8 Years	Spodosol 13 Years	Spodosol 7 Years	Ultisol 9 Years
Root zone 1 (cm)	0–30	0–50	0–9	0–5	0–9	0–30
Root zone 2 (cm)	30–60	50–150	9–18	5–25	9–18	30–60
Root zone 3 (cm)	60–90	150–200	18–28	25–57	18–28	60–90
Spodic layer (cm)	-	-	56	56	56	-

Based on the root zone classification, some field and laboratory works were established to analyze the distribution of root density $(gram/cm^3)$ in the oil palm root zone, as displayed in Figure 2. The root density represents the mass of oil palm root for each bulk soil volume. As shown in Figure 2, the value of root density varies between 0 and 0.1 $gram/cm^3$; it also varies with the soil type and crop age.

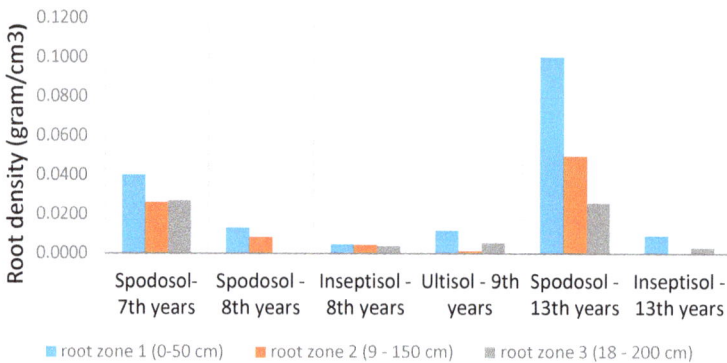

Figure 2. The distribution of oil palm root density under varying soil type and crop age.

Among all the plants studied, the 13-yearsold oil palm in spodosol has the highest root density, with root densities of 0.1001, 0.0497, and 0.0257 $gram/cm^3$ for the first, second, and third root zone, respectively. This was followed by the 7-years-old oil palm fruit in spodosol, which had root densities of 0.0400, 0.0262, and 0.0270 $gram/cm^3$ for the first, second, and third root zone, respectively. The lowest root density was obtained in the 8-years-old oil palm in inceptisol, which had root densities of 0.0046, 0.0045, and 0.0038 $gram/cm^3$ for the first, second, and third root zone, respectively. The root density was subsequently used to determine the contribution of root water uptake in the oil palm root zone.

According to the results of this study, the highest root density was in the first root zone, followed by the second root zone. Among the soil types, the spodosol contained higher root densities than the inceptisol and ultisol. As shown in Table 1, the spodosol consists of sands (89%), silt (3%), and loam (7%), while inceptisol and ultisol contain higher compositions of loam and silt. It could also be seen that the root density decreased gradually from the older to the younger oil palm plants in the same soil type.

In this study, the root water uptake is calculated based on the standard of reference evapotranspiration and crop coefficient. The root water uptake value illustrates the actual condition for adjusting the Kc value obtained by calibrating the result obtained by the soil moisture change model using Richards' equation [22] and the sensor technique. A variety of Kc values are obtained which are between 0.68 and 0.7.

Figure 3 shows the distribution of root water uptake in the oil palm root zone. Based on this study, the average root water uptake in the observation area varies between 3.07 and 3.73 mm/day, although this depends on the crop age and soil type. The highest water consumption was by the 13-years-old oil palm in spodosol, with an average daily rate of 3.73 mm/day, followed by the 8-year-old oil palm in spodosol and the 13-years-old oil palm in inceptisol, with a value of 3.51 mm/day. The lowest measured evapotranspiration, 3.07 mm/day, was for the 7-years-old oil palm in spodosol.

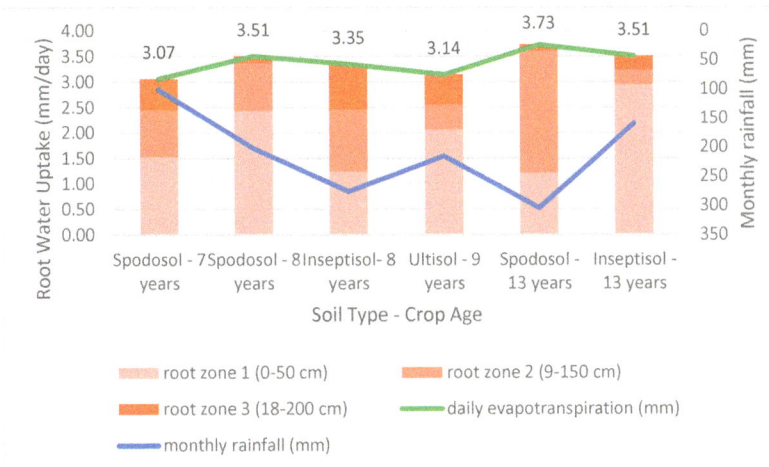

Figure 3. The distribution of root water uptake in oil palm root zone.

The distribution of root water uptake shown in Figure 3 demonstrates that the first oil palm root layer contributes more water than the second and third layers. For example, for the 13-years-old oil palm planted in inceptisol, the root water uptake from the first root layer reached almost 85%; this was followed by the 8-years-old oil palm planted in spodosol, whose root water uptake reached 69%, the 9-years-old oil palm planted in ultisol, whose root water uptake reached 65%, and the 7-years-old oil palm planted in spodosol, whose root water uptake reached 50%. The root water uptake of the 8-years-old oil palm planted in inceptisol seemed to be distributed on the root zone layer (38%, 36%, and 26% for first, second, and third root zone, respectively), while the 13-years-old oil palm planted in spodosol had a highest contribution of root water uptake 65% from second root zone, followed by 32% and 3% from the first and third root zone.

The analysis of oil palm root water for a variety of soils and plant ages could also describe the relationship and influence between the parameters. Table 3 shows the result of the correlation test between the variables derived from soil types, such as Ks (saturated hydraulic conductivity), the total available water (TAW), plant ages, yields, and climatic factors such as rainfall. From Table 3, it can clearly be seen that there are some strong relations between the parameters and the root water uptake. The total root water uptake value has a positive correlation of 0.730 with the crop age. Inverse correlations were found between the root water uptake in zone 3 and the yields.

Table 3. The variable correlations with root water uptake (RWU).

Variable	RWU total Cor_est	RWU_z1 Cor_est	RWU_z2 Cor_est	RWU_z3 Cor_est
Precipitation	0.607	−0.476	0.678	−0.069
Sat. hydraulic conductivity (Ks)	0.206	−0.279	0.547	−0.527
Crop age	0.730	0.242	0.257	−0.591
Yields	0.408	0.093	0.383	−0.816
Total available water (TAW)	−0.466	0.160	−0.442	0.333

3.3. Oil Palm Water Footprint under Varying Crop Ages and Soil Types

According to the root water uptake and yield production, as shown in Table 4, the value of the oil palm water footprint could be analyzed. Figure 4 denotes the oil palm water footprint (m^3/kg fresh fruit bunch [FFB]) under varying crop ages and soil types. The total water footprint of oil palm varied between 0.56 and 1.14 m^3/kg, with the highest water footprint of 1.14 m^3/kg being obtained for the 8-years-old oil palm in inceptisol. The lowest value, 0.56 m^3/kg, was obtained for the 7-years-old oil palm in spodosol.

Table 4. Yield production of the fresh fruit bunch (FFB) (kg/tree/month) in the field study.

Soil Type	Crop Age Years	Yield FFB (kg/Tree/Month)
Inceptisol	8	10.73
Spodosol	8	14.69
Spodosol	7	13.06
Ultisol	9	14.19
Inceptisol	13	12.83
Spodosol	13	15.62

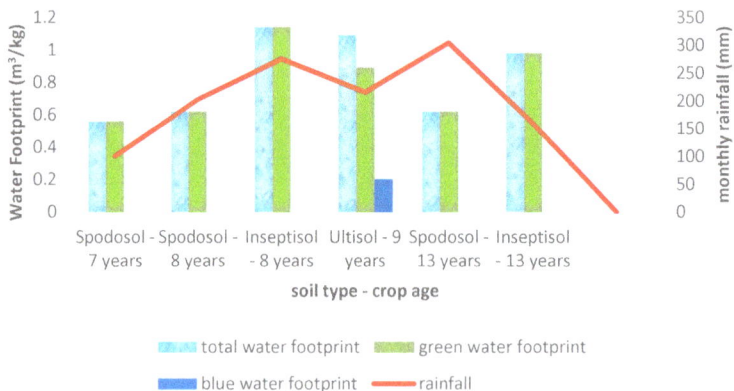

Figure 4. Oil palm water footprint (m^3/kg FFB) for varying crop ages and soil types.

Figure 4 shows that the water footprint of oil palm was mostly contributed by the green water, which was pointed out by the green water footprint for 100% contribution to the water footprint. The only case in which blue water contributed to oil palm crop water usage was for the 9-years-old oil palm in ultisol, for which blue water contributed about 28% of the total water footprint.

The contribution of groundwater was neglected in areas where the water level depth was below 10 m and where the root zone only reached a maximum of 2 m for the 8-years-old and 13-years-old oil palms in inceptisol. For the spodosol, there was a spodic layer which does not allow water to flow, both as deep percolation and capillary. The blue water indicated in the 9-years-old oil palm in ultisol was present due to the existence of the shallow water table (under 2 m under the top soil). This groundwater capillary in the third oil palm root zone contributed to the root water uptake from blue water.

The analyses of the oil palm water footprint with various soil types and ages also illustrates the influential relationship between the parameters. Table 4 shows the correlation test result between the descendant variable from various soil types, such as Ks, the total available water (TAW), and plant ages, such as crop age. This also yields some climatic factors, such as rainfall. From Table 4, it can be seen that the total value of the water footprint and the green water footprint negatively correlate to the Ks value (−0.975). The blue water footprint was positively correlated with TAW, with a value of 0.977.

4. Discussion

The reference evapotranspiration performed in this study could become the parameter of drought of an area [32–37]. Due to the absence of experimental reference evapotranspiration (ET_o) records, the modeling of reference evapotranspiration is reliable usually according to the standard FAO56 Penman Monteith equation (FAO56-PM) [38]. Based on the result, the average daily reference evapotranspiration obtained from both observation periods had an insignificantly different rate. According to the document of Food and Agricultural Organization FAO no. 56, the average value of ETo for tropical areas, particularly in humid and sub-humid zones ranges between 3 and 5 mm/day for moderate temperature and 5 and 7 mm/day for warm temperatures [17,39].

Actual evapotranspiration in this study referred to oil palm crop water usage, and the actual evapotranspiration is represented by root water uptake. Compared to the study presented in Johor, Malaysia where the annual crop evapotranspiration of oil palm, was calculated to be between 1100 and 1365 mm/year, or similar to 3 to 3.7 mm/day [40], the result showed in the same range. Additionally, several studies pointed out that the average oil palm crop evapotranspiration was 4.1 mm/day (between 3.5 and 5.5 mm/day) [41].

These results can also be compared with other types of plant. For example, the maximum value of daily evapotranspiration varies between 3.3 and 5.6 mm/day for rain-fed sunflower crops and between 6 and 7 mm/day for sunflower crops with optimal irrigation [42]. Similarly, the evapotranspiration of irrigated sunflower and canola crops varied between 3.6 and 10 mm/day and 2 to 11 mm/day, respectively [43]. This is similar to the values obtained for oil palm crops, with the consumptive water use of oil palm showing a lower rate.

On the other hand, comparing crop water use with forest plants shows that the level of evapotranspiration of the oil palm is slightly higher. A one-year daily observation in the Bornean tropical rainforest determined a varied evapotranspiration between 2.7 and 2.8 mm/day [44]. From the analysis obtained, it could be concluded that oil palm is not a crop with an extreme absorption rate that could be categorized as wasteful of water. Even if it could be compared to forest plants in the same location, the water absorption rate is only slightly different.

Root water uptake increases as the plant age increases and as the root becomes denser. The oil palm in spodosol absorbs more water than those in ultisol and inceptisol. This is in line with the root density level shown in Figure 1b, where the spodosol contains a higher root density than others. This is also supported by the production data in Table 5, where production over spodosol soil type is higher than in inceptisol. We can also conclude that the highest contribution of root water uptake was in the first root zone, which correlates to the root density distribution.

Table 5. Variable correlations with water with total, green, and blue footprint. WF: water footprint.

Variable	Yields Cor_est	WF Total Cor_est	WF Green Cor_est	WF Blue Cor_est
Precipitation	0.147	0.259	0.276	0.039
Sat. hydraulic conductivity (Ks)	0.619	−0.975	−0.948	−0.404
Crop age	0.361	0.054	0.103	−0.123
Yields	-	−0.596	−0.733	0.191
Total available water (TAW)	0.026	0.646	0.383	0.978

The findings of this study are also similar to those of one which showed that the root length density and the potential rate of root extraction decreased with the depth of the oil palm root zone [29]. The soil moisture extraction efficiency (SMEE) value increased with depth and distance from the palm. With regard to other plants, corn field extracted moisture mainly from the upper root dense soil profile when water content was in an optimal range [30]. Additionally, the distribution and density of wheat roots increases the water uptake [31].

Oil palms are often regarded as a plant capable of absorbing a large amount of water, thereby threatening the availability of ground water. From the results obtained in this study, it can be seen that oil palms display a low level of root water uptake when compared to other oil-producing plants, such as sunflower and canola. The distribution of the root water uptake of oil palm plants is mainly from the upper root zone layer. In the first layer, soil moisture comes from rainfall, while in the second and third layers, it comes from either rainwater with deep percolation or the capillary from ground water from a shallow water level with a maximum depth of roots.

The amount of crop water used by the root water uptake is used to analyze the oil palm water footprint with the supported production data listed in Table 4. The water footprint analysis in this research study is based on a specific location and a partial temporal climate. Studies related to crop water footprints mostly provide the global annual result by ignoring temporal aspects and other influential factors. However, in some areas, such as the Kalimantan region, the temporal aspects and local climate data vary greatly and affect the consumption and use of water as a major factor in the crop water footprint.

The pattern of the crop water footprint changes considerably with higher temporal resolution [6]. These changes are also shown to be sensitive to crop types due to different growth patterns leading to an increasing or decreasing water footprint. In line with this opinion, the results of this study show that there are variations in water footprint values for various conditions that represent the differences in rainfall, soil type, and growth of oil palm plantations.

The water footprint of the oil palm fresh fruit bunch for the spodosol soil type is lower than that for the inceptisol and ultisol soil types. With the same type of soil, younger plants have a higher water footprint, as shown in Figure 4. The crop water footprint is mainly driven by yield trends, while evapotranspiration plays a minor role in the annual water crop analysis for the wheat, rice, and maize and soybean footprint. Apart from correlations with yield and irrigation volume, the water footprint values are not correlated to soil properties [45]; however, it can be seen in Figure 3 that root water uptake varies.

Therefore, if drawn on the annual global scale, there will be a huge significant difference between these variables. The process of root water uptake analysis itself is strongly influenced by climatic factors, soil physical properties, and plant coefficient factors. In this analysis, the discovery of variations in root water uptake and water footprint values at the local and temporal scale could enrich the understanding of the water footprint of oil palm plants in particular, as well as other types of plants.

Another interesting fact worthy of discussion in this research study is the percentage contributions from each element of water (green, blue, and gray) to the total water footprint value of oil palm with variations in age and soil type. As shown in Figure 4, assuming no fertilization occurred during the observation process, the gray water footprint contribution would be 0%, while the green temporal water footprint reached would be 100% of the total water footprint for almost all variations.

In this oil palm water footprint analysis, the range of blue water footprint is relatively small. In its annual scope, the green, blue, and gray water footprints were 876.6, 35.9, and 91 m^3 ton^{-1}, respectively, and the contributions were 87.3%, 3.6%, and 9% for the case study in the oil palm plantation in Pundu, Central Borneo [46]. Additionally, the composition of the green, blue, and gray water footprint to be 68%, 18%, and 14% of the total average of the water footprint from several provinces in Thailand, respectively [47].

The crop water footprint for tomato cultivation showed the highest variability found in the water footprint green component, which ranged from 5% to 45.2% [45]. The blue water footprint ranged from 14.3% to 63.6% and the gray water footprint from 23.8% to 46.5% of the total water footprint. Therefore, it could be said that the range of groundwater use in oil palm plants in this study is relatively small. With no irrigation activities in the field, the possibility of using blue water only comes from the capillarity of groundwater.

The total value of the water footprint, the use of green and blue water, and the distribution of root water uptake in the rooting layers of the oil palm could be described as an indication of environmental sustainability. The various negative issues associated with the absorption of water by oil palm plants are inversely proportional to the results obtained in this study. The root water uptake of the oil palm is relatively low compared to that of other food crops. Additionally, the maximum level of water absorbed in the upper root zone also shows that oil palm plants absorb a lot of rainwater (green water), which is fast circle, compared to ground water (blue water), which is long circle.

5. Conclusions

1. Oil palm water usage in the observation area varies within different crop ages and soil types from 3.07 to 3.73 mm/day. The highest water usage was contributed by the 13-years-old oil palm in spodosol soil, with an average daily water usage of 3.73 mm/day. This was followed by the 8-years-old oil palm in spodosol soil and the 13-years-old oil palm in inceptisol soil, with a value of 3.51 mm/day. The lowest evapotranspiration was represented by the 7-years-old oil palm in spodosol soil, with a value of 3.07 mm/day. At the same soil type, the root water uptake of the oil palm increases as the plant age increases and as the root becomes denser, but there is no soil type parameter that showed a significant correlation with the root water uptake.

2. The water usage of the oil palm is distributed along the root zone in line with the root density. The upper zone of the oil palm root zone contributes more root water uptake (more than 50% of total) than the middle and the lower root zone. It can be concluded that the highest contribution of oil palm water usage was in the first root zone, which correlates to the root density distribution. The distribution of root water uptake in the rooting layers of the oil palm could be described as an indication of environmental sustainability.

3. The total water footprint of the oil palm fresh fruit bunch ranged from 0.56 to 1.14 m^3/kg for various plant ages and soil types. With higher yields, it can be concluded that the water footprint value of the oil palm fresh fruit bunch for spodosol soil types is lower than the inceptisol and ultisol soil types. It can also be stated that the plants with younger ages have relatively higher water footprint values for the same soil types. The water footprint value illustrates the efficiency of water use by plants; the higher the productivity, the larger the amount of water used. The variations in the water footprint values at the local and temporal scale could enrich the understanding of the water footprint of oil palm plants in particular, as well as other types of plants.

4. Green water contribution reached 82–100% to the total water footprint, while the blue water reached 0–28% of the total water footprint. The green water footprint reached 100% for all observed variations except for the 9-years-old oil palm in ultisol (25% blue water contribution rate out of the total water footprint due to the presence of shallow ground water with a depth of <2 m). This study showed that the source of green water from the upper oil palm root zone delivers the highest contribution for oil palm root water uptake than the blue water. The detailed description of the water footprint value could be a parameter for assessing environmental sustainability as an implication of oil palm plantations in certain regions.

Author Contributions: Conceptualization, L.S. and H.H.; Data curation, V.K.; Formal analysis, V.K.; Project administration, A.K.; Supervision, S.P.; Writing–original draft, L.S.; Writing–review and editing, S.K.S.

Funding: This research was funded by Oil Palm Plantation Fund Management Agency (BPDPKS), Indonesia grant number [PRJ-44/DPKS/2016].

Acknowledgments: This research and publication was fully supported by the Oil Palm Plantation Fund Management Agency (BPDPKS), Indonesia. We also thank our colleagues from PT Bumitama Gunajaya Agro, Indonesia, who provided the location, accommodation, and labors that greatly assisted the research. We thank our colleagues Rudyanto and Andiko Putro Suryotomo for assistance with the running of data analysis using R in finite different method and Willy Bayuardi Suwarno for his reviews and suggestions while preparing this manuscript.

Conflicts of Interest: The authors declare no conflict of interest. The founding sponsors had no role in the design of the study; in the collection, analyses, or interpretation of data; in the writing of the manuscript, and in the decision to publish the results.

References

1. Directorate General of Estate Crop Republik of Indonesia. *The Statisitic of Indonesia Plantation*; Indonesia Ministry of Agriculture: Jakarta, Indonesia, 2018; p. 5.
2. Hoekstra, A.Y.; Chapagain, A.K.; Aldaya, M.M.; Mekonnen, M.M. *Water Footprint Manual State of the Art*; Water Footprint Network: Enschede, The Netherlands, 2009.
3. Hoekstra, A.Y.; Chapagain, A.K.; Aldaya, M.M.; Mekonnen, M.M. *The Water Footprint Assessment Manual*; Water Footprint Network: Enschede, The Netherlands, 2011.
4. Jeswani, H.K. Azapagic Water footprint: Methodologies and a case study for assessing the impacts of water use. *J. Clean. Prod.* **2011**, *11*, 1288–1299. [CrossRef]
5. Cosentino, C.; Adduci, F.; Musto, M.; Paolino, R.; Freschi, P.; Pecora, G.; D'adamo, C.; Valentini, V. Low vs high "water footprint assessment" diet in milk production: A comparison between triticale and corn silage based diets. *Emirates J. Food Agric.* **2015**, *27*, 312–317. [CrossRef]
6. Pfister, S.; Bayer, P. Monthly water stress: Spatially and temporally explicit consumptive water footprint of global crop production. *J. Clean. Prod.* **2014**, *73*, 52–62. [CrossRef]
7. Mekonnen, M.M.; Hoekstra, A.Y. *The Green, Blue and Gray Water Footprint of Crops and Derived Crop Products*; Volume 1: Main Repport Value of Water Research Report Series No. 47; Twente Water Centre; University of Twente: Enschede, The Netherlands, 2010.
8. Jourdan, C.; Ferrière, N.M.; Perbal, G. Root System Architecture and Gravitropism in the Oil Palm. *Ann. Bot.* **2000**, *85*, 861–868. [CrossRef] [PubMed]
9. Jourdan, C.; Rey, H. Architecture and development of the oil-palm (Elaeis guineensis Jacq.) Root System. *Plant Soil* **1997**, *189*, 33–48. [CrossRef]
10. Lynch, J. Root architecture and plant productivity. *Plant Physiol.* **1995**, *109*, 7–13. [CrossRef] [PubMed]
11. Wigena, I.D.P.; Sudrajat Sitorus, S.R.P.; Siregar, H. Soil and Climate Characterization and Its Suitability for Nucleus Smallholder Oil Palm at Sei Pagar, Kampar District, Riau Province. *J. Soil Clim.* **2009**, *30*, 1–16.
12. Kasno, A.; Subardja, D. Soil fertility and nutrient management on spodosol for oil palm. *Agrivita* **2010**, *32*, 285–292.
13. Lubis, A.R. *Kelapa Sawit di Indonesia*; Pusat Penelitian Bandar Kuala Marihat: Pematang Siantar, Sumatera Utara, Indonesia, 2008.
14. Wiratmoko; Darlan, N.H.; Winarna Purba, A.R. Teknologi Pengelolaan Lahan Sub Optimal untuk Optimalisasi Produksi Kelapa Sawit. In , Proceedings of Seminar Optimalisasi Pemanfaatan Lahan Marginal untuk Usaha Perkebunan, Surabaya, Indonesia, 29–30 April 2015.
15. Nasrul, B.; Hamzah, A.; Anom, E. Klasifikasi tanah dan evaluasi kesesuaian lahan Kebun Percobaan Fakultas Pertanian Universitas Riau. *Jurnal Sagu* **2002**, *2*, 16–26.
16. Safitri, L.; Suryanti, S.; Kautsar, V.; Kurniawan, A.; Santiabudi, F. Study of oil palm root architecture with variation of crop stage and soil type vulnerable to drought. In *Conference Series: Earth and Environmental Science, Bogor, Indonesia*; IOP: Bogor, Indonesia, 2018.
17. Allen, R.G.; Pereira, L.S.; Raes, D.; Smith, M. *Crop Evapotranspiration-Guidelines for Computing Crop Water Requirements-FAO Irrigation and Drainage Paper 56*; FAO: Rome, Italy, 1998; Volume 300, p. D05109.
18. Walter, I.A.; Allen, R.G.; Elliott, R.; Jensen, M.E.; Itenfisu, D.; Mecham, B.; Spofford, T. ASCE's standardized reference evapotranspiration Equation. In *Watershed Management and Operations Management*; ASCE Press: Reston, VA, USA, 2000; pp. 1–11.
19. FAO. *Irrigation and Drainage Paper: Crop Evapotranspiration*; Water Resources; Development and Management Service: Rome, Italy, 2006.
20. Pereira, L.S.; Allen, R.G.; Smith, M.; Raes, D. Crop evapotranspiration estimation with FAO56, Past and future. *Agric. Water Manag.* **2015**, *147*, 4–20. [CrossRef]
21. Jensen, M.E.; Allen, R.G. (Eds.) *Evaporation, Evapotranspiration, and Irrigation Water Requirements*; American Society of Civil Engineers: Reston, VA, USA, 2016.

22. Richards, L.A. Capillary Conduction of Liquids through Porous Mediums. *Physics* **1931**, *1*, 318–333. [CrossRef]

23. Mualem, Y. A new model for predicting the hydraulic conductivity of unsaturated porous media. *Water Resour. Res.* **1976**, *12*, 513–522. [CrossRef]

24. Van Genuchten, M.T. A closed-form Equation for predicting the hydraulic conductivity of unsaturated soils. *Soil Sci. Soc. Am. J.* **1980**, *44*, 892–898. [CrossRef]

25. Van Genuchten, M.T.; Nielsen, D.R. On describing and predicting the hydraulic propertis of unsaturated soils. *Ann. Geophys.* **1985**, *3*, 615–628.

26. Feddes, R.A.; Raats, P.A.C. Parameterizing the soil-water-plant root system. In *Unsaturated Zone Modeling: Progress, Challenges and Applications*; Feddes, R.A., de Rooij, G.H., van Dam, J.C., Eds.; Wageningen UR Frontis Series; Kluwer Academic Publ.: Dordrecht, The Netherlands, 2004; pp. 95–141.

27. Huang, R.Q.; Wu, L.Z. Analytical solutions to 1-D horizontal and vertical water infiltration in saturated/unsaturated soils considering time-varying rainfall. *Comput. Geotech.* **2012**, *39*, 66–72. [CrossRef]

28. Safitri, L.; Hermantoro Purboseno, S.; Kautsar, V.; Wijayanti, Y.; Ardiyanto, A. Development of oil palm water balance tool for predicting water content distribution in root zone. *IJETS* **2018**, *5*, 38–45.

29. Nodichao, L.; Chopart, J.L.; Roupsard, O.; Vauclin, M.; Aké, S.; Jourdan, C. Genotypic variability of oil palm root system distribution in the field. Consequences for water uptake. *Plant Soil* **2011**, *341*, 505–520. [CrossRef]

30. Yadav, B.K.; Mathur, S. Modeling soil water uptake by plants using nonlinear dynamic root density distribution function. *J. Irrig. Drain. Eng.* **2008**, *134*, 430–436. [CrossRef]

31. White, R.G.; Kirkegaard, J.A. The distribution and abundance of wheat roots in a dense, structured subsoil–implications for water uptake. *Plant Cell Environ.* **2010**, *33*, 133–148. [CrossRef]

32. Ortega-Farias, S.; Irmak, S.; Cuenca, R.H. Special issue on evapotranspiration measurement and modeling. *Irrig. Sci.* **2009**, *28*, 1–3. [CrossRef]

33. Croitoru, A.E.; Piticar, A.; Dragotă, C.S.; Burada, D.C. Recent changes in reference evapotranspiration in Romania. *Glob. Planet. Chang.* **2013**, *111*, 127–136. [CrossRef]

34. Vicente-Serrano, S.M.; Beguería, S.; Lorenzo-Lacruz, J.; Camarero, J.J.; López-Moreno, J.I.; Azorin-Molina, C.; Revuelto, J.; Morán-Tejeda, E.; Sanchez-Lorenzo, A. Performance of drought indices for ecological, agricultural, and hydrological applications. *Earth Interact.* **2012**, *16*, 1–27. [CrossRef]

35. Vicente-Serrano, S.M.; Van der Schrier, G.; Beguería, S.; Azorin-Molina, C.; Lopez-Moreno, J.I. Contribution of precipitation and reference evapotranspiration to drought indices under different climates. *J. Hydrol.* **2015**, *526*, 42–54. [CrossRef]

36. Cook, B.I.; Smerdon, J.E.; Seager, R.; Coats, S. Global warming and 21 st century drying. *Clim. Dyn.* **2014**, *43*, 2607–2627. [CrossRef]

37. Senay, G.B.; Verdin, J.P.; Lietzow, R.; Melesse, A.M. Global daily reference evapotranspiration modeling and evaluation. *JAWRA J. Am. Water Resour. Assoc.* **2008**, *44*, 969–979. [CrossRef]

38. Martí, P.; González-Altozano, P.; López-Urrea, R.; Mancha, L.A.; Shiri, J. Modeling reference evapotranspiration with calculated targets. Assessment and implications. *Agric. Water Manag.* **2015**, *149*, 81–90. [CrossRef]

39. Jhajharia, D.; Dinpashoh, Y.; Kahya, E.; Singh, V.P.; Fakheri-Fard, A. Trends in reference evapotranspiration in the humid region of northeast India. *Hydrol. Process.* **2012**, *26*, 421–435. [CrossRef]

40. Yusop, Z.; Chong, M.H.; Garusu James, G.; Katiomon, A. Estimation of evapotranspiration in oil palm catchment by short-time period water-budget method. *Malays. J. Civ. Eng.* **2008**, *20*, 160–174.

41. Carr, M. The Water Relations and Irrigation Requirements of Oil Palm (Elaeis Guineensis): A Review. *Exp. Agric.* **2011**, *47*, 629–652. [CrossRef]

42. Matev, A.; Petrova, R.; Kirchev, H. Evapotranspiration of sunflower crops depending on irrigation. *Agric. Sci. Technol.* **2012**, *4*, 1313–8820.

43. Sánchez, J.M.; López-Urrea, R.; Rubio, E.; González-Piqueras, J.; Caselles, V. Assessing crop coefficients of sunflower and canola using two-source energy balance and thermal radiometry. *Agric. Water Manag.* **2014**, *137*, 23–29. [CrossRef]

44. Kume, T.; Tanaka, N.; Kuraji, K.; Komatsu, H.; Yoshifuji, N.; Saitoh, T.M.; Suzuki, M.; Kumagai, T.O. Ten-year evapotranspiration estimates in a Bornean tropical rainforest. *Agric. For. Meteorol.* **2011**, *151*, 1183–1192. [CrossRef]

45. Evangelou, E.; Tsadilas, C.; Tserlikakis, N.; Tsitouras, A.; Kyritsis, A. Water footprint of industrial tomato cultivations in the Pinios river basin: Soil properties interactions. *Water* **2016**, *8*, 515. [CrossRef]
46. Safitri, L.; Kautsar, V.; Purboseno, S.; Wulandari, R.K.; Ardiyanto, A. (Forthcoming). Water Footprint Analysis of Oil Palm (Case Study of Pundu Region, Central Borneo). *Inter. J. Palm Oil.* 2019. Available online: https://ijop.id/index.php/ijop (accessed on 21 December 2018).
47. Suttayakul, P.; Aran, H.; Suksaroj, C.; Mungkalasiri, J.; Wisansuwannakorn, R.; Musikavong, C. Water footprints of products of oil palm plantations and palm oil mills in Thailand. *Sci. Total Environ.* **2016**, *542*, 521–529. [CrossRef] [PubMed]

water

MDPI

Article

Simplified Direct Water Footprint Model to Support Urban Water Management

Wieslaw Fialkiewicz [1,*], Ewa Burszta-Adamiak [1], Anna Kolonko-Wiercik [2], Alessandro Manzardo [3], Andrea Loss [3], Christian Mikovits [4] and Antonio Scipioni [3]

[1]   Institute of Environmental Engineering, Wroclaw University of Environmental and Life Sciences, pl. Grunwaldzki 24, 50-363 Wroclaw, Poland; ewa.burszta-adamiak@upwr.edu.pl

[2]   New Technologies Center, Municipal Water and Sewage Company MPWiK S.A., ul. Na Grobli 14/16, 50-421 Wrocław, Poland; anna.kolonko@mpwik.wroc.pl

[3]   Department of Industrial Engineering, University of Padova CESQA, via Marzolo 9-35131, Padova, Italy; alessandro.manzardo@unipd.it (A.M.); andrea.loss@cesqa.it (A.L.); scipioni@unipd.it (A.S.)

[4]   Unit of Environmental Engineering, University of Innsbruck, Technikerstrasse 13, A6020 Innsbruck, Austria; christian.mikovits@mailbox.org

*   Correspondence: wieslaw.fialkiewicz@upwr.edu.pl; Tel.: +48-71-3205512

Received: 3 April 2018; Accepted: 9 May 2018; Published: 12 May 2018

Abstract: Water resources conservation corresponding to urban growth is an increasing challenge for European policy makers. Water footprint (WF) is one of the methods to address this challenge. The objective of this study was to develop a simplified model to assess the WF of direct domestic and non-domestic water use within an urban area and to demonstrate its effectiveness in supporting new urban water management strategies and solutions. The new model was tested on three Central European urban areas with different characteristics i.e., Wroclaw (Poland), Innsbruck (Austria), and Vicenza (Italy). Obtained WFs varied from 291 $dm^3/(day \cdot capita)$ in Wroclaw, 551 $dm^3/(day \cdot capita)$ in Vicezna to 714 $dm^3/(day \cdot capita)$ in Innsbruck. In addition, WF obtained with the proposed model for the city of Vicenza was compared with a more complex approach. The results proved the model to be robust in providing reasonable results using a small amount of data.

Keywords: Central Europe; modelling; urban area; water footprint; water management

---

1. Introduction

Europe is one of the most urbanized continents in the world. More than two-thirds of the European population lives in urban areas and this share continues to grow [1]. Besides the urbanization, climate change as well as demand for goods and services may influence water demand. In different cities, this impact will be different. Part of water is delivered by public water supply (public or private systems with public access). Although the share of the households water demand in total water abstraction can be relatively small, it is nevertheless often the focus of public interest, as it comprises the water volumes that are directly used by the population. The way in which water is managed in cities has consequences both for city dwellers and for the wider community and hence dictates water availability (in both quantity and quality) for other users. It thus also influences the environmental, economic, and social development of regions and countries. For those reasons sustainable, efficient and equitable management of water in cities has never been as important as in today's world. Looking forward to the next few decades, it seems likely that there will be a significant expansion in urban water infrastructure. Additionally, urban development, especially the sealing of surfaces and land use change, put pressure on urban infrastructure and quality of water discharged to the water bodies [2]. The lack of interaction between heterogeneous users, decision-makers, and isolated water managers has caused serious degradation of water resources and increased the risks to all the developmental sectors that depend upon them [3]. The traditional methods for the analysis and assessment of water availability as well as quality are not sufficient to evaluate the equitable utilization of available water and sustainable water management due to different ambient conditions as well as efficiency of use which differs between cities. One of the relevant approaches recognized by the EU to contribute positively to water management is the water footprint assessment [4]. The water footprint (WF) concept was introduced by Hoekstra as an indicator of freshwater use [5]. For years, the approach was continuously developed and now there are two methodologies (water footprint developed by The Water Footprint Network and Life Cycle Assessment developed by the Life Cycle Assessment community) used to calculate WF. Debate on their potential common grounds and differences is still ongoing [6]. WF was introduced to support better water management, however the experience at an urban level is limited. Most work on the water footprint has focused on agriculture and food production [7]. However, growing concern about water scarcity makes the concept of the water footprint potentially useful to other sectors, such as water utilities as well as with politicians, planners, and other stakeholders who have an influence on the investments and policies associated with water management at urban areas. For these reasons, the usage of tools to promote and encourage relevant measures, solutions, and technologies at a local (urban) scale is one of the key challenges for water footprint analyses, as well as the assessment and prediction of the influence of local policy on urban water [8]. Water should be managed both from a qualitative and quantitative perspective. Urban utilities frequently and independently assess water availability and vulnerability as a part of their planning processes which influence residential, commercial, and industrial development and land use patterns [6,9]. Changes in water quality pose a risk to aquatic ecosystems, but also involve the need to modify the water treatment technologies which significantly boosts the cost of its production and distribution [10]. The water footprint concept applied at an urban scale can be used as a measure to improve the communication with customers about their impact on the water environment which eventually influences conservation behavior [8]. It can be used as an awareness raising tool in decision making and in public debate by linking water supply, water use, as well as quality and quantity of sewage discharged to the receiving water [11].

Current urban studies have employed approaches for single cities and have adopted the water footprint accounting approach [12]. This kind of water footprint studies have been performed for Berlin (Germany), Delhi (India), Lagos (Nigeria) [13], Leshan (China) [14], Beijing (China) [15], Milan (Italy) [16], and Wroclaw (Poland) [17]. The majority of the cities WF studies set emphasis on evaluating virtual water (VW) which is mostly a focus on food consumption [18,19]. However, municipalities and water managers have limited influence on indirect (virtual) water use in the cities (e.g., by water saving campaigns regarding virtual water) but their impact on direct water use is much higher (e.g., applying tariffs, modernizing infrastructure, implementing water saving technologies, and organizing water saving campaigns). As the first has already been elaborated in many publications [15,20], the latter was not supported by WF analysis. In order to manage water resources in effective, efficient, and consistent way, decision making companies require access to appropriate data. A detailed assessment of the water footprint for urban areas would require the collection of a large amount of data and application of complex and sophisticated models [6]. This could restrict a wide application of the WF approach in the management of water in cities.

Therefore, the primary objective of this study is to propose a model which simplifies water footprint accounting of direct water use in urban area by adapting the approach proposed by Hoekstra et al. [12]. The proposed model is intended to support urban water managers and, as such, it can include additional aspects usually disregarded in previous water footprint of cities. To clarify this, the relevant equations to calculate components of urban WF are presented in the Methodology section. The model is tested through the application in three case studies (cities) with different characteristics. The results are presented using different metrics. To prove the effectiveness of a less data-intensive model, it is compared with the results of Manzardo et al. [11]. The authors of this paper propose to use the WF to solely investigate the direct water use in urban areas because it is the one directly managed by the local municipality. This paper is therefore a new contribution in applying WF. The discussion elaborates the secondary objective of this paper, which is the demonstration of model usefulness in supporting the definition of urban water management strategies and solutions. Differences between the proposed model and the one of Manzardo et al. [11] are further clarified in the discussion section.

2. Materials and Methods

The scope of this study focuses on the WF of direct water use which after Hoekstra et al. [12] refers to the freshwater consumption and pollution associated to the water use within city boundaries considering only urban area defined as locations with over 50% constructed surfaces [11]. Agricultural use within the city is excluded in this research. Considering that WF assessment in complex environments, such as urban areas, can be very challenging and resource consuming (time and money) a simplified method is needed. The novelty of proposed approach is not in the method itself but in improving applicability of the method in this specific context.

The simplified approach developed for urban areas and its application in urban water management is presented in Figure 1. The whole process starts with dividing the urban area into generic categories such as: impermeable area, permeable area, and water area. These categories can be subdivided further into surfaces characterized with similar water use pattern, e.g., impermeable area can be represented by paved area, roof surface, and transportation area, permeable area can consist of public and private green surfaces. The number of surfaces will depend on the local representation of urban area used by a municipality and the objectives set-up in urban water management. During the data acquisition phase, parameters characterizing all surfaces (area, evaporation coefficients), as well as water inflows and outflows (including mean annual precipitation), wastewater discharge and the concentration of pollutants are collected. The sources of the data can be found in municipalities, local water companies, legal regulations, and publicly accessible databases.

The calculation phase requires to perform a water balance for the urban area. In order to reduce the calculation effort it is recommended to use simple models [21]. The following paragraph describes how the green, blue, and grey components of WF are calculated for the urban area. This phase is complementary with the assignment of water quantity and quality in the urban area.

Calculated WF is evaluated during analysis of results phase and finally its findings are used to support creating or modifying urban water management strategies and plans. They also allow selecting activities aimed to reduce the urban WF which could stimulate policy development and create sustainable urban systems.

2.1. Urban Water Footprint Accounting Formulation

The green water footprint (WF$_{green}$) refers to the total rainwater evapotranspiration (from fields and plantations) plus the water incorporated into the harvested crop or wood [12]. In the urban environment, Manzardo et al. [11] proposed to limit accounting of WF$_{green}$ to green areas, such as private (gardens) or recreational land (lawns, public parks). According to this definition, WF$_{green}$ depends directly on the area with permeable surface covered by private and public vegetation

$$WF_{green} = PREC \times (A_{pubg} \times K_{pubg} + A_{privg} \times K_{privg}) \tag{1}$$

where the coefficients K_{pubg} and K_{privg} represent fraction of precipitation PREC (mm/a) which evapotranspirates from public green area A_{pubg} (m^2) and private green area A_{privg} (m^2), respectively. As the urban area does not include agricultural land, it is assumed that water used for agricultural activities, which might be present within city boundaries, is excluded from calculating urban WF.

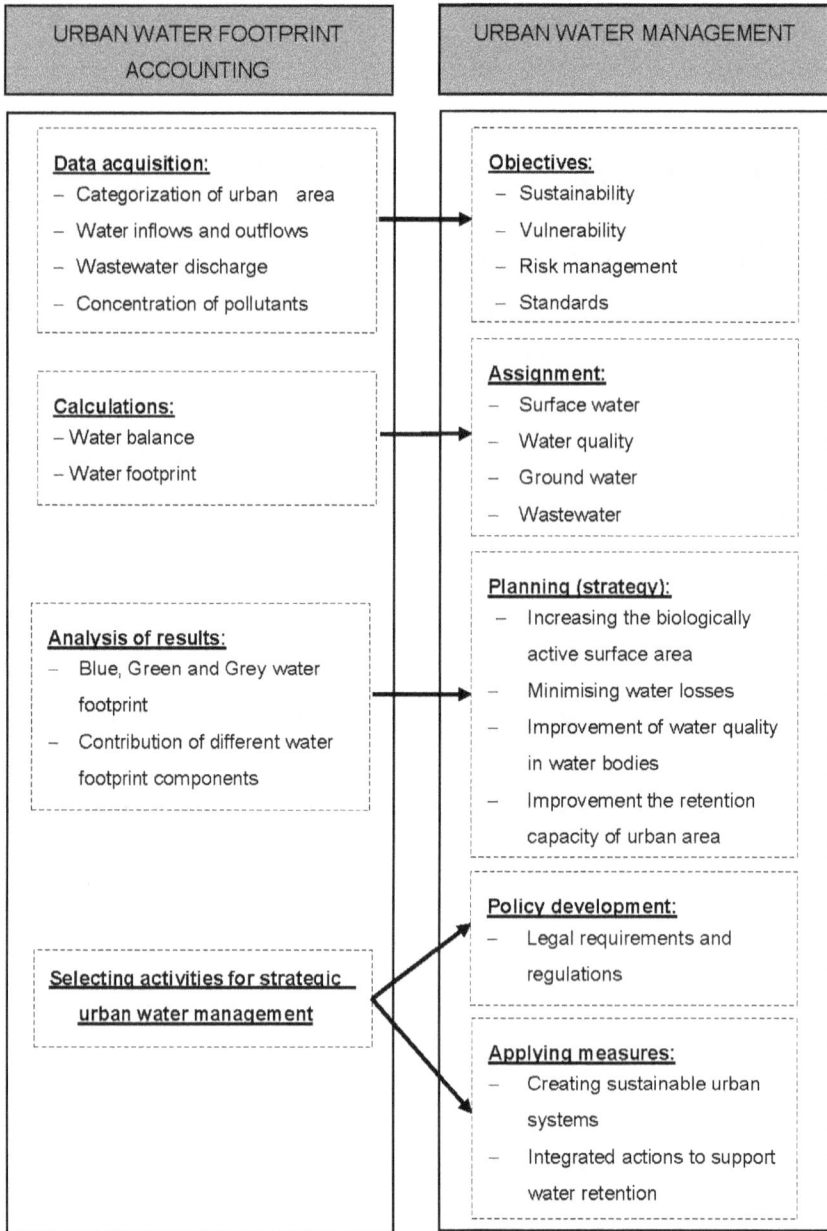

Figure 1. Interaction of the water footprint accounting with urban water management.

The blue water footprint (WF$_{blue}$) is the consumption of blue water resources, i.e., surface and groundwater withdrawn and not returned to the same water body [12]. According to this definition and its adaptation done by Manzardo et al. [11] it is proposed that WF$_{blue}$ in urban area accounts for the part of rainwater that evaporates from impervious surfaces Q$_{imperm}$ (such as roads and car parks) (m^3/a) and from water surfaces (rivers, ponds) Q$_{water}$ (m^3/a), water that is lost due to heating and cooling processes Q_{therm} (heating plants) (m^3/a), water exported outside the city boundary Q_{exp} (m^3/a), loss of supply water during transportation Q_{tl} (m^3/a), water consumed by the citizens and services and stored for a long term usage Q_{del} (m^3/a)

$$WF_{blue} = Q_{imperm} + Q_{water} + Q_{therm} + Q_{exp} + Q_{tl} + Q_{del} \qquad (2)$$

If the impermeable area is further subdivided into transportation area A_{transp} (m^2), roof area A_{roof} (m^2), and paved area A_{paved} (m^2) the volume of water evaporated from impervious surfaces can be calculated using the following formula

$$Q_{imperm} = PREC\,(A_{transp} \times K_{transp} + A_{roof} \times K_{roof} + A_{paved} \times K_{paved}) \qquad (3)$$

where K_{transp}, K_{roof}, and K_{paved} (unitless) represent fractions of precipitation PREC (mm/a) which evaporates from transportation, roof and paved surfaces respectively.

The volume of water which evaporates from the area covered by water A_{water} (m^2) is expressed as

$$Q_{water} = PREC \times A_{water} \times K_{water} \qquad (4)$$

where K_{water} (unitless) is the fraction of precipitation which evaporates from water surfaces.

The volume of water lost due to heating and cooling processes is assessed based on input–output water balance

$$Q_{therm} = Q_{cool} - Q_{heat} \qquad (5)$$

where Q_{cool} is the volume of water withdrawn from the water body (m^3/a) by a thermal power plant and Q_{heat} is the volume of water which is discharged after use to the water body (m^3/a).

The most ambitious term to assess in Equation (2) is the volume of water consumed and stored Q_{del} (m^3/a). To avoid laborious activities in collecting data about citizens water usage, a simple water balance of an urban catchment can be applied [22]

$$Q_{del} = (PREC \times A_{urban} + Q_{imp}) - (Q_{evap} + Q_{runoff} + Q_{waste}) \qquad (6)$$

where A_{urban} is the total urban area in the city, Q_{imp} the volume of water imported to the city (m^3/a), Q_{evap} the total volume of water evaporated (m^3/a), Q_{runoff} the loss of water due to surface runoff (m^3/a), and Q_{waste} wastewater discharge (m^3/a).

The grey water footprint (WF_{grey}) is defined as the volume of freshwater that is required to assimilate the load of pollutants discharged into a receiving water body based on natural background concentrations and existing ambient water quality standards [12]. In the urban environment, the pollution of water can be of chemical or thermal nature. In the case of pollution by chemicals, the WF_{grey} is calculated as

$$WF_{grey,\,chem} = (c_{sewage} \times Q_{sewage} - c_{act} \times Q_{abstr})/(c_{max} - c_{nat})] \qquad (7)$$

where c_{sewage} is the concentration of a pollutant in treated sewage discharged into receiving water body(g/m^3), Q_{sewage} the volume of sewage discharged into receiving water body by the sewage treatment plant (m^3/a), c_{act} is the actual concentration of a pollutant in water abstracted for consumption (g/m^3), Q_{abstr} the volume of abstraction by the water treatment plant (m^3/a), c_{max} the ambient water quality standard for a pollutant (the maximum acceptable concentration) (g/m^3), and c_{nat} the natural concentration of a pollutant in the receiving water body(g/m^3). In the case of separate sewage systems, WF_{grey} should be calculated separately for the treated and untreated wastewater.

When water is used for cooling—e.g., in thermal power plants—the processed water is discharged into the receiving water body, causing thermal pollution producing WF_{grey} which can be calculated as

$$WF_{grey,\,therm} = (T_{heat} - T_{act}) \times Q_{heat}/(T_{max} - T_{nat}) \qquad (8)$$

where T_{heat} is the temperature of the heated water discharged into the receiving water body (°C), T_{act} the actual temperature of water in a receiving water body (°C), T_{max} the maximum acceptable temperature in a receiving water body (°C), T_{nat} the natural temperature in a receiving water body (°C), and Q_{heat} the volume of water which was discharged after use (m³/a).

The final value of WF_{grey} is the maximum of the chemical and thermal WFs

$$WF_{grey} = \max (WF_{grey, chem}, WF_{grey, therm}) \tag{9}$$

Equation (9) is valid if the water for heating and cooling processes is released to the same water body as the water contaminated by chemical pollution. If the thermal and chemical pollutions are discharged to different water bodies, the final value of WF_{grey} should be the sum of $WF_{grey, chem}$ and $WF_{grey, therm}$.

The total value of the urban WF is the sum of green, blue, and grey WF

$$WF_{urban} = WF_{green} + WF_{blue} + WF_{grey} \tag{10}$$

2.2. Study Area Description

The assessment of urban water footprint was applied for three central European cities: Wroclaw (Poland), Vicenza (Italy), and Innsbruck (Austria). The cities assessed represent a diversity of geographical, climatic, and infrastructural aspects as presented in Table 1. The data on demographics, area, hydrology, infrastructure, and water usage were collected from the municipal authorities, sewage and water companies, law regulations, publicly accessible databases, and literature (for details see footer of the Table 1).

Table 1. General characteristics of the cities.

City	Wroclaw [1,4,5]	Innsbruck [2,4,5]	Vicenza [3,4,5]
Population, 10^3	632	125	115
Area, km²	293	105	80
Urbanized area, %	54	56	46
Arable and forest area, %	46	44	54
Paved area, ha	5487	29	1322
Roof surface area, ha	1727	423	396
Transportation area, ha	3745	482	534
Public green area, ha	1952	1026	453
Other permeable area, ha	1934	3805	818
Water area, ha	964	157	145
Average annual precipitation, mm	573	905	1889
Average annual temperature, °C	11.2	8.1	12.8
Latitude, m a.s.l.	105–155	565–2638	26–183
Major water supply	Surface water	Spring water	Ground water
Evaporation coefficients, %/100			
Public green surface	0.40	0.40	0.35
Private green surface	0.30	0.25	0.35
Water surface	0.10	0.10	0.10
Roads	0.20	0.10	0.15
Roofs	0.15	0.10	0.10
Oother impervious surface	0.20	0.10	0.15
Nitrogen concentration, g/m³			
Legal limit	10.0	15.6	30.0
In treated sewage	9.0	14.1	9.5
In the receiving water body	3.9	0.7	-

[1] [23,24]; [2] [25]; [3] [26]; [4] [27]; [5] [28].

Wroclaw is situated in the southwestern part of Poland on the Lower Silesian Lowlands. The city has two main water treatment plants in which surface and infiltration water, originating from the Sudetes mountains, are treated. The water supply system in Wroclaw connects 99% of inhabitants and is characterized by a great variance in age and material. The waste water is transported through the sewage system to one main mechanical–biological treatment plant. The sewage system in Wroclaw collects sewage from 98% of population and is comprised of two system types: combined and separate (sanitary and storm water) systems. The urbanized area in Wroclaw (54%), especially in the city center, and the large parts of the industrial area are mostly impermeable, hence most of the rainwater enters into the sewage system. Related with an increase in sealed surfaces is the lack of natural water retention for drier periods. Another important factor influencing the operation of water companies is water loss within the network, which amounts to over 10%.

Vicenza is located in the northeast part of Italy, on the Veneto Plain. Water from 18 artesian wells is treated in five plants, while one-third of the water consumed in the city is withdrawn from around 700 private wells. Currently, 97% of the population is connected to the water system. The waste water is treated in three plants. Around 92% of the population is connected to the sewerage system which consists of combined and separate systems. The latter is characteristic rather for new housing areas [29]. The annual rainfall in Vicenza is descending based on the data from the last two decades especially in winter season, which is characteristic for the whole Veneto region [26]. The yearly mean temperature is increasing which also causes an increase in evaporation leading to reduction in water reserves. At the same time, one of the main environmental issues in Vicenza is flooding, which happened a few times within recent years as a consequence of intensive rainfalls in autumn. Another reason for flooding is an overbuilt area and thus reduced ground permeability limiting water absorption. Even though the old water pipes are renovated systematically, the water losses reach up to 25%.

Innsbruck is located in Western Austria, surrounded by mountain ranges in the north and south. Only 32% (southern part) of the city is available for permanent settlement. Due to the alpine orography of the region, rainfall varies heavily in space, even within the municipality. The flow regime is influenced by snow and glacier melt in upstream regions and high precipitation during summer. The variations throughout the year and over the years according to the meteorological conditions are significant. Additionally, there is an influence of hydropower reservoirs [30]. As water flows rapidly through Innsbruck, the groundwater interaction is minimal [31]. Water intake to the distribution network relies mainly on a single spring in the mountains north of the city. All buildings are connected to the water (100%) and combined sewerage systems (99%). The major constraint influencing population density is topography, with mountain ranges north and south of the city. Both, heavy rainfalls and increasing temperatures cause accelerated glacier melting leading to higher risk of flooding [32]. The water loses in the water network are relatively small—below 10% is assumed, which might be due to the fact that about 1% of the network is rehabilitated each year.

3. Results

The data presented in Table 1 has been used to calculate three WF components: WF_{blue}, WF_{green}, and WF_{grey} for three cities in central Europe. Calculations for all cities were made on the basis of data from 2014 year except precipitation for which ten years average annual value was used. It is obvious that the total WF_{urban} is proportional to the urban area and the number of inhabitants. In order to compare cities of different size it is proposed to expressed WF per unit of area and per capita. Therefore, three different units ($Mm^3/year$, $m^3/(year \cdot ha)$, $dm^3/(day \cdot capita)$) were used to analyze obtained results as illustrated in Figure 2.

a)

b)

c)

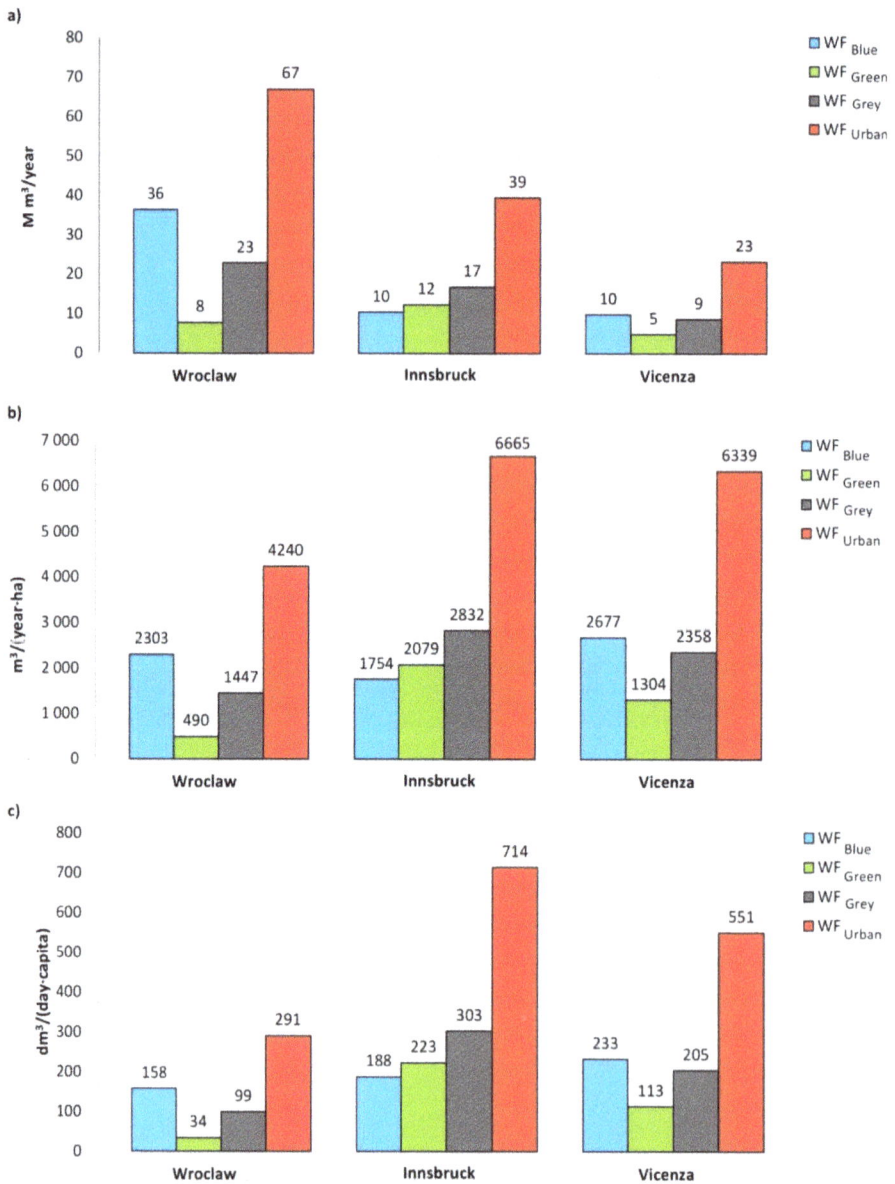

Figure 2. The urban WF for three cities expressed in: (**a**) Mm^3/year; (**b**) m^3/(year·ha); (**c**) dm^3/(day·capita).

The highest total WF$_{urban}$ of 67 Mm^3/year was obtained by Wroclaw, a 41% lower value for Innsbruck (39 Mm^3/year) and a 65% lower value for Vicenza (23 Mm^3/year) (Figure 2a). These values are in accordance to the size of the cities. For Wroclaw, WF$_{blue}$ is dominating (36 Mm^3/year), the second value is reached for WF$_{grey}$ (23 Mm^3/year), and the lowest value for WF$_{green}$ (8 Mm^3/year), due to the relatively small share of the permeable area (24%) in the urbanized area and local climate. For Innsbruck, all three WFs are similar, with the smallest value for WF$_{blue}$ (10 Mm^3/year), middle value for WF$_{green}$ (12 Mm^3/year), and highest value for WF$_{grey}$ (17 Mm^3/year). In Innsbruck, relatively high WF_{green} in proportion to other WFs in comparison to other cities is associated with a very high percentage of permeable area (ca. 82%) which consists of a green area. In Vicenza blue and grey WFs are comparable (10 and 9 Mm^3/year respectively) and the least significant is WF_{green} (5 Mm^3/year), reflecting small percentage of permeable area (ca. 35%).

Relating WF with the areas of Wroclaw, Innsbruck, and Vicenza which are 293 km^2, 105 km^2, and 80 km^2 respectively, it appears that the largest total WF of 6665 m^3/year·ha is reached for Innsbruck (Figure 2b). A very close value of 6339 m^3/year·ha was obtained for Vicenza and a relatively small value of 4240 m^3/year·ha is reached for Wroclaw. These results imply that the total urban WF is inverse proportional to the population density. In the cases of Vicenza and Wroclaw, WF_{blue} is the major component of WF_{urban} which is the result of a high share of the impermeable area in the urbanized area at 61 and 69% respectively. In Innsbruck, WF_{green} dominates over WF_{blue} which correlates with the relation of permeable (green) 82% to impermeable 16% area. However, the grey WF is greatly influencing WF_{urban} which could be explained with the very high dilution factor of 0.865 reported for Innsbruck, while Wroclaw and Vicenza have 0.51 and 0.32, respectively. In general, it should be beneficial for a city, when the WF_{green} reaches a high value as this reflects a great percentage of permeable area in the city and its retention capacity of rain water.

The comparison of WF_{urban} expressed per day and capita (Figure 2c) is especially relevant for blue and grey WFs determined to a large extent by the number of inhabitants having an impact on the volume of water used and contaminated. The results show that even though the number of citizens is the greatest in Wroclaw (632,000), the WF_{blue} per capita is the smallest (158 dm^3/d·ca), with the second value (188 dm^3/d·ca) reached in Innsbruck which is five time less populated (125,000), and the greatest value reached for Vicenza (233 dm^3/d·ca) consisting of only 115,000 citizens. The significantly high value for Vicenza is a result of a high groundwater withdrawal from private wells and high water losses in public water distribution system. Looking at WF_{green} the highest value was calculated for Innsbruck (223 dm^3/d·ca) which reflects the highest percentage of permeable green area in the city (ca. 82%) and the smallest population density (1190 inhabitants/km^2). The value of Vicenza is about half the value of Innsbruck (113 dm^3/d·ca) and Wroclaw is approximately seven times smaller (34 dm^3/d·ca). This is due to the smallest share of the green area in the urbanized area (ca. 25%) and the highest population density (2157 inhabitants/km^2). Similar relationship among the cities is observed for WF_{grey} which is also the highest in Innsbruck (303 dm^3/d·ca) while the value for Wroclaw (99 dm^3/ca·d) is approximately three times lower. Regarding the volume of produced sewage and number of inhabitants the values for Vicenza and Innsbruck are comparable thus the WF_{grey} values for these cities should be comparable. In practice, the value for Vicenza (205 dm^3/d·ca) is about one-third lower than for Innsbruck. For a better understanding of this phenomenon we need to take a close look at the Equation (7) for WF_{grey} calculation. The dilution factors which multiply the volumes of produced sewage, for Innsbruck, Wroclaw, and Vicenza are 0.865, 0.507, and 0.377 respectively. The highest dilution factor for Innsbruck determines the highest value of WF_{grey} per capita. Even though the dilution factor for Vicenza is almost 26% smaller than for Wroclaw the WF_{grey} per capita is over twice greater. This can be explained by the fact that the number of inhabitants is five times higher in Wroclaw than in Vicenza while the volume of waste water produced in Wroclaw is only higher by the factor 2.5. It is also worth mentioning that, in Vicenza, the nitrogen concentration in the treated effluent is three times lower than the legal limit (30 mg/L) while in Wroclaw (and Innsbruck) the nitrogen concentration is only 10% lower than the legal limit.

To see what contributes to specific WF_{urban} values in each city, the specific components are shown in Figure 3. The value of WF_{blue} in Wroclaw is determined mostly by water usage (16.7% of total WF_{urban}), losses in distribution system (10.1%), and evaporation from the paved area (9.3%). Water loss for heat production and cooling, as well as evaporation from transportation area, also contribute significantly (6.9% and 6.4%, respectively). In Innsbruck WF_{blue} is mostly associated with water usage (21.3% of total WF_{urban}) with other components being insignificant. In Vicenza, water loss from the water distribution system (18.5%) is dominating WF_{blue} value with evaporation from the paved area and water usage giving similar shares (9.2% and 9.0%, respectively). The high share of the public and private green areas in Innsbruck lead to high values of water evaporated from the permeable area of the city which accounts for 31.2% of total WF_{urban}. It is almost three times higher than in Wroclaw and 1.5 times higher than in Vicenza. The sewage discharged into the receiving body of the sewage treatment plant results in a significant share of WF_{grey} in WF_{urban} in all three cities, of which Innsbruck has the highest value (42.5%). The shares in Vicenza and Wroclaw are a bit lower with 37.2% and 34.1%, respectively. It has to be noted that climatic conditions (e.g., precipitation, average yearly/monthly temperature) influence WF results. This is of course particularly relevant for warmer climates such as the one in Vicenza.

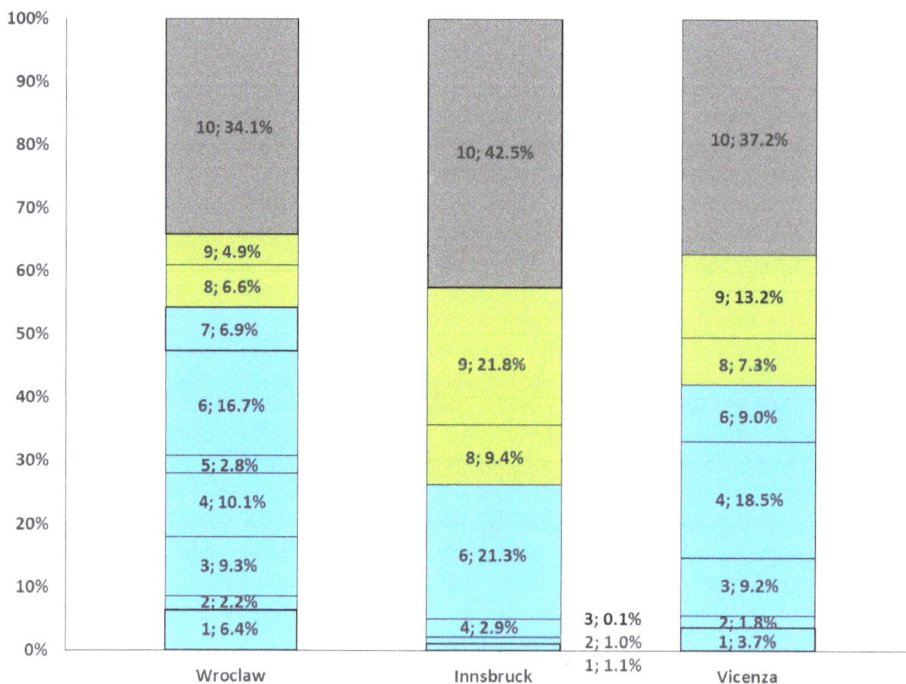

Figure 3. The WF_{urban} data for three cities with particular components specified: (1) evaporation from transportation area; (2) evaporation from roof surface; (3) evaporation from paved area; (4) water losses at transport; (5) water exported to another basin; (6) water used and stored; (7) water loss for heat production and cooling; (8) evapotranspiration from public green area; (9) wvapotranspiration from private green area; (10) treated sewage.

The simplified approach described in this paper has been compared with the more complex approach introduced by Manzardo et al. [11]. This approach assumes that the urban area is divided into basic modules with consistent characteristics which consist of building blocks with similar functions, needs, and behavior. In the accounting phase, a representative sample of building blocks for each module is identified, relevant quantitative and qualitative water data is collected and the average blue, green, and grey WF are calculated for each module—which are multiplied by the number of building blocks, providing the total WF. The flow of this methodology is that it relies on building blocks for which many parameters need to be provided to formulate water mass balance for each building block. This has been overcome in simplified approach by using the surfaces to represent the urban area. This requires less data as the water mass balance is performed for the whole city represented with homogenous surfaces and the necessary data is easily available from municipality and water and sewage companies. The two approaches have been applied to the city of Vicenza and the results of WF accounting are presented in Table 2.

Table 2. Comparison of the total WF accounting for the city of Vicenza (Italy).

Water Footprint Component	Modular Approach m³/year, [11]	Simplified Approach m³/year (This Study)	Difference %
Green water footprint	6.60×10^6	4.78×10^6	−27.6%
Blue water footprint	9.14×10^6	9.82×10^6	7.4%
Grey water footprint	8.18×10^6	8.65×10^6	5.7%
Urban water footprint	2.39×10^7	2.33×10^7	−2.8%

It is worth noticing that the simplified approach yields very close results for blue and grey WF which are overestimated with a few percent compared to modular approach. The highest difference of 27.6% was obtained for WF_{green} which might be the result of considering private green area differently. In a modular approach, private green area is included in building blocks but in a simplified approach it is a separated surface. Due to the fact that green WF was underestimated, the total WF_{urban} differs only by 2.8%. These results prove that the new simplified approach is robust and provides reliable results.

4. Discussion

Looking at the results the question arises: which city does a good job in water management? Assuming the one with the lowest water footprint might be an unequivocal answer. From the three cities analyzed, Vicenza has the lowest WF_{urban} expressed as total volume of water per year. If we relate the value of total WF_{urban} to the number of inhabitants or urban area in the city then it turned out that Wroclaw has the lowest WF per year per capita or per hectare. The answer becomes even more difficult if we consider the three components of WF: green, blue, and grey WFs. This is the merit of WF indicator as it enables to analyze different aspects of water management. In practice the urban water footprint results may be useful for decision-makers who have an influence on the investments and policies associated with water consumption, usage, and treatment. It turns out that the improvement in efficiency of water use by 40% or more is possible by implementing available technological solutions [33]. Therefore, it is important to raise the awareness of decision-makers about water scarcity and motivate them to choose environmentally friendly and sustainable solutions. In this case, the water footprint indicator can be used as a measure to improve communication.

This paper shows that each urban area is very specific regarding climatic and hydrogeological conditions and each city has a potential to improve the water and sewage management. In the cases of Vicenza and Wroclaw WF_{blue} is the major component of WF_{urban}. This may lead to a potential water scarcity issue in the future. Local problems have been noticed with droughts in Vicenza and Wroclaw occasionally, leading to withering of plants and also to water shortages during hot summers. The climate observations and prognoses indicate that the water resources might be threatened at some point in the future due to the temperature increase in recent decades, elongation of antecedent dry weather period, as well as increased frequency and intensity of heavy rainfall events, both in Wroclaw [34] and in Vicenza [26].

The efficiency of water distribution system management is also measured by the loss of water and the associated failure of the water system. High and rising water losses will increase the WF_{blue} and inform about inefficient water supply management, inadequate strategic planning or poor technical condition of the network. Results show that in Vicenza losses of supply water during transportation (18.5%) is determining WF_{blue} value. Relatively high losses are also in Wroclaw (10.1% of WF_{blue}). Investing in improvement of water supply system e.g., by means of general rehabilitation of aging water infrastructure, replacing inefficient components such as valves, pumps, pipes, and meters, monitoring domestic water use or leakage to rapidly repair leakage can reduce direct urban water use which in turn will reduce WF_{blue}.

The green water footprint (WF$_{green}$) is a good measure for assessment of natural retention capacity of urban area. In Wroclaw and Vicenza, the share of permeable area is relatively small (24% and 35%, respectively). Unlike Innsbruck with the share of permeable area of 82%. Based on the obtained results, it is recommended, especially for Wroclaw and Vicenza to incorporate more and more permeable and green spaces in the urban landscape. This can be done by building houses with green roofs, car parks, and pavements (especially walkways and squares) with permeable surface and rainwater harvesting facilities as described in Manzardo et al. [35]. Constructed wetlands, which are artificially created wetland ecosystems to treat—e.g., collected rainwater or wastewater, similarly to ponds and creeks—are also a possible solution for enhancing ecology and aesthetic value, enabling water retention for reuse for irrigation. The idea of linking water body and other open green spaces in a "blue-green infrastructure" is now recognized as part of cities planning strategy [36,37]. Local spatial management plans determine the indicators, forms, and functions of development, primarily the details of land use (including in areas excluded from construction) and the required percentage share of biologically active surfaces, providing opportunities to influence water management and mitigate the effects of flooding. Based on results of WF$_{green}$ calculations for urban areas, local governance can modify land use patterns, and thus affect water quantity and quality changes. The current trends in urban planning should highlight the need to shape compact and user-friendly cities while at the same time emphasizing the wise use of natural resources. This is evidenced by the increasingly frequent implementation of concepts based on ecological trends such as sustainable urban drainage systems, water sensitive urban design, or low impact development. Rainwater harvesting and retention is especially needed during heavy rainfall and melting snow when the sewage system is overloaded. This would help to minimize the problem with flooding noticed in Wroclaw and Vicenza and inundation of basements of buildings and streets, especially after heavy rainfalls. Such changes require promotion and might also be stimulated by the incentives and appropriate local regulations.

From an environmental point of view, it would be very helpful if not only quantitative but also qualitative requirements would be considered. The highest WF$_{grey}$ was in Wroclaw, then in Innsbruck, and finally in Vicenza. However, if we considered conversion to unit of area and capita these relationships would change. The highest value of WF$_{grey}$ was indicated for Innsbruck, Vicenza is at only 75% of the Innsbruck value and Wroclaw shows an approximately three-times smaller value than Innsbruck. Water quality changes can be significantly affected by the local governance structures, since local authorities largely influence the behaviors of inhabitants, private agents including developers, businesses, and many other stakeholders. In the case of urban areas with bigger WF$_{grey}$ value, the water and sewage companies should concentrate on potential process changes and investments that improve the contaminants removal from sewage (e.g., change of the operational scheme at the treatment stage). Reduction of the rainwater entering the sewage system will also result in reduction of the volume of treated sewage and thus WF$_{grey}$. The reduction in the treated effluent will limit the human influence on the receiving water body and maintain the river condition closer to natural. Communities further downstream may benefit especially, as well the ecosystem in general. The enhancement of awareness by means of improvement people's knowledge on water use in order to reduce wastewater generation and to facilitate the return of water that is not affected by our use to the environment is the further step to improvement of grey water footprint in the urban areas.

From a methodological perspective, in this paper a direct water footprint accounting method at urban level is presented. As such, it includes water balances at local level to support water management without addressing the consequences of water use in a more comprehensive water footprint sustainability assessment [12]. To better support informed decisions, recent scientific developments recommend adopting additional assessment such as the water scarcity or availability assessment [2,12,38–40]. For example, Bayart et al. [38] has presented the water impact index that allows the integration of consumptive and degradative water use of a process unit. The results are then characterized using a water scarcity index such as the one of Pfister et al. [41]. Moreover, Berger et al. [39] has presented the WAVE model considering atmospheric evaporation recycling and the risk of freshwater depletion. Recently, Boulay et al. [40] presented the AWARE method resulting from a consensus process lead by the UNEP SETAC Life Cycle Initiative. The outcomes of the accounting method presented in this paper can support the application of such methods by providing and organizing urban inventory data in a simplified manner when compared to previous experiences at urban level [13–17].

With reference to the design of WF accounting, indicator assumptions of the proposed method are based on the work of Manzardo et al. [11]. In the specific case of blue water, it is important to note that the consideration of rainwater evaporation is lively debated in the literature [39]. Therefore, the formulation of blue WF according to Equations (1) and (2) could be revised once consensus on this issue is found.

5. Conclusions

In this paper, a simplified model for water footprint accounting of direct water use in urban area was presented to support the definition of urban water management strategies and solutions. It was applied to three central Europe urban areas i.e., Wroclaw (Poland), Innsbruck (Austria), and Vicenza (Italy). The three cities under study represent a diversity of geographical, climatic and infrastructure aspects. This is directly reflected in three WF components: WF_{blue}, WF_{green}, and WF_{grey}. In addition, proposed model was compared with the modular approach applied to the city of Vicenza [11] and proved to be robust in providing reasonable results. The results obtained for the three cities could be the base for drawing up water management plans or strategies. For example, to assess the efficiency of water use, one should look at the blue WF per capita. Here, Vicenza shows the highest values which is a consequence of uncontrolled water intake from private wells and a large share of impermeable area. Green WF is a good measure of rainwater consumption and its low value indicate vulnerability of urban area to floods as is the case in Wroclaw which has the smallest value per hectare. WF_{grey} could help to assess the impact of the cities on water environment. The highest value observed in Wroclaw is mainly due to the largest city area and population. Even though the value is justified, it still results in the highest contamination of the receiving water body by the treated wastewater discharged in comparison with the other cities.

Though the WF directly depends on location and time, the results obtained suggest that Vicenza and Wroclaw need most modifications in the area of water management and infrastructure which should lead to restoration of natural water cycle and forming water reserves in the cities. Potential identified measures to improve local water management in analyzed cities include reduction of leakage from the drinking water network, introduction of water saving technologies, local rainwater management, education of citizens on water saving, and reduction of soil sealing in the cities.

The experience of the presented cities shows that each urban area is very specific regarding climatic and hydrogeological conditions (which cannot be changed) and each city has a great potential to improve the water and waste water management. The WF tool developed and adopted to specific city needs could be a useful tool allowing for evaluation of current water management state of the city, city area, or even single building. On the other hand, the tool could be used to compare, favor, and possibly also subsidize the best solutions proposed by the city planners, developers, and other stakeholders responsible for water management in the city. The success of using WF in water management will depend on its widespread application. The proposed simplified approach is a small contribution to achieve this goal.

Considering the outcomes of this study, future research can be planned as: (1) the development of a simplified water footprint sustainability assessment method to take into consideration also local water scarcity and availability as well as social and economic aspects [12,38]; (2) the application and possible adaptation of the proposed method at different levels, such as the regional one [42].

Author Contributions: Wieslaw Fialkiewicz, Ewa Burszta-Adamiak and Anna Kolonko-Wiercik conceived, designed and wrote the manuscript; Wieslaw Fialkiewicz and Alessandro Manzardo planned and designed the methodology; Wieslaw Fialkiewicz, Andrea Loss and Christian Mikovits performed the calculations; Antonio Scipioni supervised the whole process. All authors read and approved the final version of the manuscript.

Funding: This research was funded by European Regional Development Fund within the framework of the Central Europe Program, grant number 4CE439P3, as part of the URBAN_WFTP project, "Introduction of Water Footprint (WFTP) approach in urban areas to monitor, evaluate and improve the water use".

Acknowledgments: This work was supported by the Central Europe Program and co-financed by European Regional Development Fund (ERDF). The paper presents part of the results of the project URBAN_WFTP—"Introduction of Water Footprint (WFTP) approach in urban areas to monitor, evaluate and improve the water use" (4CE439P3).

Conflicts of Interest: The authors declare no conflict of interest.

References

1. United Nations, Department of Economic and Social Affairs, Population Division. World Urbanization Prospects: The 2014 Revision, (ST/ESA/SER.A/366). 2015. Available online: https://esa.un.org/unpd/wup/publications/files/wup2014-report.pdf (accessed on 29 April 2018).

2. Kleidorfer, M.; Mikovits, C.; Jasper-Tönnies, A.; Huttenlau, M.; Einfalt, T.; Rauch, W. Impact of a Changing Environment on Drainage System Performance. *Procedia Eng.* **2014**, *70*, 943–950. [CrossRef]

3. WWAP (World Water Assessment Programme). *The United Nations World Water Development Report 4: Managing Water under Uncertainty and Risk*; UNESCO: Paris, France, 2012.

4. European Commission. *A Blueprint to Safeguard Europe's Water Resources*; COM(2012) 673 Final; European Commission: Brussels, Belgium, 2012.

5. Hoekstra, A.Y.; Hung, P.Q. *Virtual Water Trade: A Quantification of Virtual Water Flows between Nations in Relation to International Crop Trade*; Value of Water Research Report Series No. 11; UNESCO-IHE: Delft, The Netherlands, 2002; Available online: http://www.waterfootprint.org/Reports/Report11.pdf (accessed on 26 March 2018).

6. Manzardo, A.; Mazzi, A.; Loss, A.; Butler, M.; Williamson, A.; Scipioni, A. Lessons learned from the application of different water footprint approaches to compare different food packaging alternatives. *J. Clean. Prod.* **2016**, *112*, 4657–4666. [CrossRef]

7. Silva, V.; Oliveira, S.; Braga, C.C.; Brito, J.I.; Sousa, F.; Holanda, R.; Campos, J.; Souza, E.P.; Braga, A.C.; Almeida, R.; et al. Virtual water and water self-sufficiency in agricultural and livestock products in Brazil. *J. Environ. Manag.* **2016**, *184*, 465–472. [CrossRef] [PubMed]

8. Paterson, W.; Rushforth, R.; Ruddell, B.L.; Konar, M.; Ahams, I.C.; Gironás, J.; Mijic, A.; Mejia, A. Water Footprint of Cities: A Review and Suggestions for Future Research. *Sustainability* **2015**, *7*, 8461–8490. [CrossRef]

9. Viessman, W.; Feather, T.D. *State Water Resources Planning in the United States*; Am. Soc. of Civil Eng.: Reston, VA, USA, 2006.

10. Schirmer, M.; Reinstorf, F.; Leschik, S.; Musolff, A.; Krieg, R.; Strauch, G.; Molson, J.; Martienssen, M.; Schirmer, K. Mass fluxes of xenobiotics below cities: Challenges in urban hydrogeology. *Environ. Earth Sci.* **2011**, *64*, 607–617. [CrossRef]

11. Manzardo, A.; Loss, A.; Fialkiewicz, W.; Rauch, W.; Scipioni, A. Methodological proposal to assess the water footprint accounting of direct water use at an urban level: A case study of the Municipality of Vicenza. *Ecol. Indic.* **2016**, *69*, 165–175. [CrossRef]

12. Hoekstra, A.Y.; Chapagain, A.K.; Aldaya, M.M.; Mekonnen, M.M. *The Water Footprint Assessment Manual, Setting the Global Standard*; Earthscan: London, UK, 2011; 228p, ISBN 9781849712798.

13. Hoff, H.; Döll, P.; Fader, M.; Gerten, D.; Hauser, S.; Siebert, S. Water footprints of cities; indicators for sustainable consumption and production. *Hydrol. Earth Syst. Sci.* **2013**, *10*, 2601–2639. [CrossRef]

14. Zhao, R.; He, H.; Zhang, N. Regional Water Footprint Assessment: A Case Study of Leshan City. *Sustainability* **2015**, *7*, 16532–16547. [CrossRef]

15. Wang, Z.; Huang, K.; Yang, S.; Yu, Y. An input–output approach to evaluate the water footprint and virtual water trade of Beijing, China. *J. Clean. Prod.* **2013**, *42*, 172–179. [CrossRef]

16. Vanham, D.; Bidoglio, G. The water footprint of Milan. *Water Sci. Technol.* **2014**, *69*, 789–795. [CrossRef] [PubMed]

17. Fialkiewicz, W.; Burszta-Adamiak, E.; Malinowski, P.; Kolonko, A. Urban Water Footprint—System monitorowania i oceny gospodarowania wodą w miastach. *Ochrona Środowiska* **2013**, *35*, 9–12. (In Polish)

18. Gobin, A.; Kersebaum, K.C.; Eitzinger, J.; Trnka, M.; Hlavinka, P.; Kroes, J.; Takac, J.; Ventrella, D.; Natali, F.; Dallamarta, A.; et al. Variability in the water footprint of arable crop production across European regions. *Water* **2017**, *9*, 93. [CrossRef]

19. Schyns, J.F.; Hamaideh, A.; Hoekstra, A.Y.; Mekonnen, M.M.; Schyns, M. Mitigating the risk of extreme water scarcity and dependency: The case of Jordan. *Water* **2015**, *7*, 5705–5730. [CrossRef]

20. Vanham, D.; Gawlik, B.M.; Bidoglio, G. Food consumption and related water resources in Nordic cities. *Ecol. Indic.* **2017**, *74*, 119–129. [CrossRef]

21. Mitchell, V.G.; McMahon, T.A.; Mein, R.G. Components of the total water balance of an urban catchment. *Environ. Manag.* **2003**, *32*, 735–746. [CrossRef] [PubMed]

22. Fialkiewicz, W.; Czaban, S.; Kolonko, A.; Konieczny, T.; Malinowski, P.; Manzardo, A.; Loss, A.; Scipioni, A.; Leonhardt, G.; Rauch, W.; et al. Water footprint as a new approach to water management in the urban areas. In *Water Supply and Water Quality*; Dymaczewski, Z., Jez-Walkowiak, J., Nowak, M., Eds.; PZITS: Poznan, Poland, 2014; pp. 431–439, ISBN 9788389696932.

23. *Statistical Yearbook of Poland*; Central Statistical Office: Warsaw, Poland, 2015.

24. Ziemiański, M.; Ośródka, L. *Zmiany Klimatu a Monitoring i Prognozowanie Stanu Środowiska Atmosferycznego*; The Institute of Meteorology and Water Management—National Research Institute (IMGW-PIB): Warszawa, Poland, 2012; 315p. (In Polish)

25. Stadt Innsbruck. Available online: https://www.innsbruck.gv.at/page.cfm?vpath=verwaltung/statistiken--zahlen/klima (accessed on 10 February 2018).

26. *ARPA Veneto*; Regional Environmental Protection Agency: Veneto, Italy, 2015.

27. ASCE. *Hydrology Handbook*, 2nd ed.; Task Committee on Hydrology Handbook of Management Group D of the American Society of Civil Engineers; Manual of Practice No. 28; American Society of Civil Engineers: Reston, VA, USA, 2013; pp. 1–784, ISBN 978-0-7844-7014-5.

28. EPA: Reducing Urban Heat Islands: Compendium of Strategies. Urban Heat Island Basics; 2008. Available online: https://www.epa.gov/heat-islands/heat-island-compendium (accessed on 16 January 2018).

29. Burson, M. *Strategy Report Aqueduct Territorial*; Ato Bacchiglione: Municipality of Vicenza, Italy, 2006.

30. Aulitzky, H. Sommerhochwässer 1987 in Tirol-Naturkatastrophen oder fehlende Vorbeugung. *Österreichische Wasserwirtschaft* **1988**, *40*, 122–128. (In German)

31. Lebensministerium. Green Jobs Sind Krisensichere und Klimaschützende Arbeitsplätze der Zukunft. 2010. Available online: http://minister.lebensministerium.at/article/articleview/81659/1/8111 (accessed on 20 March 2018).

32. Vanham, D.; Fleischhacker, E.; Rauch, W. Seasonality in alpine water resources management—A regional assessment. *Hydrol. Earth. Syst. Sci.* **2008**, *12*, 91–100. [CrossRef]

33. EU Final Report Water Saving Potential (Part 1) ENV.D.2/ETU/2007/0001. 2007. Available online: http://ec.europa.eu/environment/water/quantity/pdf/water_saving_1.pdf (accessed on 26 January 2018).

34. Dubicka, M.; Pyka, J.L. Wybrane zagadnienia klimatu Wrocławia w XX wieku. *Pr. Stud. Geogr.* **2001**, *29*, 101–112. (In Polish)

35. Manzardo, A.; Mazzi, A.; Rettore, L.; Scipioni, A. Water use performance of water technologies: The cumulative water demand and water payback time indicators. *J. Clean. Prod.* **2014**, *70*, 251–258. [CrossRef]

36. Howe, C.; Butterworth, J.; Smout, I.; Duffy, A.; Vairavamoorthy, K. *Sustainable Water Management in the City of the Future: Findings from the SWITCH Project 2006–2011*; UNESCO-IHE: Delft, The Netherlands, 2011; Available online: http://www.switchurbanwater.eu/outputs/pdfs/Switch_-_Final_Report.pdf (accessed on 26 March 2018).

37. Wagner, I.; Breil, P. The role of ecohydrology in creating more resilient cities. *Ecohydrol. Hydrobiol.* **2013**, *13*, 113–134. [CrossRef]

38. Bayart, J.B.; Worbe, S.; Grimaud, J.; Austin, E. The Water Impact Index: A simplified single-indicator approach for water footprinting. *Int. J. Life Cycle Assess.* **2014**, *19*, 1336–1344. [CrossRef]

39. Berger, M.; Ent van der, R.; Eisner, S.; Bach, V.; Matthias Finkbeiner, M. Water Accounting and Vulnerability Evaluation (WAVE): Considering Atmospheric Evaporation Recycling and the Risk of Freshwater Depletion in Water Footprinting. *Environ. Sci. Technol.* **2014**, *48*, 4521–4528. [CrossRef] [PubMed]

40. Boulay, A.M.; Bare, J.; Benini, L.; Berger, M.; Lathuillière, M.J.; Manzardo, A.; Margni, M.; Motoshita, M.; Núñez, M.; Pastor, A.V.; et al. The WULCA consensus characterization model for water scarcity footprints: Assessing impacts of water consumption based on available water remaining (AWARE). *Int. J. Life Cycle Assess.* **2018**, *23*, 368–378. [CrossRef]

41. Pfister, S.; Koehler, A.; Hellweg, S. Assessing the environmental impacts of freshwater consumption in LCA. *Environ. Sci. Technol.* **2009**, *43*, 4098–4104. [CrossRef] [PubMed]

42. Wang, W.; Gao, L.; Liu, P.; Hailu, A. Relationships between regional economic sectors and water use in a water-scare area in China: A quantitative analysis. *J. Hydrol.* **2014**, *515*, 180–190. [CrossRef]

water

MDPI

Article

Assessing the Water Footprint of Wheat and Maize in Haihe River Basin, Northern China (1956–2015)

Yuping Han [1,2,3], Dongdong Jia [1,*], La Zhuo [4,5], Sabine Sauvage [6], José-Miguel Sánchez-Pérez [6], Huiping Huang [1] and Chunying Wang [1,2,3]

[1] North China University of Water Resources and Electric Power, Zhengzhou 450046, China; han0118@163.com (Y.H.); 13526882916@163.com (H.H.); wangchunying1987@yahoo.com (C.W.)

[2] Collaborative Innovation Center of Water Resources Efficient Utilization and Support Engineering, Zhengzhou 450046, China

[3] Henan Key Laboratory of Water Environment Simulation and Treatment, Zhengzhou 450046, China

[4] Institute of Soil and Water Conservation, Northwest A & F University, Yangling 712100, China; zhuola@nwafu.edu.cn

[5] Institute of Soil and Water Conservation, CAS & MWR, Yangling 712100, China

[6] EcoLab, Université de Toulouse, CNRS, INPT, UPS, Toulouse, France, Avenue de l'Agrobiopole, 31326 Castanet Tolosan CEDEX, France; sabine.sauvage@univ-tlse3.fr (S.S.); jose-miguel.sanchez-perez@univ-tlse3.fr (J.-M.S.-P.)

* Correspondence: dongdongjia06@163.com; Tel.: +86-185-0085-4530

Received: 17 May 2018; Accepted: 21 June 2018; Published: 29 June 2018

Abstract: Assessing the water footprint (WF) of crops is key to understanding the agricultural water consumption and improving water use efficiency. This study assessed the WF of wheat and maize in the Haihe River Basin (HRB) of Northern China over the period1956–2015, including rain-fed, sufficient, and insufficient irrigation conditions by different irrigation intensity to understand the agricultural water use status. The major findings are as follows: (1) The annual average total WF of wheat and maize production is 20.1 (52% green, 29% blue, and 19% grey) and 15.1 (73% green, 3% blue, and 24% grey) billion m^3 $year^{-1}$, respectively. The proportion of grey WF is much larger than the world average; (2) Wheat has larger unit WF (1580 m^3 t^{-1}) than maize (1275 m^3 t^{-1}). The unit WF of both wheat and maize shows exponentially decreasing trends, indicating that water use efficiency has been improved. The unit WF is heterogeneous in space, which is larger in Tianjin and Huanghua and smaller in the Southern HRB; (3) Rain-fed crops have the largest unit WF, followed by crops under insufficient and sufficient irrigation conditions for both wheat and maize. To improve the sustainability of water resources, the application of fertilizer must be reduced, and irrigation is an effective way to improve water use efficiency in water-abundant areas.

Keywords: water footprint; irrigation intensity; wheat; maize; Haihe River Basin

1. Introduction

Water scarcity has been a growing concern worldwide [1–3]. Agriculture consumes 70% of the global freshwater withdrawal [4]. With growing populations and expanding irrigated acreage, the water demand of agriculture continues to increase. Meanwhile, extensive application of fertilizer has caused severe, diffuse agricultural water pollution, which increases the competition for freshwater [5]. In some river basins, due to limited water supply facilities and high water prices, crops are irrigated with inadequate water supply under field conditions. A comprehensive and accurate assessment of the volume and structure of agricultural water consumption under those conditions is key to improving water use efficiency and effectively managing water resources.

The water resources can be divided into green and blue water resources during water resource planning and management [6,7]. The concept of the "water footprint (WF)" was introduced by

Hoekstra [8] and it provides a tool to assist with water resource management and deals with water scarcity, such as changing consumption patterns or improving the water efficiency of production [9–12]. The WF of a product refers to the sum of the water volume consumed to produce the product [13]. The blue WF refers to the volume of surface and groundwater consumed (evaporated) as a result of the production of a good. The green water footprint refers to the rainwater consumed. For crops, this refers to the portion of rainfall that infiltrates the soil and is accessible by plants to generate vapor flow in support of biomass growth [9]. The grey WF of a product refers to the volume of freshwater that is required to assimilate the load of pollutants based on existing ambient water quality standards [13]. The WF of unit production, which is also recognized as the virtual water content [14,15] when assessing virtual water flows among regions, reflects the regional water productivity or water use efficiency.

Within the agricultural sector, WF has been intensively studied from global levels to regional levels. Mekonnen and Hoekstra [16,17] estimated the green, blue, and grey WF of global wheat and quantified the green, blue, and grey WF of global crop production for the period 1996–2005. Siebert and Döll [18] quantified the green and blue WF in global crop production, as well as potential production losses without irrigation. At the national level, Zhuo et al. [19,20] set up benchmark levels of consumptive WF of winter wheat and assessed the green and blue WF and virtual water trade in China under alternative future scenarios. Cao et al. [21] assessed the blue and green water utilization in wheat production of China. Zoumides et al. [22] employed a supply utilization approach along with two indicators, economic productivity of crop use and the blue water scarcity index, to assess the WF for the semi-arid island of Cyprus. Schyns and Hoekstra [23] demonstrated the added value of the detailed analysis of the human water footprint within Morocco and thoroughly assessed the virtual water flows. At the regional level, Bulsink et al. [24] analyzed the WF of an Indonesian province related to the consumption of crop products. Duan et al. [25] explored the spatial variations of the WF and their relationships with agricultural inputs in Northeast China. Gobin et al. [26] calibrated crop yield for a water balance model, "Aquacrop" at the field level and analyzed variability in the WF of arable crop production across European regions.

At the river basin level, Aldaya and Llamas [27] analyzed the WF and virtual water in the semiarid Guadiana Basin. Yin et al. [28] calculated the total WF and the net external WF of consumption in the Yellow River Basin of China. Zeng et al. [29] quantified the WF in the Heihe River Basin of China during 2004–2006. Zang et al. [30] reported on spatial and temporal patterns of both green and blue water flows, also in the Heihe River Basin. Zhuo et al. [31] estimated the inter- and intra-annual WF of crop production in the Yellow River Basin for the period 1961–2009. Assessing WF at the river basin level is an important step to understanding how human activities influence the water cycle and is a basis for integrated water resource management and sustainable water uses within the basin [29].

Prior studies analyzed or assessed the WF of crops by dividing them into pure rain-fed crops and irrigated crops with sufficient water. However, because of a lack of detailed long-term irrigation data, few studies assessed the WF with insufficient water supply restricted by water volume, water cost, and water supply facilities. Assessing the WF under those conditions can effectively improve our understanding regarding the agricultural water use status to improve agricultural water use efficiency. Additionally, few studies have investigated the spatial and temporal characteristics within the basin under the influence of many factors, such as climate, geography, soil property, and management practice (e.g., irrigation, fertilizer application). In order to effectively understand the agricultural water use status and reasonably allocate water resources within the basin, it is necessary to assess the spatial and temporal WF by dividing the basin into small regions according to administrative divisions which have their own record, climate, and geographical conditions. Among the above influences on spatial and temporal variations of the WF, irrigation is a key factor controlling the accuracy of WF assessment, especially in river basins facing water scarcity [22,23,32,33]. The irrigation quota is recommended by the local government to guide the farmers' irrigation practice. Furthermore, in the process, many factors such as climate, geography, soil property, and manner of irrigation are considered.

It is close to the actual scene for irrigated crops [34]. Hence, the green, blue, and grey WF can be quantified with an irrigation quota to improve the accuracy of the WF assessment.

The Haihe River Basin (HRB), the political, economic, and cultural center of China, has 146 million inhabitants [35] and is also a main grain producing area, with more than 10% of the national production. However, it is a historical water scarcity basin. The amount of water resources is 305 m^3 per capita, which is approximately 1/7 of the Chinese average (2200 m^3) and also 1/27 of the world average [36–38]. Restricted by limited water resources, high water prices, water supply facilities, and different climate conditions, crops are irrigated with different intensity in different regions within the HRB. There are great differences in the WF accounting between insufficient irrigation conditions and traditional rain-fed and sufficient irrigation conditions. However, the WF assessment under these conditions and the subsequent spatiotemporal patterns are lacking for the HRB.

The specific objectives of this study are: (1) to take account of the WF of both wheat and maize within the HRB; (2) to analyze the temporal trends and spatial variations of the WF in the entire HRB during the period 1956–2015; and (3) to allocate the WF of wheat and maize based on administrative districts within the HRB.

2. Methods and Data

2.1. Study Area

The Haihe River Basin (HRB) is located between 112° E–120° E and 35° N–43° N, with a drainage area of 318,200 km^2 (Figure 1). It encompasses Beijing, Tianjin, and 23 other large and medium cities. The basin is in a continental monsoon climate zone with annual mean temperatures between −4.9 and 15 °C, and the annual precipitation ranges from 380 to 580 mm. The precipitation in the flood season (June–September) generally accounts for 70–85% of the annual precipitation. The observed average groundwater table of the entire HRB is 6–9 m and has a decreasing trend due to overexploitation [39,40]. The most widely distributed soils in HRB are cinnamon soil and fluvo-aquic soil, with two main soil textures, sandy clay loam and sandy loam, respectively [41]. Wheat and maize are widely planted in the basin. The planting areas of wheat and maize were 3.9 and 5.1 million hectares in 2015, accounting for 27% and 36%, respectively, of the total planting area [42]. The total production of wheat and maize were 24.7 and 28.1 million tons in 2015, accounting for 20% and 13% of the nation, respectively [42].

In this study, the HRB is divided into 11 regions to illustrate the spatial variations. It is firstly divided into six administrative regions, including Beijing, Tianjin, Hebei, Shandong, Shanxi, and Henan. Since the areas of Liaoning and Inner Mongolia within HRB are small, they are incorporated into Hebei and Shanxi according to climate and geographical conditions. Among them, Shanxi Province is further divided into two regions according to the different planting systems, climate, and geographical conditions, which are also the irrigation management divisions, as recommended by the government [43]. In the southern part of Shanxi, there is a traditional rotation of winter wheat and summer maize, while in the northern part the major crops are spring maize and no wheat planted. Hebei province is divided into five regions according to the different geographical conditions, which are also the irrigation management divisions, as recommended by the government [43]. In Zhangjiakou and Chengde of Hebei, the major planting crop (spring maize) is different from that of other regions (traditional rotation of winter wheat and summer maize). A corresponding weather station was selected in each region (Table 1); the locations of the stations are shown in Figure 1.

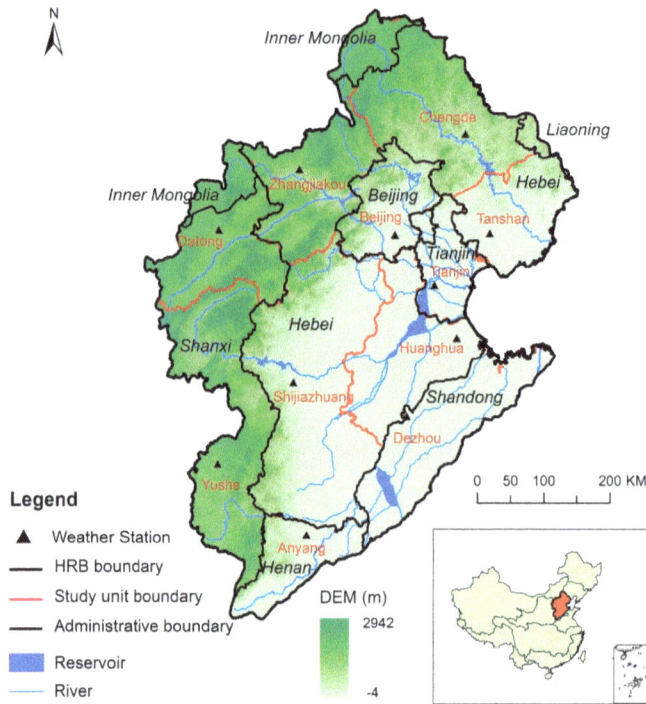

Figure 1. Locations of the study area and weather stations.

Table 1. Division of the study area and corresponding weather stations.

Code	Station	Area ($\times 10^3$ km^2)	Data Period	Weather Station Parameters			Geographical Characteristics
				Longitude (°C)	Latitude (°C)	Altitude (m)	
1	Beijing	16.4	1956–2015	116.47	39.80	31.3	Beijing municipality
2	Tianjin	11.5	1956–2015	117.07	39.08	2.5	Tianjin municipality
3	Shijiazhuang	60.7	1956–2015	114.42	38.03	81.0	Piedmont plain of Taihang
4	Tangshan	23.6	1957–2015	118.15	39.67	27.8	Hilly and plain area of Yanshan
5	Huanghua	28.0	1956–2015	117.35	38.37	6.6	Low plain of Heilonggang
6	Zhangjiakou	26.0	1956–2015	114.88	40.78	724.2	Northwestern Hebei mountains
7	Chengde	42.0	1956–2015	117.95	40.98	385.9	Mountainous area of Yanshan
8	Datong	27.3	1956–2015	113.33	40.10	1067.2	North of Shanxi
9	Yushe	38.0	1957–2015	112.98	37.07	1041.4	Middle part of Shanxi
10	Anyang	14.9	1956–2015	114.40	36.05	62.9	Plain area of northern Henan
11	Dezhou	29.8	1956–2013	116.32	37.43	21.2	North of Shandong

2.2. Methods

The green, blue, and grey WFs are quantified following the framework of Hoekstra et al. [44]. To distinguish the spatial discrepancy, the HRB is divided into 11 regions according to irrigation intensity, which refers to the irrigation quota recommended by the government of each province within the basin. In each region, a corresponding weather station was selected to represent the regional meteorological conditions.

In each region, the growing conditions of crops are divided into rain-fed and irrigated conditions. The proportion of irrigated crops is obtained by dividing the cultivated areas by irrigation areas in the statistical yearbook of each province or municipality.

For rain-fed crops, the blue WF is zero and the green WF is quantified by summing up the daily actual crop evapotranspiration (ET_a) without irrigation. For irrigated crops, the consumptive WF (green plus blue) is quantified by summing up daily actual crop evapotranspiration under different irrigation intensities. The green WF is assumed to be equal to the (ET_a) as calculated in the rain-fed scenario. The blue WF is equal to the consumptive WF minus the green WF.

To further analyze the structure of WF, the green water coefficient is defined as the ratio of green WF to the consumptive green and blue WF [45].

The grey WF is calculated by quantifying the volume of water needed to assimilate the nitrogen fertilizers that enter into the groundwater or surface water because nitrogen is the most used fertilizer in the HRB. The grey WF is calculated as:

$$WF_{grey} = \frac{\alpha \times AR}{(c_{max} - c_{nat})} \tag{1}$$

where WF_{grey} is the grey water (m^3); α is the leaching-runoff fraction (%), which is assumed to be 10%; AR is the concentration of pollutants per hectare (g ha^{-1}); c_{max} is the maximum allowable concentration of pollutants in water bodies (10 mg L^{-1}) [46]; and c_{nat} is the natural concentration of nitrogen in the receiving water body (mg L^{-1}), and is assumed to zero.

The unit WF refers to the WF for per ton of wheat or maize, which is obtained by dividing the total WF by production.

The actual crop evapotranspiration (ET_a), which depends on reference evapotranspiration, crop factor, and soil water availability [47], is calculated as:

$$ET_a[t] = K_s[t] \times K_c[t] \times ET_o[t] \tag{2}$$

where $K_s[t]$ is a dimensionless transpiration reduction factor dependent on available soil water with a value between zero and one; K_c is the crop coefficient, which varies in time as a function of the growth stage of crops, the length of the growing stage, and the crop coefficient of wheat and maize (Table 2); and ET_0 is the daily reference evapotranspiration (mm day^{-1}), which is calculated by the Penman–Monteith equation recommended by the Food and Agriculture Organization of the United Nations (FAO) [47].

The transpiration reduction factor, $K_s[t]$, is calculated based on a function of the maximum and actual available soil moisture in the root zone at daily time steps following Allen et al. [47]:

$$K_s[t] = \begin{cases} \frac{S[t]}{(1-p)S_{max}[t]} & S[t] < (1-p) \times S_{max}[t] \\ 1 & otherwise \end{cases} \tag{3}$$

where $S[t]$ is the actual available soil water in the root zone at time t (mm), which is simulated with a dynamic daily soil water balance method [16,19,48]. In this method, the irrigation quota is used as the irrigated water volume, which can be seen in Table 3. $S_{max}[t]$ is the maximum available soil water in the root zone (mm), and p is the fraction of S_{max} that a crop can extract from the root zone without suffering water stress. It is a function of crop type and potential crop evapotranspiration [47]:

$$p = p_{std} + 0.04(5 - ET_c) \qquad (4)$$

where p_{std} is a crop-specific depletion fraction when the evapotranspiration is 5 mm day^{-1}, a value of 0.55 is used for both wheat and maize in this study [47].

Table 2. Crop characteristics for winter wheat and maize in the Haihe River Basin.

	Planting Date	Growing Period (d)	Relative Length of Crop Growing Stage (–)				Crop Coefficients (–)		
			L_ini	L_dev	L_mid	L_late	K_c_ini	K_c_mid	K_c_end
Wheat	1 October	253	0.40	0.30	0.20	0.10	0.55	1.15	0.4
Maize	11 June	112	0.20	0.27	0.33	0.2	0.3	1.2	0.4
Maize *	1 May	140	0.20	0.27	0.33	0.2	0.3	1.2	0.4

Notes: L_ini, L_dev, L_mid, and L_late refer to the length of crop growing stages for initial, crop development, mid-season, and late season, respectively, as a fraction of the whole growing period; K_c_ini, K_c_mid, and K_c_end refer to crop coefficients for initial period, mid-season, and at the end of the season, respectively; Maize * refers to spring maize planted in Zhangjiakou, Chengde, and Datong; data references from Allen et al. [47]; Chen et al. [49]; and Kang et al. [50].

Table 3. Irrigation intensity [43] of wheat and maize within the HRB.

Code	Station	Irrigation Intensity (mm year^{-1})		Code	Station	Irrigation Intensity (mm year^{-1})	
		Wheat	Maize			Wheat	Maize
1	Beijing	428	75	7	Chengde	-	135
2	Tianjin	300	120	8	Datong	-	150
3	Shijiazhuang	210	68	9	Yushe	250	165
4	Tangshan	240	68	10	Anyang	180	68
5	Huanghua	248	75	11	Dezhou	270	105
6	Zhangjiakou	–	135				

2.3. Data

The weather data are obtained from the China Meteorological Administration [51], including daily maximum and minimum air temperatures, wind speed at 2 m height, average relative humidity, and daily sunshine duration from 1955 to 2015. The provincial agricultural data, including actual yield, planting area, irrigation area, fertilizer, and production from 1956 to 2015 are available at the Department of Plantation Management of the China Agriculture Ministry [42]. The production, planting area, and yield data for cities are available from the statistical yearbooks for Beijing, Tianjin, Hebei, Shanxi, Shandong, and Henan from 1983 to 2015 [52]. The yield and production data are checked and revised. The default values of yield and production for cities are calculated by multiplying a regional affecting factor by the provincial data. The regional irrigation schedules are obtained from the norm of water intake for Beijing, Tianjin, Hebei, Shanxi, Shandong, and Henan [43].

3. Results

3.1. Total Water Footprint of Wheat and Maize in the HRB

The total WF of wheat and maize over the period 1956–2015 in the HRB was calculated, and the results are shown in Table 4. The total WF of wheat is 20.1 billion m^3 year^{-1} on average. The major portion of this water (52%) comes from green water, about 29% comes from blue water, and the remaining 19% is grey water. The total WF of maize is 15.1 billion m^3 year^{-1} on average. The major portion of this water (73%) comes from green water, about 3% comes from blue water, and the remaining 24% is grey water, on average. Per hectare of cultivated land, wheat (4900 m^3 ha^{-1}) requires more water (including grey water) than maize (4580 m^3 ha^{-1}) on average. In the last ten years, the average blue WF of wheat (6.3 Gm^3 year^{-1}) and maize (0.6 Gm^3 year^{-1}) accounts for 26% and 2%, respectively, of the total agricultural water withdrawal of the HRB (24.04 Gm^3 year^{-1}) [53].

Table 4. Total water footprint (WF) of wheat and maize in the HRB, 1956–2015.

Crops	Period	Planting Area * (10^6 ha year^{-1})	Total WF * (Gm^3 year^{-1})				GWC * (%)
			Green	Blue	Grey	Total	
Wheat	1956–1965	3.7 ± 0.2	9.9 ± 1.1	4.4 ± 0.7	0.1 ± 0.0	14.4 ± 0.9	69
	1966–1975	4.0 ± 0.2	9.9 ± 0.7	5.6 ± 0.4	0.9 ± 0.3	16.4 ± 1.1	64
	1976–1985	4.4 ± 0.2	11.1 ± 1.1	6.4 ± 0.4	3.1 ± 0.3	20.6 ± 1.2	63
	1986–1995	4.4 ± 0.1	11.6 ± 1.1	6.1 ± 0.6	5.2 ± 0.6	22.9 ± 0.7	65
	1996–2005	4.2 ± 0.3	10.8 ± 0.7	5.7 ± 0.7	7.0 ± 0.4	23.5 ±1.5	66
	2006-2015	4.0 ± 0.0	9.8 ± 0.6	6.3 ± 0.4	6.7 ± 0.2	22.8 ± 0.7	61
	Average	4.1	10.5	5.8	3.8	20.1	65
Maize	1956–1965	2.1 ± 0.2	7.2 ± 0.7	0.3 ± 0.2	0.1 ± 0.0	7.6 ± 0.7	96
	1966–1975	2.6 ± 0.1	8.9 ± 0.6	0.4 ± 0.2	0.6 ± 0.2	9.9 ± 0.7	96
	1976–1985	3.3 ± 0.1	10.9 ± 0.5	0.5 ± 0.2	2.3 ± 0.3	13.7 ± 0.7	96
	1986–1995	3.3 ± 0.1	10.8 ± 0.4	0.6 ± 0.3	3.9 ± 0.5	15.3 ± 0.8	95
	1996–2005	3.9 ± 0.1	12.5 ± 0.9	0.7 ± 0.4	6.5 ± 0.2	19.7 ± 0.9	95
	2006–2015	4.8 ± 0.1	15.9 ± 0.6	0.6 ± 0.3	8.0 ± 0.1	24.5 ± 0.6	96
	Average	3.3	11.0	0.5	3.6	15.1	

* Data are the mean ± SD for every decade. GWC refers to the green water coefficient.

To further analyze the structure of WF, the green water coefficient is defined as the ratio of green WF to the consumptive green and blue WF [45]. As shown in Table 4, the green water accounts for 65% and 96% of the consumptive WF for wheat and maize, respectively. For maize, 96% of the consumptive water comes from green water, because most parts of the HRB are planted with summer maize, which mainly grows in the flood season (June to September), with 70–85% of the annual rainfall. The green water coefficients estimated in this study are very close to the previous studies by Mekonnen and Hoekstra (2010) [16] and Liu et al. [45] (80% for all crops).

The total WF has different temporal variation trends for wheat and maize. For wheat, it increased (by 64%) from 1956 to 1997 and then decreased (by 3%) following the changing trends of planting areas. For maize, it continually increased (up to 144%) over the study period due to the continual increase of planting areas. For both wheat and maize, the grey water increased before 2001 due to the increased application of the nitrogen fertilizer. The growth rate of the nitrogen fertilizer application was faster than the growth rate of the production, which reversed after 2002.

3.2. Unit Water Footprint of Wheat and Maize in the HRB

The unit WF refers to WF per ton of crop production, which is the converse of the crop water productivity, and can reflect the water use efficiency of crops [45]. Lower unit WF implies higher water use efficiency. Wheat (1580 m^3 t^{-1}) has a larger unit WF than maize (1275 m^3 t^{-1}), on average. In 2006–2015, the unit WF for wheat and maize production was 1022 m^3 t^{-1} and 934 m^3 t^{-1}, respectively.

The unit WF for both wheat and maize have exponentially decreasing trends along with the increasing production, indicating that water use efficiency has improved (Figures 2 and 3). The yield increased significantly due to the agricultural technology development, such as the large application of fertilizer, and innovation in agriculture management practices, such as the household contract responsibility system in the 1980s across China, which raised farmer's enthusiasm and increased the yield.

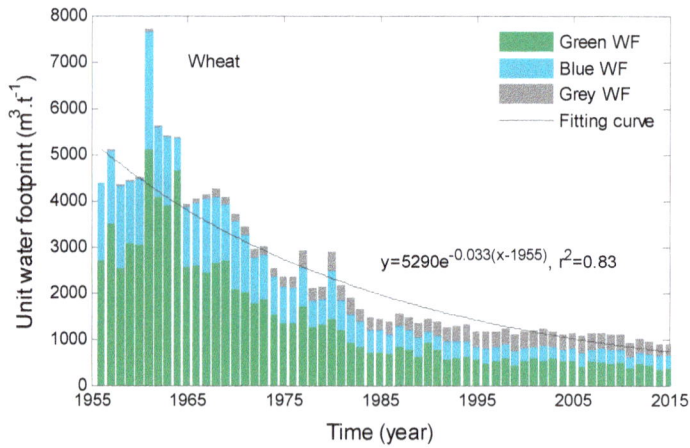

Figure 2. Unit WF and historical trend for wheat in the HRB over the period 1956–2015.

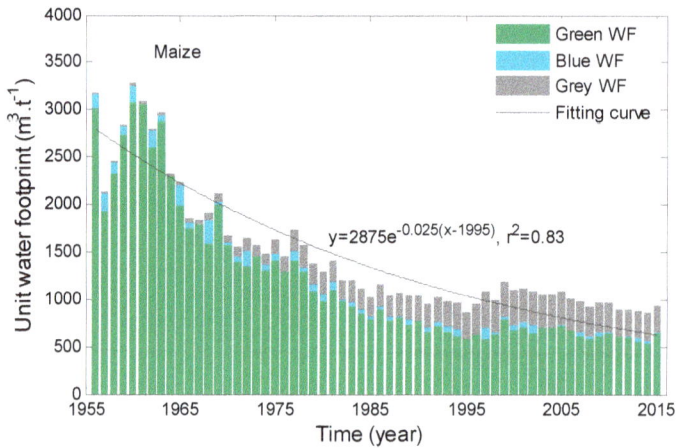

Figure 3. Unit WF and historical trend for maize in the HRB over the period 1956–2015.

The temporal variation trends of unit WF are fitted well by an exponential function with a non-linear least square method. It can be described as $y = 5290e^{-0.033(x-1955)}$ ($R^2 = 0.83$) (Figure 2) and $y = 2784e^{-0.026(x-1955)}$ ($R^2 = 0.79$) (Figure 3) for wheat and maize, respectively. The unit WF value was significantly high in 1961, with a value of 7533 m^3 t^{-1}, and that was because China experienced severe drought; at that time the production (1.6 million tons) was nearly half of the national average. In addition, in 1960 maize production decreased largely due to the severe drought, which resulted in a larger WF for maize (3254 m^3 t^{-1}).

The grey WF for unit wheat production increased significantly from 26 m^3 t^{-1} to 366 m^3 t^{-1} over the period 1956–2001 due to the increasing application of fertilizer (from 2 kg ha^{-1} to 170 kg ha^{-1}) and then decreased 30% (276 m^3 t^{-1}) in 2015, mainly because the yield increased while the fertilizer application did not change much. For maize, it increased from 17 m^3 t^{-1} to 384 m^3 t^{-1} from 1956 to 1997, and then decreased 25% (276 m^3 t^{-1}) in 2015. In 2006–2015, the grey WF was 302 m^3 t^{-1} and 304 m^3 t^{-1} for unit wheat and maize production, respectively. It was 45% and 48% larger, respectively, than the world average estimated by Mekonnen and Hoekstra [17] (207 m^3 t^{-1}). This indicates that agricultural water pollution is more severe than in other regions in the world, so the application of fertilizer should be reduced to assimilate the agricultural water pollution in the HRB.

The unit WF can be described and fitted well with a power function of crop yield (Figure 4). It is $y = 4045x^{-0.816}$ ($R^2 = 0.99$) and $y = 3384x^{-0.795}$ ($R^2 = 0.98$) for wheat and maize, respectively. A similar relationship was studied by Mekonnen and Hoekstra. [17], who argued that the trend between unit WF and the yield of cereals follows a logarithmic function. This indicated that the WF of crops is largely influenced by agricultural management rather than by climate conditions. Crop variety improvement, mechanization technologies, the rational combination of irrigation and fertilizer, and the change from family-oriented to farm management could increase the yields of wheat and maize. Further, improving crop production is an effective way to reduce the unit WF and improve water use efficiency. This can also be used to estimate the unit WF in the HRB when lacking information or to estimate the crop water use in the future.

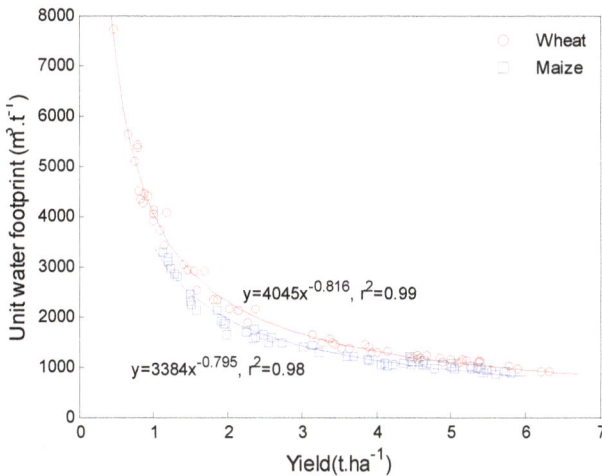

Figure 4. The relationship of unit WF and yield of wheat and maize in the HRB over the period 1956–2015.

3.3. Water Footprint Allocation among Administrative Units

Reasonable allocation of water resources within a river basin can reduce the competition for limited water resources among different regions and alleviate the intensified situation of water scarcity. The WF at the province (or municipality, which is the basic administrative district within the HRB) level was analyzed in the period of 2011–2015, and the results are shown in Table 5. Note that only the region located within the basin is calculated for each province or municipality.

The sum of the WF of wheat and maize is 47.39 billion m^3, which is much more than the water withdrawal of agriculture (24.76 Gm^3 year^{-1}). This is because the water withdrawal of agriculture excluded the green and grey water. The blue water of wheat and maize accounts for 28% of the total water withdrawal of agriculture. The total WF is 22.11 and 25.28 Gm^3 year^{-1} for wheat and maize, respectively. The largest WF for wheat was found in Hebei Province, with a value of 13.12 Gm^3 year^{-1} (43% green, 26% blue, and 31% grey), which accounts for 59% of the total WF of wheat in the basin. This is because Hebei has the largest arable land and crop area. The planting area of Hebei Province (2.4 Mha) occupies 60% of the HRB. The order for wheat WF is Hebei (59%) > Shandong (19%) > Henan (12%) > Shanxi (5%) > Tianjin (3%) > Beijing (2%) (Figure 5). For maize, the largest WF is also found in Hebei province, with a value of 15.93 Gm^3 year^{-1} (65% green, 1% blue, and 34% grey), which accounts for 63% of the total WF of maize in the HRB. The order for maize WF is Hebei (63%) > Shandong

(13%) > Shanxi (10%) > Henan (6%) > Tianjin (5%) > Beijing (3%) (Figure 5), which is slightly different for Henan and Shanxi. The proportion of WF for maize in Henan (6%) is much smaller than that of wheat (12%) because, in Henan, wheat has a larger planting area (480 kha) than maize (290 kha) within the HRB. With wheat and maize WF combined together, Hebei province has the largest WF, which accounts for 61% of the total, followed by Shanxi (16%). Beijing and Tianjin account for 2% and 4% of the total WF, respectively.

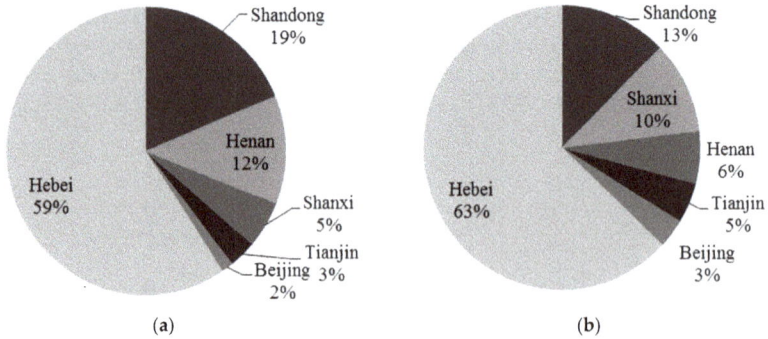

Figure 5. Allocation of WF between administrative districts within the HRB: (a) wheat; and (b) maize.

Table 5. WF for different administrative districts, 2011–2015.

Crops	Province	WF (Gm³ year⁻¹)				Yield (t ha⁻¹)	Unit WF (m³ t⁻¹)			
		Green	Blue	Grey	Total		Green	Blue	Grey	Total
Wheat	Beijing	0.07	0.14	0.09	0.30	5168	348	691	484	1523
	Tianjin	0.24	0.23	0.23	0.70	5211	458	440	442	1340
	Hebei	5.59	3.46	4.07	13.12	5807	408	252	298	958
	Shanxi	0.63	0.27	0.26	1.16	3866	627	268	256	1151
	Henan	1.12	0.73	0.82	2.67	6697	349	225	254	828
	Shandong	1.70	1.41	1.05	4.16	6630	347	289	215	851
	Total	9.35	6.24	6.52	22.11					
Maize	Beijing	0.41	0.01	0.41	0.83	6291	585	27	400	1012
	Tianjin	0.65	0.09	0.44	1.18	5213	652	89	440	1181
	Hebei	10.32	0.23	5.38	15.93	5341	620	13	323	956
	Shanxi	2.02	0.01	0.62	2.65	5685	561	4	173	738
	Henan	0.92	0.09	0.49	1.50	6088	526	50	279	854
	Shandong	2.18	0.13	0.88	3.19	7142	497	30	200	727
	Total	16.50	0.56	8.22	25.28					

The unit WF, which reflects the water use efficiency, was significantly different between different administrative districts in 2011–2015. The largest unit WF was found in Beijing and Tianjin for wheat and maize, respectively. The blue WF in Beijing (691 m³ t⁻¹) was the largest for unit wheat production due to the large amount of irrigation (428 mm year⁻¹). Water-saving irrigation systems could be used to reduce the amount of irrigation water and improve the efficiency in the future.

3.4. Spatial Distribution of Unit Water Footprint

The unit WF of wheat and maize varies largely across regions, as shown in Figures 6 and 7. For unit wheat production, Tianjin has the largest WF on average (1956–2015). The second largest group is the region surrounding Tianjin, containing Beijing, Tangshan, and Huanghua. The remaining areas, containing Yushe, Shijiazhuang, Anyang, and Dezhou, have relatively smaller WF, because these areas are mainly grain-producing areas, especially Anyang and Dezhou, which have larger yields than the others due to efficient and centralized management.

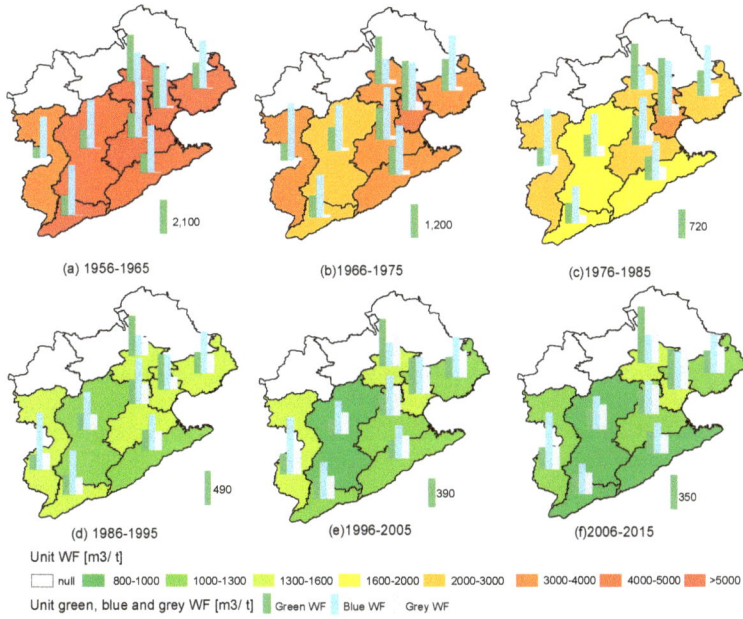

Figure 6. Unit WF of wheat in the HRB over 1956–2015 (m³ t⁻¹).

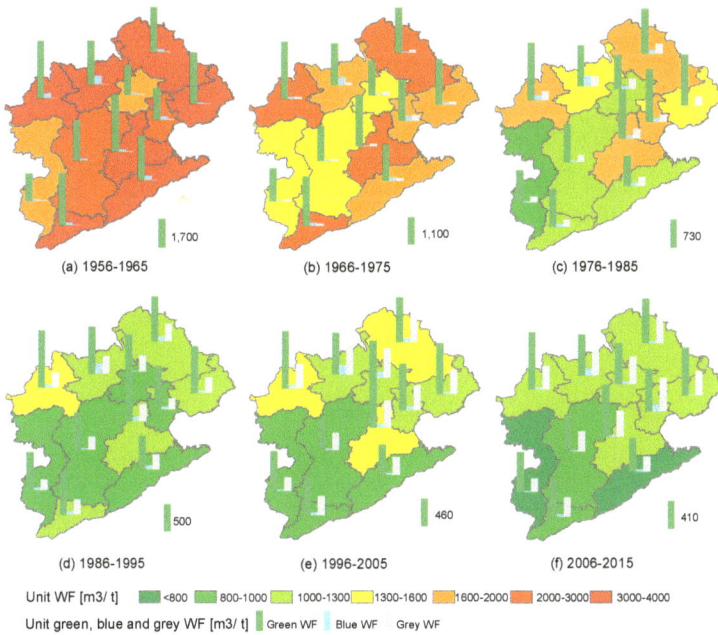

Figure 7. Unit WF of maize in the HRB over 1956–2015 (m³ t⁻¹).

The spatial distribution of unit WF for maize is different from wheat. First, the unit WF of spring maize planted in Northwestern HRB (Zhangjiakou, Chengde, and Datong) is larger than for summer maize in other regions. This is because the growing period of spring maize is longer than that of summer maize. In the summer maize-planted areas, Huanghua has the largest unit WF (1860 m^3 t^{-1}) on average. This is because the yield of maize in Huanghua is much smaller than in other regions. In contrast, relatively lower unit WF is found in the south of the HRB (e.g., Dezhou) because of the relatively higher yield.

The variation in space could be attributed to the different climate conditions, geography, soil properties, and management practices among these regions. Tianjin has the largest unit WF because of the lower yield and large application of fertilizer. Many factors might cause a lower crop yield, such as soil physiochemical properties and management practices. These factors should be further studied to improve crop productivity.

3.5. Water Footprint under Different Irrigation Conditions

Irrigation is a key factor affecting the accuracy of WF assessment. In this study, a comparison was made of the WF of wheat and maize under rain-fed and irrigated conditions with sufficient and insufficient water. The crops suffered water stress under conditions of rain-fed and insufficient irrigation, and the yields were simulated by a yield reduction fraction caused by the reduction of crop evapotranspiration proposed by Doorenbos and Kassam [54].

Maize production per hectare requires more water than wheat under all conditions. This is because maize has a much shorter growing period than wheat. The yield of wheat under sufficient water conditions is 93% larger than for rain-fed. This indicates that irrigation plays a vital role in the wheat growing period. For maize, the yield is relatively good even without irrigation, because it mainly grows in the rainy season and there is sufficient water during the crop-growing period.

For both wheat and maize, the unit WF under sufficient irrigation conditions was lower than rain-fed and crops under insufficient irrigation conditions (Table 6). This is because irrigation can significantly improve the crop yield, though more blue water is required. The result is close to other studies [17].

Table 6. Comparison of WF under different irrigation conditions, 2006–2015.

Crops	Irrigation Conditions	WF (m^3 ha^{-1})	Yield (t ha^{-1})	Unit WF (m^3 t^{-1})
Wheat	Rain-fed	4029	4.21	957
	Sufficient irrigation	7427	8.11	916
	Insufficient irrigation	6096	6.58	926
Maize	Rain-fed	4952	5.5	900
	Sufficient irrigation	5198	5.85	889
	Insufficient irrigation	5116	5.73	893

4. Discussion

4.1. Comparison of Unit Water Footprint with Other Studies

Results of unit WF in this study are compared to studies by Mekonnen and Hoekstra [17] during 1996–2005 and Cao et al. [21] in 2010 who estimated unit WF in each province of China, and their results were calculated in the HRB (Table 7). This can illustrate the discrepancies between large-scale datasets and what happens on a more local level [55].

Table 7. The comparison of unit WF with other studies in the HRB.

Study	Period	Crops	Unit WF in the HRB ($m^3\ t^{-1}$)			
			Green	Blue	Grey	Total
Mekonnen and Hoekstra [17]	1996–2005	Wheat	650	608	436	1694
This study			536	281	347	1164
Cao et al. [21]	2010	Wheat	597	329	-	926
This study			491	287	315	1093
Mekonnen and Hoekstra [17]	1996–2005	Maize	791	115	293	1199
This study			676	41	354	1071

For wheat, the unit WF in this study is 31% lower than that estimated by Mekonnen and Hoekstra [17] in 1996–2005 in the HRB. The green, blue, and grey WF in this study are 18%, 54%, and 20% lower than their estimates, respectively. That might be due to the different yield and irrigation areas. The yield estimated by Mekonnon and Hoekstra [17] was simulated by water stress proposed by Doorenbos and Kassam [54]. The maximum yield values were obtained by multiplying the corresponding national average yield values (4.0 t ha^{-1} in China in 1996–2005) by a factor of 1.2. The calculated yields were scaled to fit the national average FAO yield data [56] while, in this study, the yield of each province (4.8 t ha^{-1} in the HRB in 1996–2005) was obtained from the statistical data [42], which is 20% larger than the national average. The difference of yield might explain the discrepancy of green and grey WF, while the discrepancy of blue WF might be explained by different irrigation areas. Mekonnon and Hoekstra [17] used irrigation areas from the MICRA2000 grid database [57], which is larger than the irrigation areas from the statistical data [42] in this study.

To compare with Cao et al. [21], this study estimated the unit consumptive WF (green WF plus blue WF) for wheat in 2010, which was 778 $m^3\ t^{-1}$, 16% lower than Cao et al. [21]. The green and blue WF are 18% and 13% lower than their estimates, respectively. In their study, the yield was obtained from the statistical yearbook of each province, which was the same as this study. The green WF being higher than this study could mainly be due to different calculation methods. Cao et al. [21] estimated green WF by effective precipitation on a 10-day time step, while in this study a daily water balance model was used, referring to Allan et al. [47]. The blue WF is higher than this study, because in their study it is the sum of the evaporated surface water from the water intake point to the field and the field evapotranspiration, while, in this study, only the field evapotranspiration is taken into account.

For maize, the unit WF in this study is 11% lower than the estimate by Mekonnen and Hoekstra [17] in 1996–2005 in the HRB. This can be explained by the discrepancy of green WF. Green WF takes a large proportion of evapotranspiration (the green water coefficient is 95% in this study) in the maize growing period. The green WF estimated by Mekonnen and Hoekstra [17] in the HRB was 15% larger than this study; this could mainly be due to the different growing period, since the yield estimated by Mekonnen and Hoekstra [17] is close to this study. The growing period estimated in this study (112 days) is based on the observed values provided by the China Meteorological Administration [51], and that estimated by Mekonnen and Hoekstra [17] was based on the FAO [56], which is at least 125 days for the growing period of maize. The longer growing period could result in higher evapotranspiration. The grey WF estimated by Mekonnen and Hoekstra [17] is lower than this study, which could be because that study assumed that crops receive the same amount of nitrogen fertilizer per hectare in all grid cells in a country, while in this study the application amount of nitrogen fertilizer of each province published by CAM [42] was used.

4.2. Management of Green and Blue Water

Green water accounts for a large proportion of consumptive WF, which is 65% and 96% for wheat and maize, respectively. This indicates that climate (precipitation) contributes more to WF than human activities (irrigation) in the HRB. Attention should be paid to the utilization of rain water in water-stressed arid and semi-arid environments in the future, which is an important way to alleviate the stress from water scarcity. For example, water ponds for rainwater harvesting and utilization could be built to promote rainwater utilization. Terraces could be built in mountainous areas of the Western HRB to improve rainwater use efficiency. Additionally, planting and harvesting dates should be carefully planned for utmost utilization of rainwater, though the precipitation is highly variable.

Blue water plays a vital role in the growing period of wheat and maize in the HRB [32,58]. Irrigation can improve crop production to feed more people, and could also improve water use efficiency (irrigated crops have smaller unit water footprint). Hence, we can build some water supply facilities in some water scarce regions for irrigation. Additionally, we can develop water-saving technologies, like drip irrigation, which can improve blue water use efficiency [59]. Considering that using blue water costs much more than using green water, especially in many water-scarce regions, we should comprehensively consider the degree of water shortage, water costs, economic levels, and the requirement of food in future agricultural water management.

4.3. Limitations in This Study

There were a number of limitations and uncertainties during the assessment of WF in this study. First, the limited data influences the accuracy of WF accounting and spatial distribution. In this study, only 11 weather stations were selected for WF assessment. Spatial and temporal climate variability could not be clearly shown over such large areas. The proportion of irrigated crops is assumed to be the same in each province (or municipality). The soil conditions are assumed to be uniform and the effect of terrain is not taken into account. Second, the same planting dates and length of growing period are assumed during the study period of 60 years. The temporal differences of planting dates and the length of growing period under climate variability were not taken into account, which could affect the crop WF calculation. Third, the irrigation data were chosen based on a series of norms of the water intake in each province. Though it is close to the real irrigation situation, there are still some discrepancies in different hydrological years. In addition, the crop parameters are assumed to be the same for both irrigated and rain-fed crops, following Mekonnen and Hoekstra [17]. However, in rain-fed crops, roots are deeper than those in irrigated crops, and that affects the soil water balance in the root zone. More accurate irrigation amounts and crop parameters for irrigated and rain-fed crops should be used in future studies to enhance the accuracy of agricultural WF assessment.

5. Summary and Conclusions

The spatial and temporal characteristics of WF of wheat and maize are analyzed in the period 1956–2015 in the HRB. The major portion of total WF comes from green water, especially for maize production, indicating that we should pay more attention to the management of rain water in the future. In all, 19% and 24% of total WF are required to eliminate agricultural water pollution for wheat and maize, respectively. Those are much higher percentages than the world average, indicating that fertilizer use efficiency should be improved in the future. The total WF of wheat and maize varied largely in 1956–2015, mainly following the changing planting areas.

Per ton of crop, wheat ($1581 \ m^3 \ t^{-1}$) required more water than maize ($1275 \ m^3 \ t^{-1}$). The unit WF of wheat and maize both have exponentially decreasing trends due to increasing production, indicating that water use efficiency has improved. However, increased production was mainly caused by increased fertilizer use, resulting in increased grey water.

Considering the total WF of crop production allocation based on administrative districts (only the area located in the HRB was considered), Hebei Province has the largest WF of both wheat and maize, which accounts for 61% of the total WF, followed by Shanxi (16%), and Beijing and Tianjin account for 2% and 4% of the total WF, respectively. The allocation of WF between administrative districts within the HRB provides an effective way to reduce the conflict among different regions over competition for limited water resources.

The WF varies largely in space. Spring maize has relatively larger unit WF than summer maize. Tianjin has the largest unit WF for wheat because of the poor yield and large application of fertilizer. Other factors might also cause poor crop yield, such as soil physicochemical properties and management practices. These factors should be further studied to improve water productivity. Spring maize has relatively larger unit WF than summer maize due to the longer growing period.

A comparison was made of the WF of wheat and maize under different irrigation conditions. Maize production per hectare requires more water than wheat for all conditions. For both wheat and maize, the unit WF under sufficient irrigation conditions is lower than rain-fed and crops under insufficient irrigation conditions, though much blue water is consumed.

Overall, this study assessed the WF of wheat and maize in HRB of Northern China over the period 1956–2015. The WF analysis for wheat and maize in the HRB shows very large spatial and temporal variations. Analyzing the spatial and temporal characteristics of WF is helpful for basin agencies to make proper water management decisions to improve agricultural water use efficiency and control diffuse agricultural water pollution.

In future studies, in order to improve the accuracy of WF assessment, high-resolution climate datasets and detailed soil datasets should be considered. Meanwhile, in order to obtain more accurate WF and calibrate and validate the model parameters, the soil water conditions and crop growing status in the HRB should be monitored by field experiments.

Author Contributions: Y.H. and D.J. conceived, designed, and drafted the manuscript; D.J. and H.H. planned and designed the methodology; L.Z., S.S., and J.-M.S.-P. revised the manuscript; Y.H. and C.W. guided and supervised the whole process; and all authors read and approved the final manuscript.

Funding: This research received no external funding.

Acknowledgments: This study was funded by the National Key Research and Development Program of China (2016YFC0401402), the National Natural Science Foundation of China (51279063; 51679089; 51609084; 51709107), and the Innovation Foundation of North China University of Water Resources and Electric Power for PhD Graduates. The authors also thank the editor and two anonymous reviewers for their insightful comments and constructive suggestions.

Conflicts of Interest: No conflict of interest exists in the submission of this manuscript, and the manuscript is approved by all authors for publication.

References

1. Oki, T.; Kanae, S. Global hydrological cycles and world water resources. *Science* **2006**, *313*, 1068–1072. [CrossRef] [PubMed]
2. Vörösmarty, C.J.; Green, P.; Salisbury, J.; Lammers, R.B. Global water resources: Vulnerability from climate change and population growth. *Science* **2000**, *289*, 284–288. [CrossRef] [PubMed]
3. Vörösmarty, C.J.; McIntyre, P.; Gessner, M.O.; Dudgeon, D.; Pusevich, A.; Green, P.; Glidden, S.; Bunn, S.E.; Sullivan, C.A.; Liermann, C.R. Global threats to human water security and river biodiversity. *Nature* **2010**, *467*, 555–561. [CrossRef] [PubMed]
4. FAO. *Coping with Water Scarcity: An Action Framework for Agriculture and Food Security*; Food and Agriculture Organization of United States (FAO): Rome, Italy, 2012.
5. Pimentel, D.; Berger, B.; Filiberto, D.; Newton, M.; Wolfe, B.; Karabinakis, E.; Clark, S.; Poon, E.; Abbett, E.; Nandagopal, S. Water resources: Agricultural and environmental issues. *BioScience* **2004**, *54*, 909–918. [CrossRef]
6. Falkenmark, M.; Rockström, M. *Balancing Water for Man and Nature: The New Approach to Ecohydrology*; Routledge: London, UK, 2004.
7. Falkenmark, M.; Rockström, J. The new blue and green water paradigm: Breaking new ground for water resources planning and management. *J. Water Resour. Plan. Manag.* **2006**, *132*, 129–132. [CrossRef]
8. Hoekstra, A.Y. Virtual Water Trade. In *Proceedings of the International Expert Meeting on Virtual Water Trade, Delft, The Netherlands, 12–13 December 2002*; Value of Water Research Report Series No. 12; UNESCO-IHE: Delft, The Netherlands, 2003.

9. Chenoweth, J.; Hadjikakou, M.; Zoumides, C. Review article: Quantifying the human impact on water resources: A critical review of the water footprint concept. *Hydrol. Earth Syst. Sci. Discuss.* **2013**, *10*, 9389–9433. [CrossRef]

10. Hoekstra, A.Y.; Mekonnen, M.M. The water footprint of humanity. *Proc. Natl. Acad. Sci. USA* **2012**, *109*, 3233–3237. [CrossRef] [PubMed]

11. Hoekstra, A.Y.; Chapagain, A.K. Water footprint of nations: Water use by people as a function of their consumption pattern. *Water Resour. Manag.* **2007**, *21*, 35–38. [CrossRef]

12. Hoekstra, A.Y. Water footprint assessment: Evolvement of a new research field. *Water Resour. Manag.* **2017**, *31*, 3061–3081. [CrossRef]

13. Hoekstra, A.Y. Human appropriation of natural capital: Comparing ecological footprint and water footprint analysis. *Ecol. Econ.* **2007**, *68*, 1963–1974. [CrossRef]

14. Allan, J.A. Overall perspectives on countries and regions. In *Water in the Arab World: Perspectives and Prognoses*; Rogers, P., Lydon, P., Eds.; Harvard University Press: Cambridge, MA, USA, 1994; pp. 65–100.

15. Allan, J.A. Virtual water: A strategic resources global solutions to regional deficits. *Ground Water* **1998**, *36*, 545–546. [CrossRef]

16. Mekonnen, M.M.; Hoekstra, A.Y. A global and high-resolution assessment of the green, blue and grey water footprint of wheat. *Hydrol. Earth Syst. Sci.* **2010**, *14*, 1259–1276. [CrossRef]

17. Mekonnen, M.M.; Hoekstra, A.Y. The green, blue and grey water footprint of crops and derived crop products. *Hydrol. Earth Syst. Sci.* **2011**, *15*, 1577–1600. [CrossRef]

18. Siebert, S.; Döll, P. Quantifying blue and green virtual water contents in global crop production as well as potential production losses without irrigation. *J. Hydrol.* **2010**, *384*, 198–217. [CrossRef]

19. Zhuo, L.; Mekonnen, M.M.; Hoekstra, A.Y. Benchmark levels for the consumptive water footprint of crop production for different environmental conditions: A case study for winter wheat in China. *Hydrol. Earth Syst. Sci.* **2016**, *20*, 4547–4559. [CrossRef]

20. Zhuo, L.; Mekonnen, M.M.; Hoekstra, A.Y. Consumptive water footprint and virtual water trade scenarios for China with a focus on crop production, consumption and trade. *Environ. Int.* **2016**, *94*, 211–223. [CrossRef] [PubMed]

21. Cao, C.X.; Wu, P.T.; Wang, Y.B.; Zhao, X.N. Assessing blue and green water utilization in wheat production of China from the perspectives of water footprint and total water use. *Hydrol. Earth Syst. Sci.* **2014**, *18*, 3165–3178. [CrossRef]

22. Zoumides, C.; Bruggeman, A.; Hadjikakou, M.; Zachariadis, T. Policy-relevant indicators for semi-arid nations: The water footprint of crop production and supply utilization of Cyprus. *Ecol. Indic.* **2014**, *43*, 205–214. [CrossRef]

23. Schyns, J.F.; Hoekstra, A.Y. The added value of water footprint assessment for national water policy: A case study for Morocco. *PLoS ONE* **2014**, *9*, e99705. [CrossRef] [PubMed]

24. Bulsink, F.; Hoekstra, A.Y.; Booij, M.J. The water footprint of Indonesian provinces related to the consumption of crop products. *Hydrol. Earth Syst. Sci.* **2010**, *14*, 119–128. [CrossRef]

25. Duan, P.; Qin, L.; Wang, Y.; He, H. Spatiotemporal correlations between water footprint and agricultural inputs: A case study of maize production in Northwest China. *Water* **2015**, *7*, 4026–4040. [CrossRef]

26. Gobin, A.; Kersebaum, K.C.; Eitzinger, J.; Trnka, M.; Hlavinka, P.; Takáč, J.; Kroes, J.; Ventrella, D.; Marta, A.D.; Deelstra, J.; et al. Variability in the water footprint of arable crop production across European regions. *Water* **2017**, *9*, 93. [CrossRef]

27. Aldaya, M.M.; Llamas, M.R. *Water Footprint Analysis for the Guadiana River Basin*; Research Report Series No. 35; UNESCO-IHE Institute for Water Education: Delft, The Netherlands, 2008.

28. Yin, J.; Wang, H.; Cai, Y. Water footprint calculation on the basis of input-output analysis and a biproportional algorithm: A case study for the Yellow River basin, China. *Water* **2016**, *8*, 363. [CrossRef]

29. Zeng, Z.; Liu, J.; Koeneman, P.H.; Zarate, E.; Hoekstra, A.Y. Assessing water footprint at river basin level: A case study for the Heihe River Basin in northwest China. *Hydrol. Earth Syst. Sci.* **2012**, *16*, 2771–2781. [CrossRef]

30. Zang, C.F.; Liu, J.; van der Velde, M.; Kraxner, F. Assessment of spatial and temporal patterns of green and blue water flows under natural conditions in inland river basins in Northwest China. *Hydrol. Earth Syst. Sci.* **2012**, *16*, 2859–2870. [CrossRef]

31. Zhuo, L.; Mekonnen, M.M.; Hoekstra, A.Y.; Wada, Y. Inter- and intra-annual variation of water footprint of crops and blue water scarcity in the Yellow River Basin (1961–2009). *Adv. Water Resour.* **2016**, *87*, 29–41. [CrossRef]

32. Sun, S.K.; Wu, P.T.; Wang, Y.B.; Zhao, X.N.; Liu, J.; Zhang, X.H. The impacts of inter-annual climate variability and agricultural inputs on water footprint of crop production in an irrigation district of China. *Sci. Total Environ.* **2013**, *444*, 498–507. [CrossRef] [PubMed]

33. Chukalla, A.D.; Krol, M.S.; Hoekstra, A.V. Green and blue water footprint reduction in irrigated agriculture: Effect of irrigation techniques, irrigation strategies and mulching. *Hydrol. Earth Syst. Sci.* **2015**, *19*, 4877–4891. [CrossRef]

34. Shang, S.H.; Mao, X.M. Application of simulation based optimization model for winter wheat irrigation scheduling in North China. *Agric. Water Manag.* **2006**, *85*, 314–322. [CrossRef]

35. National Bureau of Statistics of the People's Republic of China. *Population Census of China in 2010*; National Bureau of Statistics of the People's Republic of China: Beijing, China, 2011.

36. Bao, Z.X.; Zhang, J.Y.; Wang, G.Q.; Fu, G.B.; He, R.M.; Yan, X.L.; Jin, J.L.; Liu, Y.L.; Zhang, A.J. Attribution for decreasing stream flow of the Haihe River Basin, northern China: Climate variability or human activities? *J. Hydrol.* **2012**, *460–461*, 117–129. [CrossRef]

37. Sun, S.K.; Wang, Y.B.; Wu, P.T.; Zhao, X.N. Spatial variability and attribution analysis of water footprint of wheat in China. *Trans. Chin. Soc. Agric. Eng.* **2015**, *31*, 142–148. (In Chinese)

38. CMWR (Ministry of Water Resources of China). *China Water Resources Bulletin 2011*; China Water and Power Press: Beijing, China, 2012.

39. Zou, J.; Xie, Z.H.; Zhan, C.S.; Qin, P.H.; Sun, Q.; Jia, B.H.; Xia, J. Effects of anthropogenic groundwater exploitation on land surface processes: A case study of the Haihe River Basin, northern China. *J. Hydrol.* **2015**, *524*, 625–641. [CrossRef]

40. Haihe River Water Conservancy Commission, Ministry of Water Resources. Groundwater Bulletin in Haihe River Basin. September 2013. Available online: http://www.hwcc.gov.cn/ (accessed on 15 March 2018).

41. FAO; IIASA; ISRIC; ISS-CAS; JRC. *Harmonized World Soil Database (Version 1.1)*; FAO: Rome, Italy; IIASA: Laxenburg, Austria, 2009.

42. CAM (China Agriculture Ministry). The Department of Plantation Management of China Agriculture Ministry. Available online: http://zzys.agri.gov.cn/nongqing.aspx (accessed on 23 October 2017).

43. Ministry of Water Resources of China; Department of Water Resources and Water Resources Management Center. *Norm of Water Intake for Beijing, Tianjin, Hebei, Shanxi, Shandong and Henan of China (DB13/T 1161.1–2009; DB14/T 1049.1–2015; DB41/T 958–2014; and DB12/T 159.01–2003)*; Ministry of Water Resources of China: Beijing, China, 2013.

44. Hoekstra, A.Y.; Chapagain, A.K.; Aldaya, M.M.; Mekonne, M.M. *The Water Footprint Manual: Setting the Global Standard*; Earthscan: London, UK, 2011.

45. Liu, J.; Zehnder, A.J.B.; Yang, H. Global consumptive water use for crop production: The importance of green water and virtual water. *Water Resour. Res.* **2009**, *45*, W05428. [CrossRef]

46. Chapagain, A.K.; Hoekstra, A.Y.; Savenije, H.H.G.; Gautam, R. The water footprint of cotton consumption: An assessment of the impact of worldwide consumption of cotton products on the water resources in the cottern producing countries. *Ecol. Econ.* **2006**, *60*, 186–203. [CrossRef]

47. Allen, R.G.; Pereira, L.S.; Raes, D.; Smith, M. *Crop Evapotranspiration: Guidelines for Computing Crop Water Requirements*; FAO Drainage and Irrigation Paper 56; Food and Agriculture Organization: Rome, Italy, 1998.

48. Zhuo, L.; Mekonnen, M.M.; Hoekstra, A.Y. Sensitivity and uncertainty in crop water footprint accounting: A case study for the Yellow River Basin. *Hydrol. Earth Syst. Sci.* **2014**, *18*, 2219–2234. [CrossRef]

49. Chen, Y.; Guo, G.; Wang, G.; Kang, S.; Luo, H.; Zhang, D. *Main Crop Water Requirement and Irrigation of China*; Hydraulic and Electric Press: Beijing, China, 1995.

50. Kang, S.Z.; Gu, B.J.; Du, T.S.; Zhang, J.H. Crop coefficient and ratio of transpiration to evapotranspiration of winter wheat and maize in a semi-humid region. *Agric. Water Manag.* **2003**, *59*, 239–254. [CrossRef]

51. CMA (China Meteorological Administration). China Meteorological Data Sharing Service System, Beijing, China. Available online: http://data.cma.cn/ (accessed on 29 November 2016).

52. NBSC (National Bureau of Statistics of China). *China Statistical Yearbook (1983–2015)*; China Statistical Press: Beijing, China, 1984–2016.

53. Zhou, X.; Pang, J.; Wu, Q.; Huang, H.; Cai, Y.; Chen, G. *China First Census for Water. Data Compilation of National Water Conservancy*; China Water and Power Press: Beijing, China, 2016.

54. Doorenbos, J.; Kassam, A.H. *Yield Response to Water*; Drainage and Irrigation Paper 33, FAO Irrigation and Drainage Paper N°33; FAO: Rome, Italy, 1979.

55. Zoumides, C.; Bruggeman, A.; Zachariadis, T. Global versus Local Crop Water Footprints: The Case of Cyprus. In *Solving the Water Crisis: Common Action toward a Sustainable Water Footprint*; UNESCO-IHE Institute for Water Education: Delft, The Netherlands, 2012; p. 7.

56. FAO. FAOSTAT On-Line Database, Food and Agriculture Organization of United States, Rome. 2008. Available online: http://faostat.fao.org (accessed on 10 October 2008).

57. Portmann, F.T.; Siebert, S.; Döll, P. MIRCA2000—Global Monthly Irrigated and Rainfed Crop Areas around the Year 2000: A New High-Resolution Data Set for Agricultural and Hydrological Modeling. *Global Biogeochem. Cycles* **2010**, *24*, GB1011. Available online: http://www.geo.uni-frankfurt.de/ipg/ag/dl/forschung/MIRCA/data_download/index.html (accessed on 15 October 2009). [CrossRef]

58. Sun, S.K.; Wu, P.T.; Wang, Y.B.; Zhao, X.Y. Temporal Variability of water footprint for maize production: The case of Beijing from 1978 to 2008. *Water Resour. Manag.* **2013**, *27*, 2447–2463. [CrossRef]

59. Zwart, S.J.; Bastiaanssen, W.G.M. Review of measured crop water productivity values for irrigated wheat, rice, cotton and maize. *Agric. Water Manag.* **2004**, *69*, 115–133. [CrossRef]

water

MDPI

Article

Evaluating Water Use for Agricultural Intensification in Southern Amazonia Using the Water Footprint Sustainability Assessment

Michael J. Lathuillière [1,*], Michael T. Coe [2,3], Andrea Castanho [2], Jordan Graesser [4] and Mark S. Johnson [1,5]

[1]   Institute for Resources, Environment and Sustainability, University of British Columbia, Vancouver, BC V6T 1Z4, Canada; mark.johnson@ubc.ca
[2]   Woods Hole Research Center, Falmouth, MA 02540-1644, USA; mtcoe@whrc.org (M.T.C.); acastanho@whrc.org (A.C.)
[3]   Instituto de Pesquisa Ambiental da Amazônia, Brasília DF 71503-505, Brazil
[4]   Department of Earth and Environment, Boston University, Boston, MA 02215, USA; graesser@bu.edu
[5]   Department of Earth, Ocean and Atmospheric Sciences, University of British Columbia, Vancouver, BC V6T 1Z4, Canada
*   Correspondence: mlathuilliere@alumni.ubc.ca

Received: 13 February 2018; Accepted: 19 March 2018; Published: 21 March 2018

Abstract: We performed a Water Footprint Sustainability Assessment (WFSA) in the Xingu Basin of Mato Grosso (XBMT), Brazil, with the objectives of (1) tracking blue (as surface water) and green water (as soil moisture regenerated by precipitation) consumption in recent years (2000, 2014); and (2) evaluating agricultural intensification options for future years (2030, 2050) considering the effects of deforestation and climate change on water availability in the basin. The agricultural sector was the largest consumer of water in the basin despite there being almost no irrigation of cropland or pastures. In addition to water use by crops and pasture grass, water consumption attributed to cattle production included evaporation from roughly 9463 ha of small farm reservoirs used to provide drinking water for cattle in 2014. The WFSA showed that while blue and green water consumptive uses were within sustainable limits in 2014, deforestation, cattle confinement, and the use of irrigation to increase cropping frequency could drive water use to unsustainable levels in the future. While land management policies and practices should strive for protection of the remaining natural vegetation, increased agricultural production will require reservoir and irrigation water management to reduce the potential threat of blue water scarcity in the dry season. In addition to providing general guidance for future water allocation decisions in the basin, our study offers an interpretation of blue and green water scarcities with changes in land use and climate in a rapidly evolving agricultural frontier.

Keywords: water footprint; water management; soybean; cattle; land use change; Amazon; Cerrado; Mato Grosso

1. Introduction

Southern Amazonia, Brazil, has experienced significant development since the 1990s, with agricultural production expanding rapidly through land use change in both the Amazon and Cerrado (or savanna) biomes [1]. Natural vegetation cover has been gradually replaced by pasture and soybean land use systems [2], often through a natural vegetation to pasture to cropland transition [3,4]. This increase in agricultural production has had important socio-economic and environmental implications. Socio-economic indicators suggest a growth in the tertiary sector up- and down-stream of soybean production, with evidence of local investment and financial returns [5]. At the same time, deforestation has been shown to alter local climate and water cycles, thereby pushing the Amazon towards a tipping point [6] that could significantly alter the biome. Changes to above and belowground carbon stocks have implications for global climate change [7], while land use change can affect the water cycle by increasing river discharge [8] and diminishing water vapor supply to the atmosphere with implications for regional precipitation [3,9,10]. Changes to the water cycle, in particular, affect economic activity through hydropower generation and agriculture [11–13], but can also affect aquatic and terrestrial ecosystems [14,15].

Agricultural expansion in the region has been followed by infrastructure development such as road networks [6], population growth and land use activities that trigger further deforestation. Between 1991 and 2010, the population of Mato Grosso increased from 2 to 3 million, while the animal population increased from 22 to 82 million, led mainly by cattle [16]. These increases put additional pressures on land use and local demand of natural resources, particularly water. Atmospheric feedbacks could negatively affect agricultural production when considering changes to regional climate and precipitation regimes [12], but could also trigger infrastructure investment in irrigation with additional effects on water withdrawals and feedbacks on water resources [15]. Therefore, feedbacks between agricultural production, land use change, and human and animal population growth need to be investigated in order to evaluate future development scenarios in Southern Amazonia.

This study aims to quantify these changes by carrying out a Water Footprint (WF) Sustainability Assessment (WFSA) [17] in the Xingu Basin of Mato Grosso (XBMT) located in Southern Amazonia, an area that has experienced the land use change dynamics described above. Since 2002, the WF has been increasingly used to quantify direct and indirect water use of production and consumption processes as a means to put these activities into a context of regional and global water resources, as well as potential environmental impacts [18–20]. In a WF context, water resources are typically separated into blue water (which represents the surface water and groundwater stocks), green water (which characterizes soil moisture stocks regenerated by precipitation [21]), and grey water (i.e., the amount of water required to dilute chemical or thermal pollution loads to ambient water quality standards [17]). When focusing on water quantity in a WFSA, the blue and green WF are compared to local water availability to derive local water scarcity as a step towards formulating a policy recommendation [17]. Many studies have applied WFSAs to derive blue water scarcity at a global scale (e.g., [22,23]), but only one study to date has attempted to quantify green water scarcity [24]. More studies using the concept of green water scarcity are thus needed to verify the full extent of WF assessments [17,25].

We build upon previous research results on the water cycle of XBMT to carry out a WFSA for the 2000-2015 period. We also evaluate scenarios for 2030-2031 and 2050-2051 with the objectives of formulating responses for water resources management based on past and future land and water use decisions. The combination of land use change, climate change and agricultural production scenarios within a blue and green WFSA is informative to both water resources management, and the WF community seeking to apply this assessment regionally. The XBMT represents a unique basin for such a study, given its geographic location in the so-called "arc-of-deforestation" and the importance in future land use change for agricultural production, but also because of agriculture's reliance on precipitation in the region. The combination of land use and hydrologic data with information on domestic and industrial water consumption remains mostly unexplored in Southern Amazonia. Therefore, there is an opportunity to use such information in a WFSA to provide a greater context for water resources management and inform decision-making for regional production processes.

Following a description of the XBMT and the details of the required steps for the blue and green WFSA (Section 2), we describe our results of past and future blue and green water availabilities and scarcities (Section 3). Then in Section 4, we discuss our results within the context of regional agricultural development and the effects of land and water management on water scarcity, prior to formulating a policy response for land and water resources in the basin.

2. Materials and Methods

2.1. The Xingu Basin of Mato Grosso

The XBMT (Figure 1) is a 170,000 km² basin located in Southern Amazonia, separated into the Xingu Headwaters (139,000 km²) that flow North into the Upper Xingu Basin (31,000 km²) [26,27], through the state of Pará and into the Amazon River, to constitute the greater Xingu River Basin (510,000 km²) [28]. The XBMT is located at the intersection of both the Amazon (80% of the basin) and Cerrado (20%) biomes and had 50% (85,000 km²) of its forest cover in 2010, of which about 34,000 km² was contained within conservation areas that include parts of the Xingu Indigenous Reserve [9] (Figure 1). Between 2001 and 2010, the XBMT lost 18,838 km² of forest to either cropland (3347 km²) or pasture (15,491 km²) with further evidence of conversion of 4962 km² of pasture into cropland [9]. In 2015, agricultural production for municipalities in the basin consisted of 1.3 Mha of soybean [16], about 5.4 Mha of pasture [10], and less than 12,000 ha of permanent crops (e.g., papayas, bananas, rubber trees) [16]. In addition, the XBMT contains close to 10,000 small farm reservoirs mainly used to supply drinking water for cattle [29]. In 2015, the cattle population reached about 3.5 million heads in the municipalities of the basin [16].

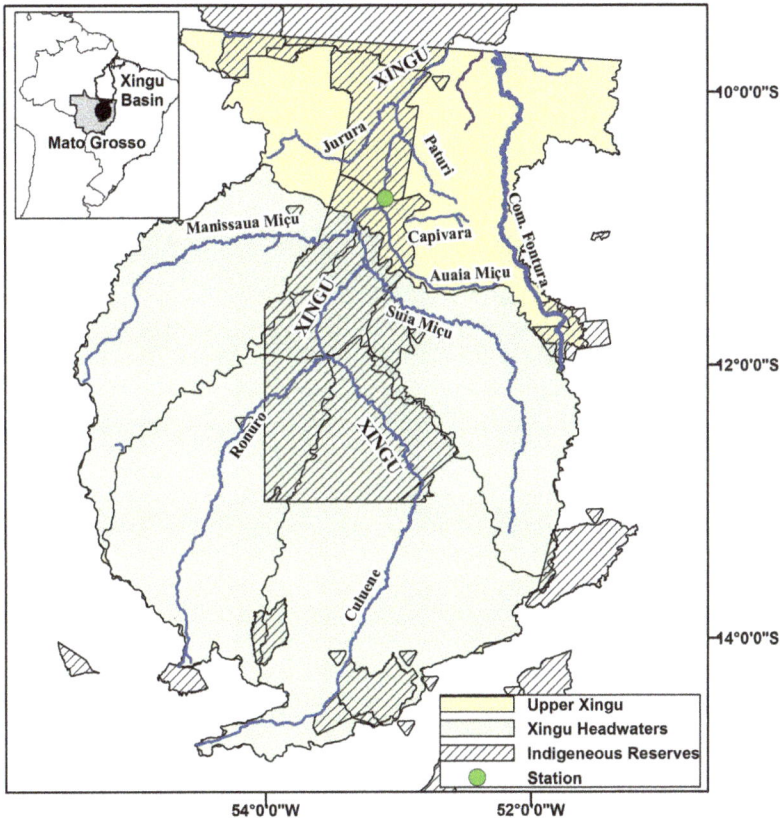

Figure 1. The Xingu Basin of Mato Grosso (XBMT) and its sub-basins: the Upper Xingu Basin (yellow) and the Xingu Headwaters (green) with the main rivers and the location of the discharge measurement station used for validation [30]. The inset shows the position of XBMT (black) in relation to the Xingu River Basin (black outline) and the state of Mato Grosso (grey).

From a total of 199,015 people living in XBMT in 2007, 125,279 made up the urban population (63%), and 73,736 represented the rural population (37%), with the portion serviced by the general water network reaching 47% and 49%, respectively [26]. Most of the drinking water for communities in the Xingu Headwaters is supplied by deep wells (60%), followed by surface water (20%), shallow wells (10%) and a mix of surface water and deep wells (10%), while 100% of the water in the Upper Xingu is supplied exclusively by deep wells [26]. Total domestic water demand was estimated at 0.0208 m^3 s^{-1} in the Xingu Headwaters and 0.1814 m^3 s^{-1} in the Upper Xingu, while industrial demand (as transformation industry) was 0.0023 m^3 s^{-1} and 0.226 m^3 s^{-1}, respectively [26]. Given the importance of the agricultural sector in the region, there is additional water demand for livestock, aquaculture with about 47.6 ha of fish tanks, and a total irrigation demand of 1.447 m^3 s^{-1} in 2006 [26].

2.2. Integrated BIosphere Simulator (IBIS)

Hydrology in the XBMT was modeled using the Integrated BIosphere Simulator (IBIS) (v.2.5), which combines ecological processes related to the water and carbon cycles with vegetation dynamics, climate, canopy and vegetation physiology, and phenology on a monthly or annual basis [31–33]. IBIS represents the soil-plant-atmosphere continuum to simulate soil moisture and evapotranspiration (ET) through six soil layers to 8 m depth (and soil temperatures), vegetation structure, stomatal conductance and photosynthetic pathways, all forced with atmospheric conditions [31,33]. The model was previously validated by Panday et al. [28] in a study of the water balance of the Xingu River Basin from 2001 to 2010 using atmospheric forcing with data from the Climate Research Unit (CRU TS v.3.2.1). Surface runoff was derived as the difference between ET and the balance of soil moisture, with the latter derived from infiltration (from the Green Ampt equation) and dynamics in the soil (from the Richards equation) [28].

Following Panday et al. [28], we combine IBIS results with land use maps to derive the monthly water balance of the XBMT for 2000-2001, 2014-2015, 2030-2031, and 2050-2051 (0.5° resolution, and hydrologic years as September to August) following two simulations: (1) considering the basin's potential natural vegetation (PNV) as defined by Ramankutty and Foley [34]; (2) considering the replacement of all natural vegetation by C4 grass (G) as a representation of complete deforestation in the basin. Hydrology for 2030-2031 and 2050-2051 was obtained from an average of 23 IPCC global climate models and considering two different Representative Concentration Pathways (RCP) of 4.5 and 8.5 W m^{-2}.

We derived total runoff in the basin through linear association of PNV and G IBIS simulations for the basin in hydrologic year t, defined from September to August of each year [28]

$$R(t) = R_{PNV}(t) \times F_f(t) + (1 - F_f(t)) \times R_G(t) \tag{1}$$

where R(t) (mm mo^{-1}) is the monthly discharge in the basin, R$_{PNV}$(t) (mm mo^{-1}) is the total runoff in the basin under a PNV simulation, F$_f$(t) (dimensionless) is the fraction of forest cover in the pixel of interest, and R$_G$(t) (m^3 mo^{-1}) is the total runoff in the G simulation. The fraction F$_f$(t) was obtained from land cover maps derived from Landsat imagery (30-m resolution) [35], while future land use in 2030 and 2050 was obtained from Soares-Filho et al. [36] based on distinct deforestation scenarios: a business-as-usual scenario (BAU) in which 1997-2002 deforestation is maintained with planned transportation infrastructure, and a governance scenario (GOV) which assumes similar deforestation rates as BAU, but in which a maximum deforested area representing 50% of each Amazonian sub-region is imposed [36]. When combining climate change with deforestation scenarios, we obtained four distinct scenarios for 2030 and 2050 (BAU$_{RCP4.5}$, BAU$_{RCP8.5}$, GOV$_{RCP4.5}$, GOV$_{RCP8.5}$). Values of R(2000), R(2014), R(2030), R(2050) were obtained for the XBMT, and R(2000) was obtained for the Xingu Headwaters and validated against monthly mean river discharge measured at Marcelândia, Mato Grosso (Passagem BR80, station 18430000, 10°46′38″ S, 53°5′44″ W) [30] with a Pearson correlation of 0.83 (see Figure S1 in the Supplemental Material). Values of R(t) were obtained annually and interannually with three-month averages for the years listed above.

Values of R(t) were then used to derive annual basin ET ($ET_T(t)$, mm y^{-1}) using the water balance equation shown in Equation (2), and assuming a change in annual storage close to 0 following findings from Panday et al. [28],

$$ET_T(t) = P(t) - R(t) \qquad (2)$$

where P(t) (mm y^{-1}) is the precipitation input to the IBIS model. Similarly, we use Equation (2) to derive $ET_{PNV}(t)$, or the annual ET of the basin under PNV, using $R_{PNV}(t)$ and the IBIS precipitation input. All values of ET were obtained for 2000, 2014, 2030 and 2050 hydrologic years.

2.3. Water Footprint Sustainability Assessment

2.3.1. Goal and Scope Definition

The goal of this study is to determine changes in blue and green water scarcities from production processes in the XBMT in recent history, and considering deforestation and climate change scenarios for 2030 and 2050, to: (1) provide a hotspot analysis of water use in the basin as guidance for future water allocation decisions; and (2) explore links between blue and green water scarcities in the basin considering land use change histories. This assessment focuses exclusively on water quantity and therefore considers blue and green WFs separately, and does not address water quality as expressed by the grey WF.

2.3.2. Water Footprint Accounting

The accounting step includes the calculation of the blue and green WFs of all processes occurring in the basin for the 2000, 2014, 2030 and 2050 hydrologic years, representing production in recent years (2000, 2014) and defined following distinct scenarios for future conditions (2030, 2050, see Section 2.3.4). The selection of the 2000 and 2014 hydrologic years was based on the intense land use change history in the basin within this time period, as attested by land use maps [9,28,35]. Long-term runoff observation in the Xingu River Basin at Marcelândia [30] showed a change in runoff of −14% (February 2001) and +23% (December 2000) compared to the mean 1975-2005 discharge. We focus exclusively on production processes, leaving out any local consumption of products that might be produced outside the basin. This assumption is reasonable given the regional focus on agricultural products for export [37], with a majority of crops grown in the region supplied as input feed for livestock. Cropland and pasture in Mato Grosso have been nearly exclusively rainfed [10], and therefore only require green water whose consumption is estimated by ET.

Green Water Footprint of Agriculture in the Context of Basin Land Use Systems

We obtain the green WF of agriculture by combining top-down and bottom-up approaches to track changes in the green WF from 2000 to 2050 hydrologic years (top-down approach) and 2000 to 2014 hydrologic years (bottom-up approach). First, we propose that total annual ET of the XBMT is equal to the sum of contributions from natural vegetation, agricultural land, and a residual term as described in Equation (3)

$$ET_T(t) = ET_{NV}(t) + ET_{AG}(t) + ET_R(t) \qquad (3)$$

where $ET_T(t)$ (m^3 y^{-1}) is the annual ET in the basin in hydrologic year t obtained from Equation (2), $ET_{NV}(t)$ (m^3 y^{-1}) is the annual ET from natural vegetation (as tropical humid or savanna forest, shrubland, etc.) in the basin, $ET_{AG}(t)$ (m^3 y^{-1}) is the annual ET from agricultural land (as cropland and pasture combined), and $ET_R(t)$ (m^3 y^{-1}) is a residual ET term, which accounts for other land use systems (e.g., forest clearance, urban areas, etc.) and water bodies (e.g., rivers, wetlands) that may or may not be included in human consumption activity. In the top-down approach, we extract $ET_{AG}(t) + ET_R(t)$ from a calculation of $ET_{NV}(t)$ in Equation (4)

$$ET_{NV}(t) = \sum_j ET_{PNV,j}(t)A_{NV,j}(t)F_{NV,j}(t) \tag{4}$$

where $ET_{NV}(t)$ $(m^3\ y^{-1})$ is the natural vegetation ET contribution in the basin, $ET_{PNV,j}(t)$ $(m\ y^{-1})$ is the ET of the IBIS PNV simulation for each IBIS raster j of area $A_{NV,j}(t)$ $(3080\ 10^6\ m^2)$ within the basin, and considering the fraction of forested land $F_{NV,j}(t)$ (dimensionless). This approach allowed for the disaggregation of ET_T into ET_{NV} and $(ET_{AG} + ET_R)$, which we use to analyze the hydrologic years between 2000 and 2050.

The bottom-up approach was applied for the 2000 and 2014 hydrologic years in which we used average pasture and cropland ET estimates from Lathuillière et al. [10,38] together with land use estimates extracted from Landsat imagery [35]. We considered single- and double-cropped soybean (with rice or maize) as the main crops in the region (Table S1). This assumption is reasonable considering that between 2000 and 2014, soybean represented 48–69% of total annual cropland in the basin, while maize and rice represented 12–23% and 33–3%, respectively [16] with an ever increasing amount of maize double cropping in Mato Grosso [39]. During the same time period, perennial crops represented less than 1% of total agricultural land [16] and were therefore not considered further in this green water accounting step. Residual ET (ET_R) was then derived using Equation (3) and, in this approach, may include ET that could be allocated to a production activity occurring in urban areas, or other landscapes with no immediate productive activity (e.g., ET following forest clearance). Differences in ET_R between the top-down and bottom-up approaches may be interpreted as a systematic error in the allocation of ET to a particular land use systems or human activity.

Blue Water Footprint of Agriculture

The blue WF of agriculture includes irrigation, but also water consumption from livestock production systems. In 2006, about 770 ha of perennial crops were irrigated within XBMT and therefore we assume that the majority of irrigation in the XBMT was not applied to soybean or pasture between 2000 and 2015. Blue water use was estimated for livestock production systems in pasture (ruminants), as well as confined facilities (chicken and swine), and includes drinking water as well as water used for washing of animal housing. Feed for all livestock production was assumed to be sourced from within the region, and is therefore already accounted for in the agricultural green WF (see above). Water consumption for cattle follows the steps described in Lathuillière et al. [40], who allocated green and blue water per kg of live weight based on sex, animal development stage and diet. Here, we consider drinking water sourced from small farm reservoirs in the basin detected by remote sensing [40]. All other animals were assumed to have their drinking water sourced by the main water system. As described in Lathuillière et al. [40], cattle population reported by agricultural production data [16] is a total animal population which does not consider the annual live population in their different stages of development. The annual live animal population for municipality i and calendar year t is the difference between the total herd population ($H_{i,t}$) and the number of animals slaughtered ($0.17H_{i,t}$). The live annual population $L_{i,t}$ can then be expressed by

$$L_{i,t} = 0.27H_{i,t} + (L_{i,t} - 0.27H_{i,t} - 0.17H_{i,t+1}) + 0.17H_{i,t+1} \tag{5}$$

where $0.27H_{i,t}$ is the sum of the calf population (15–18 month duration), $L_{i,t} - 0.27H_{i,t} - 0.17H_{i,t+1}$ is the sum of the adult population (24–27 month duration), and $0.17H_{i,t+1}$ is the animal population in the finishing stages (to be slaughtered in calendar year t + 1, 6–8 month duration) [40]. Sheep and goat annual offtake rates were assumed identical to that of cattle (17%), while horses, donkeys, and mules were not considered to be consumed and therefore their live population was equated to the total herd population reported by the Brazilian Institute of Geography and Statistics (IBGE as the Portuguese acronym) [16].

The swine and chicken development cycles were assumed to be 70 and 42 days respectively [41,42], from which we derived average swine and chicken populations following Equation (6) [43]

$$\overline{P_{k,i}(t)} = days \frac{P_{k,i}(t)}{365} \tag{6}$$

where $\overline{P_{k,i}(t)}$. (animals) is the average population of animal k, in municipality i and calendar year t, *days* is the total number of days of the animal's development cycle, and $P_{k,i}(t)$ is the population of animal k reported by national statistics [16]. To reflect animal population information available from IBGE [16] for calendar years into the hydrologic years used in this study, we take the average of the two consecutive calendar years that overlap with each hydrologic year. Similar to crops, animal population for each municipality located inside the basin was scaled based on the percent area located inside XBMT (Table S2).

vestock water consumption was derived following the National Water Agency (ANA as the Portuguese acronym) [26] and Food and Agriculture Organization (FAO) [44] which provide water demand per animal, assuming an average adult consumption. For confined swine and chicken production we assumed 0.0034 m^3 animal^{-1} used to clean animal housing after slaughter following Palhares [42] for swine, which was also assumed for chicken housing. These volumes are assumed to be entirely consumed. While our drinking water consumption estimate is based on adult animal water demand and likely constitutes an overestimation of animal blue water consumption, our water consumption estimate for cleaning is likely an underestimate given the housing turnaround for both swine (70 days) and chicken (42 days) production (Table S2). Large and small ruminants were not allocated any water for cleaning as they were assumed to spend their lifetime in pasture.

Domestic and Industrial Blue Water Footprints

We estimated domestic water consumption based on urban and rural human populations within the basin and the total population receiving municipal services based on information from ANA [26]. We assumed a constant population growth in the basin at a rate of 3.0% y^{-1} until 2014–2015 [16]. By assuming a total basin population of 199,015 in 2010 (the same as the 2007 information reported by ANA [26]), we derived total population in the remaining years, maintaining the same proportion of urban and rural population not serviced by the municipal system (47% and 49%, respectively) (Table S3). Water consumption was calculated assuming a 50% return flow to surface water, and based on a rural water demand of 70 10^{-3} m^3 d^{-1} cap^{-1} and an urban demand of 0.260 m^3 d^{-1} cap^{-1}. The 47% of the urban population that was not serviced by the municipalities was assigned a water demand equal to rural demand [26] (Table S3).

Industrial water consumption was based on the number of industrial workers in both extraction and transformation industries assuming an industrial demand of 3.5 m^3 d^{-1} cap^{-1} [26]. In 2010, the number of industrial workers was 4.1% of the total population within the basin [16], which we assumed to be of constant proportion between 2000 and 2015. Similar to domestic water consumption, we assumed a 50% return flow of industrial water (Table S3).

2.3.3. Water Scarcity Calculation

We estimate water scarcity within the XBMT in hydrologic year t following Equation (7) [17]

$$WS(t) = \frac{\sum_j WF_j(t)}{WA(t)} \tag{7}$$

where WS(t) (dimensionless) is the water scarcity, WF$_j$(t) (m^3 y^{-1}) is the WF of all activities j (determined in Section 2.3.2), and WA(t) (m^3 y^{-1}) is the water available in the basin over time t.

Values of WS(t) are defined for both blue and green water and vary from 0 (no scarcity) to 1 (extreme scarcity) to gauge how water use has evolved within the basin. For both blue and green water resources, WA(t) is defined following Equations (8) and (9) [17]

$$WA_B(t) = R(t) - EFR \tag{8}$$

$$WA_G(t) = ET_T(t) - ET_{RNV}(t) - ET_{UN}(t) \tag{9}$$

where $WA_B(t)$ ($m^3 \, y^{-1}$) is the blue water availability, $R(t)$ ($m^3 \, y^{-1}$) is the natural discharge (or discharge without human appropriation in the basin, defined in Equation (1)), and EFR ($m^3 \, y^{-1}$) is the environmental flow requirement defined for the XBMT. When considering our top-down WF accounting approach, the value of EFR was defined according to mean annual runoff following Smakhtin et al. [45] with a value of 45.9 $km^3 \, y^{-1}$ to keep natural ecosystems in a "fair" condition (see Supplemental Material). When considering our bottom-up WF accounting approach, values of EFR were defined for each 3-month mean discharge between 2000 and 2014 hydrologic years as $0.20R(t)$ following Richter et al. [46]. Green water availability in hydrologic year t, $WA_G(t)$ ($m^3 \, y^{-1}$), was obtained by subtracting from the $ET_T(t)$ ($m^3 \, y^{-1}$) the ET reserved to natural vegetation, as $ET_{RNV}(t)$ ($m^3 \, y^{-1}$), and the ET of areas agriculturally unproductive, $ET_{UN}(t)$ ($m^3 \, y^{-1}$). We interpret $ET_{UN}(t)$ as the amount of small reservoir evaporation for cattle production whose area we consider unavailable for agricultural expansion. The value of $ET_{RNV}(t)$ is interpreted as a percentage of total basin ET ($ET_T(t)$) as measured in the 2000 hydrologic year and based on the Federal Forest Code minimum requirements for natural forest cover in both the Amazon (80%), Cerrado (35%), and transition (50%) zones [47]. As a result, WA_G and WS_G were calculated using these three minimum requirements expressed in $ET_{RNV}(t)$ in Equation (9) and equal to $0.80ET_T(2000)$, $0.35ET_T(2000)$ and $0.50ET_T(2000)$.

2.3.4. Interpretation and Response Formulation through Scenarios

Blue and green water scarcities were interpreted following previously defined benchmarks. Blue water scarcity was "severe" when $WS_B > 2$, "significant" when $1.5 < WS_B < 2$, "moderate" when $1 < WS_B < 1.5$, and "low" when $WS_B < 1$ [23]. Green water scarcity was "unsustainable" when $WS_G > 1$, a "threat" when $0.5 < WS_G < 1$, "within sustainable limits" when $0.25 < WS_G < 0.5$, and sustainable when $WS_G < 0.25$ [24]. Results were then interpreted following the deforestation (BAU, GOV) and climate change (RCP 4.5 and 8.5 W m^{-2}) scenarios described above, and onto which we added population growth and agricultural production scenarios (Table 1).

First, we assumed that human population will continue to grow at current rates, or 3.0% y^{-1} until 2050, and assumed a similar breakdown in rural and urban population as in the 2000s (see Section 2.3.2), with industrial activity assumed to be proportional to population growth. Primary sector growth was based on projections made by the Brazilian Ministry of Agriculture for the 2025–2026 period focusing specifically on soybean, maize, cattle, swine and chicken production [48], assuming continuous growth in the basin between 2030 and 2050. We assumed a 35% increase in soybean production (or a 30% increase in soybean area at current yields) from 6.3 Mtons (or 2.1 Mha) in 2015 to 8.5 Mtons (or 2.8 Mha) in 2030 and an additional 35% increase (at current yields) to 11.5 Mtons (3.8 Mha) in 2050. When considered together, the total surface area for soybean and pasture were well within non-forested areas in the deforestation scenarios for 2030 and 2050 of 13 Mha and 14 Mha (BAU), and 10 Mha and 11 Mha (GOV), respectively. Cattle, pig and chicken populations were assumed to increase respectively 3.0% y^{-1}, 2.7% y^{-1} and 2.4% y^{-1} until 2050 [48] but with organizational differences in production systems based on two agricultural production options (Table 1).

We considered two agricultural production options based on increases in green water (the Green Option) and blue water (the Blue Option) resources appropriation as a means to increase agricultural output. In the BAU scenario, average livestock density for the XBMT in 2014 (0.87 live cattle ha^{-1}) was maintained to require a 0.4 Mha of additional pasture in 2030 (total of 4.4 Mha) and 3 Mha (total of 7.0 Mha) in 2050 (the Green Option). Evaporation from small farm reservoirs in 2030 and 2050 was scaled with cattle population on pasture based on 40 m^3 cattle^{-1} y^{-1} of evaporation obtained for 2014–2015 [40]. In the GOV scenario, all additional cattle in 2030 were confined on 2014–2015 pasture area to reach a livestock density of 1.3 cattle ha^{-1} (affecting 5.2 million animals). In 2050, additional cattle were confined with a total population breakdown of 5.2 million cattle on pasture and 3.1 million raised in confinement. We assumed that confined cattle did not use small farm reservoirs, but other sources that do not carry evaporation (e.g., groundwater). At the same time, we assumed that 90 mm of irrigation was applied in September–October to the entire soybean area (the Blue Option).

Table 1. Description of scenarios for 2030 and 2050 activities in the Xingu Basin of Mato Grosso following deforestation (business-as-usual (BAU) and governance (GOV) [36]), and climate change scenarios (Representative Concentration Pathways (RCP) 4.5 and 8.5 W m^{-2}). BAU and GOV scenarios also illustrate agricultural intensification options focused respectively on green water (BAU) and blue water (GOV) appropriation.

Scenario	Year	Human Population; Industrial Workers	Livestock Population	Description
BAU$_{RCP4.5}$ BAU$_{RCP8.5}$	2030	336,335; 211,722	5,233,040 cattle; 74,069 pigs; 792,674 chicken	Human population increases at historic growth rate; Industry grows proportionally to human settlement; Soybean production requires 2.8 Mha of land; Cattle population requires 4.4 Mha of pasture
BAU$_{RCP4.5}$ BAU$_{RCP8.5}$	2050	568,407; 357,809	8,372,864 cattle; 114,066 pigs; 1,173,157 chicken	Human population increases at historic growth rate; Industry grows proportionally to human settlement; Soybean production requires 3.8 Mha of land; Cattle population requires 7.0 Mha of pasture
GOV$_{RCP4.5}$ GOV$_{RCP8.5}$	2030	336,335; 211,722	5,233,040 cattle; 74,069 pigs; 792,674 chicken	Human population increases at historic growth rate; Industry grows proportionally to human settlement; Soybean production requires 2.8 Mha of land; Cattle population is requires 4 Mha of pasture
GOV$_{RCP4.5}$ GOV$_{RCP8.5}$	2050	568,407; 357,809	8,372,864 cattle; 114,066 pigs; 1,173,157 chicken	Human population increases at historic growth rate; Industry grows proportionally to human settlement; Soybean production requires 3.8 Mha of land; Cattle population is split between pasture (5.2 million) and confinement (3.1 million); soybean is irrigated 90 mm in September–October.

2.3.5. Data Processing and Sensitivity Analysis

Data processing was carried out using statistical software R (v.3.4.0) in R Studio (v.1.0.143) [49] with packages: raster (v.2.5-8) [50], sp [51,52], rgdal (v.1.2-7) [53], maptools (v.0.9-2) [54], and ncdf4 (v.1.16) [55]. Our results are provided using a series of values to highlight the extent of water scarcity in the basin, such as the use of both bottom-up (2000, 2014) and top-down (2030, 2050) approaches for allocating ET to vegetation. Our response formulation for green water resources was based on a suite of restrictions following mandatory natural vegetation cover outlined in the Federal Forest Code (35%, 50%, and 80%), which served as a sensitivity analysis for green water scarcity (WS$_G$). Blue WF values were considered to be conservative estimates, particularly for cattle production [40], as well as the high return flows (50%) attributed to withdrawals.

3. Results

3.1. Past and Future Water Footprints

Between 2000 and 2014, the sum of cropland and pasture areas increased 31% from 4.7 Mha to 6.2 Mha. Changes in the consumption of blue water expressed by the total blue WF increased from 0.153 km^3 y^{-1} in 2000 to 0.218 km^3 y^{-1} in 2014 (Figure 2). The blue WF was dominated by agriculture, representing 97% of total water use, followed by domestic and industrial uses (Table S5). Water evaporation from small farm reservoirs represented 66% of total agricultural blue WF in 2000, and 67% in 2014, followed by livestock drinking (respectively 32% and 31%) and irrigation (2% of total consumption in both years) (Figure 2). Between 2000 and 2014, the total area of small farm reservoirs increased 37% from 6914 ha to 9463 ha of water, leading to a total evaporation of 0.099 km^3 y^{-1} and 0.141 km^3 y^{-1}, respectively. Domestic blue water consumption computed here was similar to values from ANA [26], which reported 3.47 10^{-3} km^3 y^{-1} in 2007, while our industrial consumption estimates were three orders of magnitude smaller than the 3.55 10^{-2} km^3 y^{-1} reported for 2007 [26]. Differences in industrial uses are primarily attributed to our separation of confined livestock from industry, as well as our focus on extractive and transformative industries. Combining livestock and industrial water consumptive uses raised our computed values closer to those reported by ANA [26]. The total blue WF increased with larger human and livestock populations in 2030 and 2050. In 2030, agricultural water use nearly doubled to 0.258 km^3 y^{-1}, while the combined industrial and domestic uses increased to 9.90 10^{-3} km^3 y^{-1} (Table S5). In 2050, agricultural water use increased to 0.517 km^3 y^{-1} and 0.391 km^3 y^{-1} in the BAU and GOV scenarios, respectively. In the case of cattle confinement and early season soybean irrigation (GOV$_{RCP8.5}$), consumption rose to 3.81 km^3 y^{-1}.

Agricultural expansion resulted in an increase in the total green WF of agriculture (as ET$_{AG}$) from 40.6 km^3 y^{-1} in 2000 to 49.9 km^3 y^{-1} in 2014 (Table S7). This change was led by cropland ET which increased from 7 to 29% of ET$_{AG}$, while pasture dropped from 93 to 71% of ET$_{AG}$ in the same time period (Figure 3). The increase in green WF occurred at the expense of the natural vegetation whose contributions to ET dropped 11% between 2000 and 2014 due to a decrease in forest cover by roughly 1.4 Mha. Changes in ET$_{AG}$ and ET$_{NV}$ obtained through the bottom-up approach were similar to results from Silvério et al. [9] (Table S7). Further deforestation for agriculture in 2030 and 2050 increased ET of non-forested areas to 188.6 km^3 y^{-1} and 209.6 km^3 y^{-1} for the BAU scenarios in 2030 and 2050 respectively, and 147.2 km^3 y^{-1} and 147.3 for the GOV scenarios (average climate change scenarios) (Figure S4, Table S8).

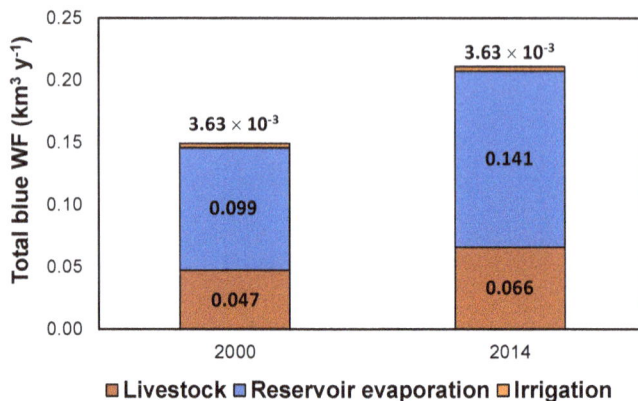

Figure 2. Total blue Water Footprint (WF) of agriculture in the Xingu Basin of Mato Grosso for the 2000 and 2014 hydrologic years.

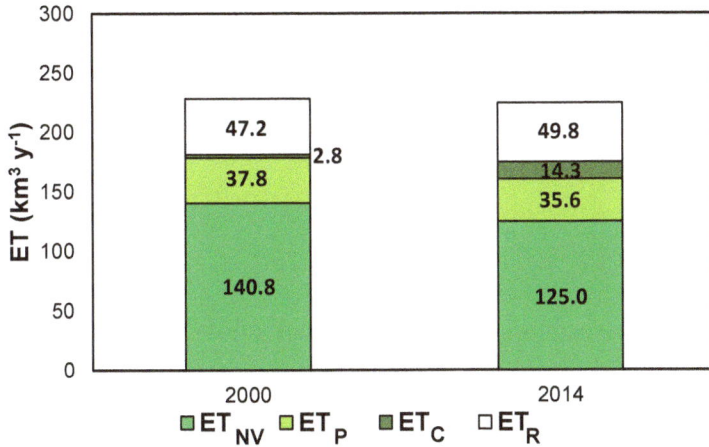

Figure 3. Changes in contributions to evapotranspiration (ET) for natural vegetation (ET_{NV}), pasture (ET_P), cropland (ET_C) and residual landscapes (ET_R) in the Xingu Basin of Mato Grosso in the 2000, and 2014 hydrologic years (September–August). Values obtained through the bottom-up approach as described in the text.

3.2. Blue and Green Water Availability and Scarcity

Annual runoff decreased from 74.9 km^3 y^{-1} to 70.4 km^3 y^{-1} between 2000 and 2014 (Table S6), which, when considering environmental flow requirements, left 43.4 km^3 y^{-1} (in 2000) and 40.8 km^3 y^{-1} (in 2014) of blue water available in the basin. The decrease in annual runoff followed the decline in precipitation from 1999 mm y^{-1} in 2000 to 1934 mm y^{-1} in 2014 (Table S6). When considering 3-month windows, the decrease in runoff was more prominent in the December–February period where values decreased from 20.7 km^3 3-month^{-1} in 2000–2001 to 14.3 km^3 3-month^{-1} in 2014–2015 (Table S6), which we relate to a reduction in September–November precipitation from 519 mm 3-months^{-1} in 2001 to 447 mm 3-months^{-1} in 2014.

The combination of deforestation and climate change in the scenarios generally increased runoff by 2% in 2030 when compared to 2000 (GOV$_{RCP4.5}$), and by 8% in 2050 (BAU$_{RCP4.5}$) despite a reduction in precipitation (Table S6). The GOV$_{RCP8.5}$ scenario was the only exception with a decrease in runoff of 1% in 2050 for a precipitation decline to 1952 mm y^{-1}. Focusing on climate change effects alone, runoff with potential natural vegetation cover in the basin decreased from 69.8 km^3 y^{-1} in 2000 to 64.1 km^3 y^{-1} in 2014, 67.9–69.1 km^3 y^{-1} in 2030 and 65.7–69.0 km^3 y^{-1} in 2050 (Table S6). Inter-annual changes in runoff were apparent when considering 3-month windows: runoff generally increased at the beginning of the wet season (September–November, +13–20%), before decreasing at the end of the wet season (December–February, -62–71%). Dry season runoff increased between 22% and 52% in the June to August periods when compared to 2000 (Table S6).

Land contributions to ET in the basin were similar between 2000 (279.0 km^3 y^{-1}) and 2014 (272.0 km^3 y^{-1}) (Table S8). In 2000, forests represented 50–69% of contributions (bottom-up and top-down estimates), while agriculture represented 15% (bottom-up estimates) (Tables S7 and S8). In 2014, these values changed to 46–63% and 18% for forests and agriculture, respectively. Total land contributions to ET dropped by up to 4% in the GOV$_{RCP4.5}$ scenario in 2030 and both BAU$_{RCP4.5}$ and GOV$_{RCP4.5}$ scenarios in 2050 (Table S8, Figure S4) with differences in contributions based on forest cover. Forests in the GOV$_{RCP4.5}$ scenario provided 122.9 km^3 y^{-1} and 122.5 km^3 y^{-1} of ET in 2030 and 2050, respectively. In contrast, BAU$_{RCP4.5}$ showed a reduction of natural vegetation ET from 82.1 km^3 y^{-1} in 2030 to 60.0 km^3 y^{-1} in 2050 as a result of reduced forest cover (Table S8, Figure S4).

Annual blue water scarcity values were less than 0.10 (Figure 4) with the largest value recorded in the GOV$_{RCP8.5}$ scenario in 2050 (0.09). Inter-annual values increased to 0.65 for the GOV$_{RCP8.5}$ scenario between September and November 2050 due to early soybean planting and irrigation (with inter-annual blue water scarcity values ≤0.03 the rest of the year). Annual green water scarcity values changed according to deforestation scenarios, but also due to restrictions placed on the allocation of natural vegetation. Between 2000 and 2014, green water scarcity was at least "within sustainable limits" (WS$_G$ < 0.50) when considering the bottom-up approach, moving closer to "threat" conditions (0.5 < WS$_G$ < 1) in the top-down approach (Figure 4). In 2030 and 2050, green water scarcity values increased to 1.1 in the BAU-2050 scenario considering 35% of natural vegetation allocated to the basin, and beyond 1.2 when allocation increased to 50% and 80%. In the same time period, the GOV scenario maintained WS$_G$ < 1 with a 50% allocation to natural vegetation, but moved to "unsustainable conditions" in both 2030 and 2050 when allocating 80% of the basin allocated to natural vegetation (Figure 4).

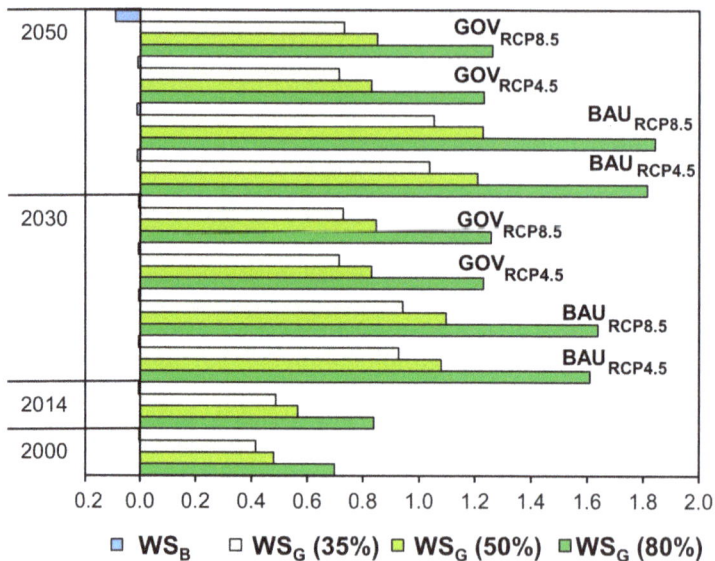

Figure 4. Annual blue (WS$_B$) and green (WS$_G$) water scarcities for the Xingu Basin of Mato Grosso in 2000 and 2014 hydrologic years, business-as-usual (BAU) and governance (GOV) deforestation scenarios considering Representative Concentration Pathways (RCP 4.5 and 8.5 W m^{-2}) (Table 1). Values of WS$_G$ were obtained assuming that 35%, 50%, and 80% natural vegetation cover in the basin was maintained as described in the text.

4. Discussion

4.1. Agricultural Development and Water Resources

Agriculture was found to be the largest contributor to the total blue WF in the basin, with livestock water consumption for drinking and from reservoir evaporation representing the largest component. Water allocated to livestock production systems in 2014 was equivalent to the consumption of 2.3 million people connected to the municipal system. Animal population in the basin was historically led by cattle, but pig and chicken production have increased in recent years [16], effectively increasing water consumption and the water supply needed for production. Chickens and pigs are typically raised in confined facilities in Mato Grosso and, therefore, rely on surface or groundwater pumped for drinking water. In contrast, cattle in Mato Grosso rely on small reservoirs whose evaporation constitutes more than half of agricultural blue water consumption. Some of these reservoirs are constructed from impoundments of small streams, which contribute to stream warming with potential effects on stream chemistry [29] and hydrologic connectivity [56]. The regional effects of these reservoirs on hydrology remain relatively unexplored in Southern Amazonia.

The replacement of natural vegetation by cropland and pasture was illustrated by an increase in green water appropriation in the basin. We report a decline in pasture area in 2014 compared to 2000, which, when combined with increasing cattle population, led to an increase in cattle density (0.57 cattle ha^{-1} in 2001 to 0.97 cattle ha^{-1} in 2015), following general trends in the state of Mato Grosso [40]. The replacement of deep rooted natural vegetation with shallow-rooted crops and pasture affects radiation partitioning by decreasing latent heat and increasing sensible heat fluxes [15]. These changes in radiation partitioning have important consequences on surface temperatures. Silvério et al. [9] showed that cropland and pasture surface temperatures in the XBMT were 6.4 °C and 4.3 °C greater than forests. As a result, deforestation between 2000 and 2010 led to an average basin temperature increase of 0.3 °C on top of the 1.7 °C increase that had occurred because of deforestation prior to 2000 [9]. The Xingu Indigenous Park located in the heart of the basin (Figure 1) showed surface temperatures 3 °C lower inside the protected area compared to the rest of the basin [57]. Such effects illustrate the importance of maintaining natural forest cover.

Water consumption in future agricultural production varied substantially based on the conditions of production, which include land expansion and intensification. Our evaluation of two agricultural expansion scenarios highlights the extent of future green water appropriation from rain-fed agriculture which carries consequences for the carbon and water cycles [37]. Agricultural intensification for both crops and livestock requires either more efficient use of green water on current land, a reallocation of green water resources for production (e.g., cropland expansion into pasture), additional blue water consumption from irrigation, or a combination of the above [15]. Under current production practices, the onset of the wet season dictates when (or if) a second crop (typically maize) could be planted [39,58]. Farmers may plant soybean earlier in the season (e.g., in September) and irrigate fields until the onset of the wet season (e.g., approximately 16 October 2007 in the basin [58]) to allow for earlier planting and harvesting of maize, and the potential success of two crops. Under this strategy, farmers could also add a third irrigated dry season crop (e.g., bean) leading to additional blue water consumption [40].

Similarly, future cattle production may include additional confinement as a strategy to free pasture for cropland expansion. A larger cattle population means greater appropriation of both green water (through feed) and blue water (through drinking, small farm reservoirs, cleaning of pens, etc.) [40]. Confinement could also move towards the use of blue water sources other than those stored in small farm reservoirs (e.g., groundwater), in which case the total blue WF of cattle could drop. However, this apparent efficiency has to be assessed considering the use of reservoirs in the long term, or their possible decommissioning or alternative use in other production systems (e.g., as irrigation). Potential water savings through efficiencies in the cattle production system (e.g., reservoir evaporation management) could also reduce the blue WF of cattle to allow greater water availability downstream [40,59].

Since 2000, the state of Mato Grosso increased meat production for both domestic consumption and international exports. The amount of water used for production is therefore virtually transferred to consumers within and outside of Brazil [60] (80% of Brazilian production is consumed within the country according to the Bovine Support Fund (FABOV as the Portuguese acronym) [61]). Between 2000 and 2014, Mato Grosso meat exports rose from 27,000 tons to 387,000 tons [62], thereby increasing the amount of water consumed regionally for foreign export, along with soybean commodities [37]. For instance, 27% of Europe's virtual water imports between 2006 and 2015 came from soybean trade [63].

The selection of future production systems proposed through our scenarios can therefore change the resource appropriation for regional production, which already carries nutrient and carbon footprints that can be allocated to consumers [37]. This connection between consumption and production centers has inspired demand-side management of water use through the supply chain. For instance, Vanham et al. [64] estimated the WF of different European diets and their implications for water resources. Supply chain interventions in the region have been motivated by deforestation and climate change implications though both the "Soybean Moratorium" or the "Cattle Agreement" [65], but could also include water resources given the close link between land and water resources management in agricultural production systems [15].

4.2. Changes in Water Scarcity with Land and Water Management

Activities in the basin through present day were found to be within blue water sustainable limits. Green water resources, however, were within sustainable limits under specific conditions only. Inter-annual blue water scarcity moved closer to "moderate" under irrigation expansion and cattle confinement, reflecting the potential vulnerability of the basin to dry season agricultural water use. A total of 234 irrigation pivots covering almost 28,000 ha were identified in the municipalities overlapping XBMT [66] and expansion could increase given the 10 Mha irrigation capacity estimated for Mato Grosso [67]. Similarly, the developed reservoir capacity for cattle is a measure to ensure continued drinking water in the dry season when animals may need more water due to meteorological conditions [68]. This water consumed for agricultural production is then unavailable for other human and ecosystem uses in the greater Xingu River Basin, and may affect wetlands or hydroelectric power production [14,28]. Water rationing has already taken place as a result of drought (e.g., 2005) and the lack of infrastructure to cope with low water levels, particularly in the Xingu Headwaters [26]. We therefore expect future water use for irrigation and cattle to also come from additional sources (e.g., surface and groundwater sources) should water become scarce in the dry season. Consequently, both intensification of soybean and cattle production should carefully observe the effects on future water scarcity in the basin in agricultural management plans.

While policies have mostly focused on maintaining forest cover to protect biodiversity and reduce greenhouse gas emissions, these policies can also play a role in maintaining sustainable water resource use. The sustainable limits that we calculated relied on our estimate of water availability (WA_B, WA_G) which depended in turn on the interaction of land and water management initiatives. We found that natural runoff (i.e., runoff without any consumption activity, affecting WA_B) would change in 2030 and 2050 as a result of deforestation and climate change, while total land ET (affecting WA_G) responded directly to the allocation of land to natural vegetation cover, with a feedback on natural runoff. Changes in the natural runoff resulting from deforestation and climate change have already been measured in the region. For example, the 15% forest cover loss between 1971 and 2010 in the Xingu River Basin led to a 6% increase in runoff, while climate variability led to a 2% decrease in precipitation and 14% decrease in runoff [28]. Groundwater is known to act as a buffer in the basin, particularly in the dry season when runoff could diminish due to an extended dry season ET [69]. Changes in water availability can therefore be affected by the amount of deforestation in the basin represented by the BAU and GOV scenarios also guided by Brazilian Federal law.

The determination of green water scarcity assumed an increasing amount of land allocated to natural vegetation in the basin based on natural vegetation cover mandated by the Federal Forest Code [47]. As such, our interpretation of green water scarcity was based on the amount of vegetation cover lost in the basin in relation to Federal thresholds, which vary by biome from 35% (Cerrado savanna) to 80% (Amazon forest). For instance, in 2014, green water scarcity was within "threat" conditions when allocating 80% of the basin to natural vegetation (based on ET in the 2000 hydrologic year as described in Section 2.3.3). These "threat" conditions mean that from the total amount of green water available in the XBMT (represented by total ET, ET_T), the amount that could be put to use for agricultural production approached the limits mandated by the retention of natural vegetation cover (80%). Even in a restrictive deforestation scenario (GOV), green water appropriation would be unsustainable unless the policy goal for natural vegetation cover were reduced from 80% to 50%, in which case the basin's green water scarcity changes from "unsustainable" to "threat" conditions. The XBMT is located within the Amazon and Cerrado biomes, which have different mandatory levels of natural vegetation cover based on whether a property was within the Cerrado (35%), Amazon (80%), or the transition zone between the two (50%). We therefore conclude, that future green water appropriation will, at best, remain under "threat" conditions considering both a restrictive deforestation scenario (GOV) and a 50% natural vegetation cover. This analysis, however, does not include potential indirect land use change that might occur outside the limits of the basin [2,39,70].

The increase in green water appropriation by cropland and pasture from natural vegetation through agricultural extensification, was previously observed in the basin [9], at the Mato Grosso state level [10], and the Cerrado [3]. These studies show that land use change can impact the water cycle by returning less water vapor to the atmosphere when compared to natural vegetation with a potential reduction on regional precipitation [71,72]. Regional precipitation is sourced from green water resources as opposed to ocean evaporation [73], such that land use change may, in turn, affect water availability within and outside the basin [11–13,74]. This so-called "moisture recycling", however, is also expected to be affected by the expansion of irrigation practices which could transfer additional water vapor to the atmosphere in the dry season when regional recycling is enhanced [75].

4.3. Response Formulation and Study Limitations

Our scenarios represent two possible agricultural production options [15] considering agriculture remains the largest water consumer in the basin. These options reflected whether agricultural intensification relied on cropland expansion into pasture (the Green Option), or whether cropping frequency and livestock confinement becomes more widespread in the future (the Blue Option) (Table 2). Further appropriation of green water from either natural vegetation or pasture depends on land use policies and incentives (e.g., Federal Forest Code, Protective Areas, etc.), while blue water use depends on water management, which has generally focused on human rather than ecosystem requirements [14]. Both options have consequences for future water availability: continued reduction in natural vegetation cover, which is accompanied by reduced water vapor supply to the atmosphere could also affect terrestrial ecosystems that rely on precipitation for ecosystem functioning [15], while dry season water consumed in intensified livestock and irrigation systems could impact aquatic ecosystems downstream.

Regional water resources planning requires that connectivity of the water cycle among basins and biomes be maintained in order to secure future water availability within the basin and beyond. Water resources management options should consider upstream rain-fed agriculture and small farm reservoirs and their effects on downstream hydroelectric power. Currently, large hydropower dams (>10 MW) require environmental licenses and impact assessment studies, while smaller dams do not [14], suggesting possible conflicts between up- and downstream water uses. As 22% and 48% of evaporation in the Xingu and Amazon Basins, respectively, return to the same basins as precipitation [76,77], land and water management in a basin should go beyond its physical boundary. So far, effects of land use change on moisture recycling has been absent in water management, in part, due to the difficulty to connect precipitation source and sink regions in governance [78].

Water management strategies should also include green and blue water resource use efficiency gains at the field level. For instance, small farm reservoir management should strive to reduce total evaporation [15,59], especially when combining livestock confinement with the widespread use of irrigation for soybean planted at the end of the dry season (as described in our Blue Option). Moreover, green water use should attempt to improve transpiration over evaporation [15], while irrigation should be used efficiently. These actions depend on each individual farmer, their production systems, and the available training for capacity building of such options. For instance, the recent increase of cattle density on the current pastureland relied on increased pasture productivity with the potential to reduce the amount of water for feed [40]. However, such an initiative has been difficult to implement in the region [79], and the financial returns of increased cattle density still depend strongly on price fluctuations in the beef market [80].

Our study focused on environmental aspects related to water quantity, not social nor economic implications of water consumption, nor the effects of water quality on scarcity through the grey WF. As the largest water consuming sector in the basin, agriculture likely carries the greatest impacts both socially and economically. Some studies have made strong connections between agricultural development and human and economic development [5,81]. The effects on water quality resulting from widespread fertilizer application in the XBMT have been inconclusive thus far with respect to eutrophication [82], while few studies have investigated the effects of pesticides on water quality in Southern Amazonia [81]. The increase in livestock confinement for both swine and chicken production suggest additional on-farm waste management, which could also affect water quality and were not considered in this study.

Results of this study relied on the accuracy of the IBIS model to represent the water cycle from land use maps. Our bottom-up approach relied on maps obtained from Landsat imagery which were used to infer runoff, and ET using average land use system values derived from previously published results. The derived runoff and ET results were used exclusively for the 2000–2001 and 2014–2015 period and were close to the observations (see Supplemental Material). Our top-down approach used for the 2030–2031 and 2050–2051 periods relied on the assumption that cropland and pasture ET were equal. Cropland and pasture ET can differ by almost 100 mm (see Table S1) suggesting a potential overestimation of agricultural land ET (Figure S4). A reduction in agricultural ET would increase the estimated runoff and decrease agricultural green water consumption. These changes would have a small effect on our annual blue water scarcity values, and limited effect on our green water scarcity values which were more sensitive to the allocation of ET to natural vegetation (ET_{RNV}).

Table 2. Summary of effects and responses for two agricultural production options focused on production intensification in the Xingu Basin of Mato Grosso.

Description	The Green Option		The Blue Option	
	Crops	Cattle	Crops	Cattle
Strategy	Increase production by increasing cropped area	Intensify production on current land	Increase crop frequency (triple cropping)	Intensify production through confinement
Land use	Expansion of crops into pastureland	Concentration of cattle on current and more productive pastureland	Cropland expansion into pastureland	Increase confinement of cattle
Water use	Reallocate water from cattle to cropland	Reduce water use for a more productive pasture; Feed sourced off-farm (virtual water transfers); Increase small reservoir capacity	Use irrigation for early soybean planting and include a dry season irrigated crop	Increase small reservoir capacity; Supplemental drinking from surface and groundwater in confined systems
Effects on blue water use and scarcity	Blue water consumption increases with animal population, reservoir evaporation and groundwater use, but remains within sustainable limits		Blue water consumption approaches sustainable limits in the dry season with potential effects on downstream water availability	
Effects on green water use and scarcity	Green water use increases for crops and decreases for pasture keeping green water scarcity constant; Green water availability may change in the long-term due to local (land use) and global (CO_2 emissions) climate change and additional evaporation from farm reservoirs increase water vapor flows to the atmosphere; Changes in precipitation affect blue and green water availability in- and outside the basin.		Green water use increases for crops and decreases for pasture keeping green water scarcity constant; Green water availability may change in the long-term due to local (land use) and global (CO_2 emissions) climate change but additional ET from crop irrigation and farm reservoirs increase water vapor flows to the atmosphere; Changes in precipitation affect blue and green water availability in- and outside the basin.	
Water management considerations	Improve efficiency of blue water use, especially the reduction of evaporation from farm reservoirs; Consider effects of land (precipitation and runoff) beyond the basin; Integrate land and water policies.		Improve efficiencies in blue water use for irrigation and confined livestock; Groundwater management or the use of old farm reservoirs could be used without affecting runoff; Consider effects of land use on water availability, especially the effects of additional water vapor supply to the atmosphere.	

Our results used IBIS to infer natural runoff under deforestation and climate change scenarios, which do not include the feedbacks of water consumption activities. First, blue water scarcity values were estimated based on the appropriation of runoff as the blue water source. The currently reported XBMT water use is made up of only 20% of surface water with the remainder coming from deep and shallow wells [26]. We therefore expect future dry season blue water scarcity limits to take longer to reach as a result of groundwater extraction in the case of soybean irrigation and cattle confinement. Groundwater in Southeastern Amazonia is deep and known to also feed streams in the Xingu Headwaters [83,84]. Therefore, the effects of extensive groundwater extraction could only partially contribute to blue water scarcity, unlike other regions [85]. Our results, however, are still expected to represent a general trend towards greater water scarcity given the large contribution of drinking water for cattle and evaporation from small farm reservoirs which was entirely attributed to surface water.

In this case we also expect groundwater storage to act as a blue water source available to alleviate agricultural water demand in cases of domestic, industrial consumption and additional demand from confined livestock and irrigated agriculture which merit further investigation. It is important to note that the inter-annual water scarcity values were based on 3-month means of natural runoff obtained from IBIS, which we found to be close to observed values between September and November when blue water scarcity was its greatest in 2050.

Moisture recycling feedbacks resulting from reduced vegetation cover and an expanding small farm reservoir network were not included in our estimate of both long-term green and blue water availability and, therefore, water scarcity indicators. A reduction in precipitation as a result of land use change would reduce green water availability in the basin and therefore increase the magnitudes of our estimates towards more unsustainable limits. Similarly, reduced precipitation in the basin can further affect runoff at the regional scale [14], thereby increasing blue water scarcity as estimated here. Both of these limitations, therefore suggest that our results represent mainly a conservative estimate of the effects described in this study.

5. Conclusions

The application of the WFSA revealed the importance of the agricultural sector for future land and water management initiatives in the XBMT. Our study has also provided an important case for estimating blue and green water scarcities in the context of land use change, climate change and agricultural production scenarios. Agricultural expansion between 2000 and 2015 led to conditions under which green water scarcity moved towards "threat" conditions, while blue water resources remained within sustainable limits. The evaluation of two water resource use options for agricultural intensification confirmed the importance of land use policies in further reducing deforestation as a driver for intensifying agricultural production in the basin. Future cropland expansion can rely on further green water appropriation by expanding onto pasture, while cattle confinement and cropland irrigation for increased cropping frequency have the potential of bringing the basin towards dry season sustainable limits. Future studies should consider the role of small farm reservoirs and irrigation in the water cycle to identify their importance for regional groundwater storage, downstream blue water availability, and also for large scale moisture recycling and the atmospheric water balance.

Supplementary Materials: The following are available online at http://www.mdpi.com/2073-4441/10/4/349/s1, Figure S1: Validation of the monthly discharge for the Xingu Headwaters, Figure S2: Modeled compared to observed 3-month mean discharge for the Xingu Headwaters, Figure S3: Exceedance probability curve for the Xingu Headwaters, Figure S4: Values of ET (top-down approach), Table S1: Cropland and pasture ET with respective area estimates, Table S2: Average livestock population, water demand and living condition assumptions, Table S3: Urban, rural, industrial worker population and domestic and industrial blue water demand, Table S4: Total forest cover obtained from land use maps, Table S5: Blue Water Footprint results, Table S6: Runoff results for the Xingu Basin of Mato Grosso from IBIS simulations and land use, Table S7: Individual land use contributions to ET, Table S8: Values of ET (top-down approach).

Acknowledgments: This research was supported by the Natural Sciences and Engineering Research Council (NSERC) through the Vanier Canada Graduate Scholarship to M.J.L. (201411DVC-347484-257696). Results constitute a contribution to the project "Integrating land use planning and water governance in Amazonia: Towards improving freshwater security in the agricultural frontier of Mato Grosso" supported by the Belmont

Forum and the G8 Research Councils Freshwater Security Grant to M.S.J. through NSERC (G8PJ-437376-2012). Additional funding was provided by The National Science Foundation (ICER-1343421) and the Gordon and Betty Moore Foundation to M.T.C. We thank Divino Silvério for his help in sharing results for this study, and Kylen Solvik, and Marcia Macedo for providing reservoir area. We also thank Trent Biggs and two anonymous reviewers for their comments to help improve the quality of this paper.

Author Contributions: M.J.L. conceived the study and wrote the paper, M.T.C. and A.C. performed the simulations and J.G. provided the land use maps. M.J.L., M.T.C., A.C., J.G., and M.S.J analyzed the data and provided feedback on drafts of the paper.

Conflicts of Interest: The authors declare no conflict of interest.

References

1. Simon, M.F.; Garagorry, F.L. The expansion of agriculture in the Brazilian Amazon. *Environ. Conserv.* **2006**, *32*, 203. [CrossRef]
2. Barona, E.; Ramankutty, N.; Hyman, G.; Coomes, O.T. The role of pasture and soybean in deforestation of the Brazilian Amazon. *Environ. Res. Lett.* **2010**, *5*, 024002. [CrossRef]
3. Spera, S.A.; Galford, G.L.; Coe, M.T.; Macedo, M.N.; Mustard, J.F. Land-use change affects water recycling in Brazil's last agricultural frontier. *Glob. Chang. Biol.* **2016**, *22*, 3405–3413. [CrossRef] [PubMed]
4. Macedo, M.N.; DeFries, R.S.; Morton, D.C.; Stickler, C.M.; Galford, G.L.; Shimabukuro, Y.E. Decoupling of deforestation and soy production in the southern Amazon during the late 2000s. *Proc. Natl. Acad. Sci. USA* **2012**, *109*, 1341–1346. [CrossRef] [PubMed]
5. Richards, P.; Pellegrina, H.; VanWey, L.; Spera, S. Soybean Development: The Impact of a Decade of Agricultural Change on Urban and Economic Growth in Mato Grosso, Brazil. *PLoS ONE* **2015**, *10*, e0122510. [CrossRef] [PubMed]
6. Davidson, E.A.; de Araújo, A.C.; Artaxo, P.; Balch, J.K.; Brown, I.F.; Bustamante, M.M.C.; Coe, M.T.; DeFries, R.S.; Keller, M.; Longo, M.; et al. The Amazon basin in transition. *Nature* **2012**, *481*, 321–328. [CrossRef] [PubMed]
7. Galford, G.L.; Melillo, J.; Mustard, J.F.; Cerri, C.E.P.; Cerri, C.C. The Amazon Frontier of Land-Use Change: Croplands and Consequences for Greenhouse Gas Emissions. *Earth Interact.* **2010**, *14*, 1–24. [CrossRef]
8. Coe, M.T.; Costa, M.H.; Soares-Filho, B.S. The influence of historical and potential future deforestation on the stream flow of the Amazon River—Land surface processes and atmospheric feedbacks. *J. Hydrol.* **2009**, *369*, 165–174. [CrossRef]
9. Silvério, D.V.; Brando, P.M.; Macedo, M.N.; Beck, P.S.A.; Bustamante, M.; Coe, M.T. Agricultural expansion dominates climate changes in southeastern Amazonia: The overlooked non-GHG forcing. *Environ. Res. Lett.* **2015**, *10*, 104015. [CrossRef]
10. Lathuillière, M.J.; Johnson, M.S.; Donner, S.D. Water use by terrestrial ecosystems: Temporal variability in rainforest and agricultural contributions to evapotranspiration in Mato Grosso, Brazil. *Environ. Res. Lett.* **2012**, *7*, 024024. [CrossRef]
11. António Sumila, T.C.; Pires, G.F.; Fontes, V.C.; Costa, M.H. Sources of Water Vapor to Economically Relevant Regions in Amazonia and the Effect of Deforestation. *J. Hydrometeorol.* **2017**, *18*, 1643–1655. [CrossRef]
12. Oliveira, L.J.C.; Costa, M.H.; Soares-Filho, B.S.; Coe, M.T. Large-scale expansion of agriculture in Amazonia may be a no-win scenario. *Environ. Res. Lett.* **2013**, *8*, 024021. [CrossRef]
13. Stickler, C.M.; Coe, M.T.; Costa, M.H.; Nepstad, D.C.; McGrath, D.G.; Dias, L.C.P.; Rodrigues, H.O.; Soares-Filho, B.S. Dependence of hydropower energy generation on forests in the Amazon Basin at local and regional scales. *Proc. Natl. Acad. Sci. USA* **2013**, *110*, 9601–9606. [CrossRef] [PubMed]
14. Castello, L.; Macedo, M.N. Large-scale degradation of Amazonian freshwater ecosystems. *Glob. Chang. Biol.* **2016**, *22*, 990–1007. [CrossRef] [PubMed]
15. Lathuillière, M.J.; Coe, M.T.; Johnson, M.S. A review of green- and blue-water resources and their trade-offs for future agricultural production in the Amazon Basin: What could irrigated agriculture mean for Amazonia? *Hydrol. Earth Syst. Sci.* **2016**, *20*, 2179–2194. [CrossRef]
16. IBGE. Banco de Dados Agregados. Available online: www.sidra.ibge.gov.br/ (accessed on 1 June 2016).
17. Hoekstra, A.Y.; Chapagain, A.K.; Aldaya, M.M.; Mekonnen, M.M. *The Water Footprint Assessment Manual*; Earthscan: London, UK, 2011; ISBN 9781849712798.

18. Chenoweth, J.; Hadjikakou, M.; Zoumides, C. Quantifying the human impact on water resources: A critical review of the water footprint concept. *Hydrol. Earth Syst. Sci.* **2014**, *18*, 2325–2342. [CrossRef]

19. Quinteiro, P.; Ridoutt, B.G.; Arroja, L.; Dias, A.C. Identification of methodological challenges remaining in the assessment of a water scarcity footprint: A review. *Int. J. Life Cycle Assess.* **2017**. [CrossRef]

20. Hoekstra, A.Y. Water Footprint Assessment: Evolvement of a New Research Field. *Water Resour. Manag.* **2017**, *31*, 3061–3081. [CrossRef]

21. Falkenmark, M.; Rockström, J. The New Blue and Green Water Paradigm: Breaking New Ground for Water Resources Planning and Management. *J. Water Resour. Plan. Manag.* **2006**, *132*, 129–132. [CrossRef]

22. Mekonnen, M.M.; Hoekstra, A.Y. Four billion people facing severe water scarcity. *Sci. Adv.* **2016**, *2*, e1500323. [CrossRef] [PubMed]

23. Hoekstra, A.Y.; Mekonnen, M.M.; Chapagain, A.K.; Mathews, R.E.; Richter, B.D. Global Monthly Water Scarcity: Blue Water Footprints versus Blue Water Availability. *PLoS ONE* **2012**, *7*, e32688. [CrossRef] [PubMed]

24. Miguel Ayala, L.; van Eupen, M.; Zhang, G.; Pérez-Soba, M.; Martorano, L.G.; Lisboa, L.S.; Beltrao, N.E. Impact of agricultural expansion on water footprint in the Amazon under climate change scenarios. *Sci. Total Environ.* **2016**, *569–570*, 1159–1173. [CrossRef] [PubMed]

25. Schyns, J.F.; Hoekstra, A.Y.; Booij, M.J. Review and classification of indicators of green water availability and scarcity. *Hydrol. Earth Syst. Sci.* **2015**, *19*, 4581–4608. [CrossRef]

26. ANA. *Plano Estratégico de Recursos Hídricos dos Afluentes da Margem Direita do Rio Amazonas*; Agência Nacional de Águas: Brasilia, Brazil, 2013.

27. Velasquez, H.Q.C.; Bernasconi, P. *Fique por Dentro: A Bacia do Rio Xingu em Mato Grosso*; Instituto Socioambiental, Instituto Centro de Vida: São Paulo, Brazil, 2010.

28. Panday, P.K.; Coe, M.T.; Macedo, M.N.; Lefebvre, P.; Castanho, A.D.d.A. Deforestation offsets water balance changes due to climate variability in the Xingu River in eastern Amazonia. *J. Hydrol.* **2015**, *523*, 822–829. [CrossRef]

29. Macedo, M.N.; Coe, M.T.; DeFries, R.; Uriarte, M.; Brando, P.M.; Neill, C.; Walker, W.S. Land-use-driven stream warming in southeastern Amazonia. *Philos. Trans. R. Soc. B Biol. Sci.* **2013**, *368*, 20120153. [CrossRef] [PubMed]

30. ANA. Hidroweb. Available online: http://hidroweb.ana.gov.br/ (accessed on 1 June 2016).

31. Foley, J.A.; Kucharik, C.J.; Polzin, D. *Integrated Biosphere Simulator Model (IBIS), Version 2.5*; ORNL DAAC: Oak Ridge, TN, USA, 2005.

32. Foley, J.A.; Prentice, I.C.; Ramankutty, N.; Levis, S.; Pollard, D.; Sitch, S.; Haxeltine, A. An integrated biosphere model of land surface processes, terrestrial carbon balance, and vegetation dynamics. *Glob. Biogeochem. Cycles* **1996**, *10*, 603–628. [CrossRef]

33. Kucharik, C.J.; Foley, J.A.; Delire, C.; Fisher, V.A.; Coe, M.T.; Lenters, J.D.; Young-Molling, C.; Ramankutty, N. Testing the performance of a dynamic global ecosystem model: Water balance, carbon balance, and vegetation structure. *Glob. Biogeochem. Cycles* **2000**, *14*, 795–825. [CrossRef]

34. Ramankutty, N.; Foley, J.A. Estimating historical changes in global land cover: Croplands from 1700 to 1992. *Glob. Biogeochem. Cycles* **1999**, *13*, 997–1027. [CrossRef]

35. Graesser, J.; Ramankutty, N. Detection of cropland field parcels from Landsat imagery. *Remote Sens. Environ.* **2017**, *201*, 165–180. [CrossRef]

36. Soares-Filho, B.S.; Nepstad, D.C.; Curran, L.M.; Voll, E.; Garcia, R.A.; Ramos, C.A.; McDonald, A.J.; Lefebvre, P.A.; Schlesinger, P. *LBA-ECO LC-14 Modeled Deforestation Scenarios, Amazon Basin: 2002–2050*; ORNL DAAC: Oak Ridge, TN, USA, 2013. [CrossRef]

37. Lathuillière, M.J.; Johnson, M.S.; Galford, G.L.; Couto, E.G. Environmental footprints show China and Europe's evolving resource appropriation for soybean production in Mato Grosso, Brazil. *Environ. Res. Lett.* **2014**, *9*, 074001. [CrossRef]

38. Lathuillière, M.J.; Bulle, C.; Johnson, M.S. Land Use in LCA: Including Regionally Altered Precipitation to Quantify Ecosystem Damage. *Environ. Sci. Technol.* **2016**, *50*, 11769–11778. [CrossRef] [PubMed]

39. Spera, S.A.; Cohn, A.S.; VanWey, L.K.; Mustard, J.F.; Rudorff, B.F.; Risso, J.; Adami, M. Recent cropping frequency, expansion, and abandonment in Mato Grosso, Brazil had selective land characteristics. *Environ. Res. Lett.* **2014**, *9*, 064010. [CrossRef]

40. Lathuillière, M.J. *Harmonizing Water Footprint Assessment for Agricultural Production in Southern Amazonia*; The University of British Columbia: Vancouver, BC, Canada, 2018.

41. Mesa, D.; Muniz, E.; Souza, A.; Geffroy, B. Broiler-Housing Conditions Affect the Performance. *Rev. Bras. Ciência Avícola* **2017**, *19*, 263–272. [CrossRef]

42. Palhares, J.C.P. Pegada hídrica dos suínos abatidos nos Estados da Região Centro-Sul do Brasil. *Acta Sci. Anim. Sci.* **2011**, *33*, 309–314. [CrossRef]

43. IPCC. *2006 IPCC Guidelines for National Greenhouse Gas Inventories, Prepared by the National Greenhouse Gas Inventories Programme*; Eggleston, S., Buendia, L., Miwa, K., Ngara, T., Tanabe, K., Eds.; IGES: Hayama, Japan, 2006.

44. FAO. *Guidelines for Water Use Assessment of Livestock Production Systems and Supply Chains*; Livestock Environmental Assessment and Performance (LEAP) Partnership: Rome, Italy, 2018.

45. Smakhtin, V.; Revenga, C.; Döll, P. A Pilot Global Assessment of Environmental Water Requirements and Scarcity. *Water Int.* **2004**, *29*, 307–317. [CrossRef]

46. Richter, B.D.; Davis, M.M.; Apse, C.; Konrad, C. A presumptive standard for environmental flow protection. *River Res. Appl.* **2012**, *28*, 1312–1321. [CrossRef]

47. Presidência da República, Casa Civil, Subchefia para Assuntos Jurídicos. *Lei n⁰ 12.651, de 25 de maio de 2012*; Presidência da República, Casa Civil, Subchefia para Assuntos Jurídicos: Brasilia, Brazil, 2012.

48. MAPA. *Projeções do Agronegócio: Brasil 2015/16 a 2025/26 Projeções de Longo Prazo*; Ministério da Agricultura, Pecuária e Abastecimento: Brasilia, Brazil, 2016.

49. R Core Team. *R: A Language and Environment for Statistical Computing*; R Foundation for Statistical Computing: Vienna, Austria, 2017.

50. Hijmans, R.J. Raster: Geographic Data Analysis and Modeling. R package Version 2.5-8. Available online: https://cran.r-project.org/package=raster (accessed on 1 June 2016).

51. Bivand, R.S.; Pebesma, E.; Gomez-Rubio, V. *Applied Spatial Data Analysis with R*, 2nd ed.; Springer: New York, NY, USA, 2013.

52. Pebesma, E.J.; Bivand, R.S. Classes and Methods for Spatial Data in R. R News 5 (2). Available online: https://cran.r-project.org/doc/Rnews/ (accessed on 1 June 2016).

53. Bivand, R.; Keitt, T.; Rowlinsgon, B. Rgdal: Bindings for the Geospatial Data Abstraction Library. R Package Version 1.2-7. Available online: https://cran.r-project.org/package=rgdal (accessed on 1 June 2016).

54. Bivand, R.; Lewin-Koh, N. Maptools: Tools for Reading and Handling Spatial Objects. R package version 0.9-2. Available online: https://cran.r-project.org/package=maptools (accessed on 1 June 2016).

55. Pierce, D. ncdf4: Interface to Unidata netCDF (Version 4 or Earlier) Format Data Files. R Package Version 1.16. Available online: https://cran.r-project.org/package=ncdf4 (accessed on 1 June 2016).

56. Callow, J.N.; Smettem, K.R.J. The effect of farm dams and constructed banks on hydrologic connectivity and runoff estimation in agricultural landscapes. *Environ. Model. Softw.* **2009**, *24*, 959–968. [CrossRef]

57. Coe, M.T.; Brando, P.M.; Deegan, L.A.; Macedo, M.N.; Neill, C.; Silvério, D.V. The Forests of the Amazon and Cerrado Moderate Regional Climate and Are the Key to the Future. *Trop. Conserv. Sci.* **2017**, *10*, 1–6. [CrossRef]

58. Arvor, D.; Dubreuil, V.; Ronchail, J.; Simões, M.; Funatsu, B.M. Spatial patterns of rainfall regimes related to levels of double cropping agriculture systems in Mato Grosso (Brazil). *Int. J. Climatol.* **2014**, *34*, 2622–2633. [CrossRef]

59. Baillie, C. Assessment of Evaporation Losses and Evaporation Mitigation Technologies for on Farm Water Storages across Australia. *Coop. Res. Cent. Irrig. Futures Irrig. Matters Ser.* **2008**, *05/08*, 1–52.

60. Da Silva, V.; de Oliveira, S.; Hoekstra, A.; Dantas Neto, J.; Campos, J.; Braga, C.; de Araújo, L.; Aleixo, D.; de Brito, J.; de Souza, M.; et al. Water Footprint and Virtual Water Trade of Brazil. *Water* **2016**, *8*, 517. [CrossRef]

61. FABOV. *Diagnostico da Cadeia Produtiva Agroindustrial da Bovinocultura de Corte do Estado de Mato Grosso*; Fundo de Apoio à Bovinocultura de Corte: Cuiabá, Mato Grosso, Brazil, 2007.

62. MAPA. Agrostat. Available online: http://indicadores.agricultura.gov.br/agrostat (accessed on 1 June 2016).

63. Ercin, A.E.; Chico, D.; Chapagain, A.K. Dependencies of Europe's Economy on Other Parts of the World in Terms of Water Resources. Available online: http://waterfootprint.org/media/downloads/Imprex-D12-1_final.pdf (accessed on 1 June 2016).

64. Vanham, D.; Mekonnen, M.M.; Hoekstra, A.Y. The water footprint of the EU for different diets. *Ecol. Indic.* **2013**, *32*, 1–8. [CrossRef]

65. Nepstad, D.; McGrath, D.; Stickler, C.; Alencar, A.; Azevedo, A.; Swette, B.; Bezerra, T.; DiGiano, M.; Shimada, J.; Seroa da Motta, R.; et al. Slowing Amazon deforestation through public policy and interventions in beef and soy supply chains. *Science* **2014**, *344*, 1118–1123. [CrossRef] [PubMed]

66. ANA & Embrapa/CNPMS. *Levantamento da Agricultura Irrigada por Pivôs Centrais no Brasil—ano 2014*; ANA and Embrapa: Brasilia, Brazil, 2016.

67. FEALQ. *Análise Territorial Para o Desenvolvimento da Agricultural Irrigada no Brasil*; Fundação de Estudos Agrários Luiz de Queiros: Piracicaba, São Paulo, Brazil, 2014.

68. Palhares, J.C.P.; Morelli, M.; Junior, C.C. Impact of roughage-concentrate ratio on the water footprints of beef feedlots. *Agric. Syst.* **2017**, *155*, 126–135. [CrossRef]

69. Pokhrel, Y.N.; Fan, Y.; Miguez-Macho, G. Potential hydrologic changes in the Amazon by the end of the 21st century and the groundwater buffer. *Environ. Res. Lett.* **2014**, *9*, 84004. [CrossRef]

70. Noojipady, P.; Morton, C.D.; Macedo, N.M.; Victoria, C.D.; Huang, C.; Gibbs, K.H.; Bolfe, L.E. Forest carbon emissions from cropland expansion in the Brazilian Cerrado biome. *Environ. Res. Lett.* **2017**, *12*, 025004. [CrossRef]

71. Khanna, J.; Medvigy, D.; Fueglistaler, S.; Walko, R. Regional dry-season climate changes due to three decades of Amazonian deforestation. *Nat. Clim. Chang.* **2017**, *7*, 200–204. [CrossRef]

72. Zemp, D.C.; Schleussner, C.-F.; Barbosa, H.M.J.; Hirota, M.; Montade, V.; Sampaio, G.; Staal, A.; Wang-Erlandsson, L.; Rammig, A. Self-amplified Amazon forest loss due to vegetation-atmosphere feedbacks. *Nat. Commun.* **2017**, *8*, 14681. [CrossRef] [PubMed]

73. Wright, J.S.; Fu, R.; Worden, J.R.; Chakraborty, S.; Clinton, N.E.; Risi, C.; Sun, Y.; Yin, L. Rainforest-initiated wet season onset over the southern Amazon. *Proc. Natl. Acad. Sci. USA* **2017**, *114*, 8481–8486. [CrossRef] [PubMed]

74. Keys, P.W.; Wang-Erlandsson, L.; Gordon, L.J. Revealing Invisible Water: Moisture Recycling as an Ecosystem Service. *PLoS ONE* **2016**, *11*, e0151993. [CrossRef] [PubMed]

75. Bagley, J.E.; Desai, A.R.; Harding, K.J.; Snyder, P.K.; Foley, J.A. Drought and Deforestation: Has Land Cover Change Influenced Recent Precipitation Extremes in the Amazon? *J. Clim.* **2014**, *27*, 345–361. [CrossRef]

76. Berger, M.; van der Ent, R.; Eisner, S.; Bach, V.; Finkbeiner, M. Water Accounting and Vulnerability Evaluation (WAVE): Considering Atmospheric Evaporation Recycling and the Risk of Freshwater Depletion in Water Footprinting. *Environ. Sci. Technol.* **2014**, *48*, 4521–4528. [CrossRef] [PubMed]

77. van der Ent, R.J.; Savenije, H.H.G.; Schaefli, B.; Steele-Dunne, S.C. Origin and fate of atmospheric moisture over continents. *Water Resour. Res.* **2010**, *46*, 1–12. [CrossRef]

78. Keys, P.; Wang-Erlandsson, L.; Gordon, L.; Galaz, V.; Ebbesson, J. Approaching moisture recycling governance. *Glob. Environ. Chang.* **2017**, *45*, 15–23. [CrossRef]

79. Latawiec, A.E.; Strassburg, B.B.N.; Silva, D.; Alves-Pinto, H.N.; Feltran-Barbieri, R.; Castro, A.; Iribarrem, A.; Rangel, M.C.; Kalif, K.A.B.; Gardner, T.; et al. Improving land management in Brazil: A perspective from producers. *Agric. Ecosyst. Environ.* **2017**, *240*, 276–286. [CrossRef]

80. Garcia, E.; Ramos Filho, F.; Mallmann, G.; Fonseca, F. Costs, Benefits and Challenges of Sustainable Livestock Intensification in a Major Deforestation Frontier in the Brazilian Amazon. *Sustainability* **2017**, *9*, 158. [CrossRef]

81. Arvor, D.; Tritsch, I.; Barcellos, C.; Jégou, N.; Dubreuil, V. Land use sustainability on the South-Eastern Amazon agricultural frontier: Recent progress and the challenges ahead. *Appl. Geogr.* **2017**, *80*, 86–97. [CrossRef]

82. Neill, C.; Jankowski, K.; Brando, P.M.; Coe, M.T.; Deegan, L.A.; Macedo, M.N.; Riskin, S.H.; Porder, S.; Elsenbeer, H.; Krusche, A.V. Surprisingly Modest Water Quality Impacts From Expansion and Intensification of Large-Sscale Commercial Agriculture in the Brazilian Amazon-Cerrado Region. *Trop. Conserv. Sci.* **2017**, *10*, 1–5. [CrossRef]

83. Hayhoe, S.J.; Neill, C.; Porder, S.; McHorney, R.; Lefebvre, P.; Coe, M.T.; Elsenbeer, H.; Krusche, A.V. Conversion to soy on the Amazonian agricultural frontier increases streamflow without affecting stormflow dynamics. *Glob. Chang. Biol.* **2011**, *17*, 1821–1833. [CrossRef]

84. Pokhrel, Y.N.; Fan, Y.; Miguez-Macho, G.; Yeh, P.J.F.; Han, S.C. The role of groundwater in the Amazon water cycle: 3. Influence on terrestrial water storage computations and comparison with GRACE. *J. Geophys. Res. Atmos.* **2013**, *118*, 3233–3244. [CrossRef]

85. Gleeson, T.; Wada, Y.; Bierkens, M.F.P.; van Beek, L.P.H. Water balance of global aquifers revealed by groundwater footprint. *Nature* **2012**, *488*, 197–200. [CrossRef] [PubMed]

water

MDPI

Article

Threshold Based Footprints (for Water)

Benjamin L. Ruddell

School of Informatics Computing and Cyber Systems, Northern Arizona University, Flagstaff, AZ 86011, USA;
Benjamin.ruddell@nau.edu; Tel.: +1-928-523-3124

Received: 4 July 2018; Accepted: 31 July 2018; Published: 3 August 2018

Abstract: Thresholds are an emergent property of complex systems and Coupled Natural Human Systems (CNH) because they indicate "tipping points" where a complicated array of social, environmental, and/or economic processes combine to substantially change a system's state. Because of the elegance of the concept, thresholds have emerged as one of the primary tools by which socio-political systems simplify, define, and especially regulate complex environmental impacts and resource scarcity considerations. This paper derives a general framework for the use of thresholds to calculate scarcity footprints, and presents a volumetric Threshold-based Water Footprint (TWF), comparing it with the Blue Water Footprint (BWF) and the Relevant for Environmental Deficiency (RED) midpoint impact indicator. Specific findings include (a) one requires all users' BWF to calculate an individual user's TWF, whereas one can calculate an individual user's BWF without other users' data; (b) local maxima appear in the Free from Environmental Deficiency (FED) efficiency of the RED metric due to its nonlinear form; and (c) it is possible to estimate the "effective" threshold that is approximately implied by the RED water use impact metric.

Keywords: blue water footprint; water scarcity footprint; threshold; embedded resource accounting; life cycle analysis; regulation

1. Introduction

The 21st century's problems are increasingly systemic and rooted in the indirect connections of a complex Coupled Natural–Human system (CNH) [1]. Decision making is confounded by indirect effects, joint effects, and unintended consequences. As a result, leaders are calling for the development of sustainability metrics that link decisions to their systemic consequences [2].

Threshold metrics indicate "tipping points" where a complicated array of social, environmental, and/or economic processes combine to substantially change a system's state. The system may be complex, but the threshold is simple and easily communicated. A threshold is measured against a single system performance index or metric. Because of the elegance of the concept, thresholds have emerged as one of the primary tools by which socio-political systems simplify, define, and especially regulate complex environmental impacts. The elegance of thresholds makes them both very useful and very dangerous, because they facilitate both simplification and oversimplification of the concept of "impact" in complex systems. It is therefore important to develop methods of characterization for human consumption and its impacts that are compatible with the ubiquitous threshold-based regulatory paradigm.

Thresholds often manifest as sharp discontinuities between system states that are sustainable versus states that are unsustainable [3–5], and between states that are affordable versus states that are expensive. For example, crops wither and regions erupt into violence when drought reaches a certain threshold of severity [6]. Marine ecosystems collapse when phytoplankton drops below a threshold [7]. Freshwater ecosystem health requires sustained environmental flows [8]. Thresholds have been used by ecological economists because they can integrate issues of cost and value in complex and diverse contexts [9], such as when the impacts of human economic activity increase beyond an acceptable level [10–13]. The frequently employed concepts of "maximum concentration levels" of pollutants [14,15], "adverse resource impact" limits [12], ecological flow requirements [16], and carrying capacities [17–19] are excellent examples of thresholds placed on resource stocks to distinguish between sustainable and unsustainable states. Western US water law speaks of groundwater use that either "is" or "is not" impacting surface flows, whereas the hydrological truth is somewhere in-between. The Colorado River Treaty's water allocations depend in part on whether water levels in Lake Mead are above or below a key elevation, which is a legal threshold separating relatively "abundant" from relatively "scarce" Colorado River water. Thresholds draw a sharp "black and white" line where the underlying science typically reflects "shades of grey", but an accurately defined threshold can be extremely useful owing to its simplicity, clarity, communicability, and legal compatibility.

Thresholds can be defined using metrics that integrate socio-ecological pressures, services, values, and impacts [20–24]. Or, threshold can be based on physical quantities such as a sustainable yield, an ecosystem flow requirement, flow variability, inputs, or planetary boundaries [11,16,25–35]. Increasingly, sustainability indices have focused on social sustainability and social capital in addition to environmental sustainability [34–36]. Empirical scientific work establishing socio-environmental thresholds of the global CNH is diverse and mature [37–41].

Thresholds have emerged from the complex socio-political system as a favored conceptual framework for environmental regulation, and are a favored approach of the U.S. Environmental Protection Agency and other environmental agencies. Regulatory thresholds are set using a complex combination of social, environmental, and economic factors and processes that cannot be reduced to a simple technocratic formula but instead include multiple considerations unique to a given context [11,28]. It is wise to recognize this emergent legal and socio-environmental concept in the construction of sustainability indices, by linking these indices to regulatory systems and participatory government [42,43]. Otherwise, our "policy suffers from a profound disconnect between science and law." [43]. In other words, regulatory thresholds tend be socio-economic, and political, in addition to environmental, in their constitution.

A footprint is a quantitative and usually volumetric (i.e., conserving mass or energy) measure of the depletion of an inventory of a natural resource stock [21,44]. Typical examples of footprints include Ecological, Carbon, and Water Footprints [21,44–46]. Footprint indices are relatives of Life Cycle Analysis methods (LCA). Whereas footprints usually emphasize carrying capacities, planetary boundaries, and straightforward units such as mass, energy, or volume, LCA methods focus on the translation of volumetric, inventory, and pressure metrics via mid-point metrics into end-point impact metrics that are a type of index for the environmental cost or price of a process or product [47,48]. The focus of this paper's discussion is on footprint indices and on LCA volumetric mid-point indicators (attributional indicators), but LCA end-point (consequential) indicators are beyond this scope and are not addressed.

This paper presents the simple mathematics of a Threshold-based Water Footprint (TWF), which is a special case of the generalized Threshold-based Footprint (FT). The implications of these mathematics are explored through a comparison with the Blue Water Footprint (BWF) [49] and the Relevant for Environmental Deficiency (RED) [50] mid-point impact metric that characterizes the context-based impact of the BWF using the Water Scarcity Index (WSI) [51]. Section 2 derives the simple mathematics of TWF and compares them with BWF and RED. Section 3 presents comparisons between BWF, TWF, and RED and argues that TWF is a simple approximation for RED. Section 4 summarizes conclusions and discusses their implications.

2. Materials and Methods

2.1. Mathematics of a Threshold-Based Footprint

During some differential time interval there is an initial (subscript zero) resource Stock capacity, S_0, before the stock is used. Then, S_0 is drawn down by a gross Withdrawal, W, due to the aggregated direct actions of all processes in the system. Stock-reducing withdrawals are positive by sign convention. The term "stock" is used generally and not strictly, and the "stock" could be one of a variety of environmental quantities: stock, flow, resource, event magnitude, population, or incidence. For instance, in the water footprint example in Section 2.2, the metric of interest is surface water flow.

The aggregate net volumetric Footprint F is some fraction of W, adjusted by the withdrawal-weighted average consumption coefficient c of the processes, such that $F = c\,W$. F is also the Consumptive Use, in water applications. The Threshold-based Footprint F^T is the nonnegative difference between F and the Threshold T (Equation (1), Figure 1). The stock's Threshold is often a limit on the sustainable consumption or degradation of that stock. For example, an environmental flow requirement R would be the difference between the flow Q and the surface water stock's threshold, such that $R = Q - T$. The Free Footprint F^f is the portion of F falling below the threshold T. This free portion of the footprint is discounted and characterized as having no impact because it has negligible environmental or economic cost (Equation (2), Figure 1). The relative Threshold $T' = T/S$ is the Threshold expressed as a fraction of the Stock. Units of W, F, and T are those of S; c is unitless. If $T = 0$ then all resource withdrawals are adverse and $F^T = F$; if $T = S$ then all impacts are discounted and $F^T = 0$. $F = F^f$ until $F > T$. Observe in Equations (1) and (2) that the inventory of the stock S does not appear in the footprint calculation, but T does appear and is presumably related somehow to S. If $F^T = 0$ the resource is "abundant" (negligible marginal value and cost), at least from the point of view of this decision making process. If $F^T > 0$ the resource is scarce from the point of view of this decision making process and has non-negligible marginal value and cost.

$$F^T = \max\left\{ \begin{array}{c} 0 \\ F - T \end{array} \right\} \tag{1}$$

$$F^f = \min\left\{ \begin{array}{c} T \\ F \end{array} \right\} \tag{2}$$

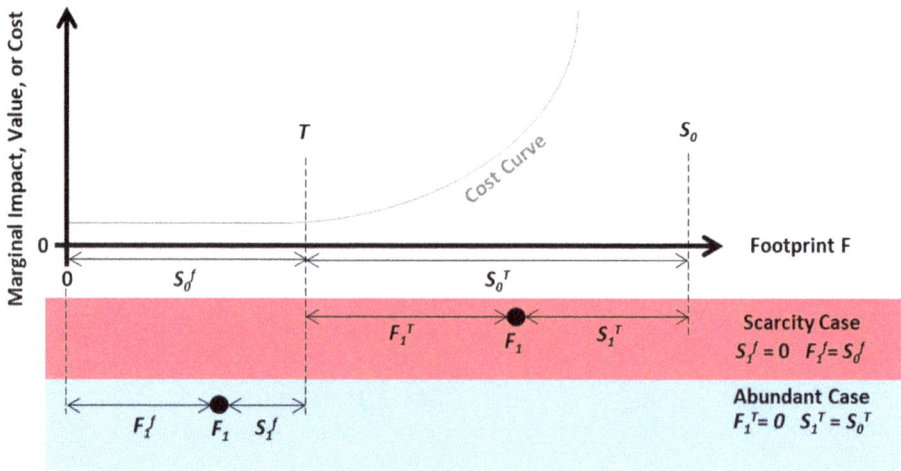

Figure 1. Schematic of the Threshold-based Footprint concept. The vertical axis gives the marginal value, impact or cost of water use, and the horizontal axis gives the Footprint. The cost of the footprint increases slowly at first as the footprint rises, but beyond a threshold it increases sharply and in unbounded fashion as the resource becomes "scarce" and the marginal value, impact, and cost begin to rise. Illustrations of footprint components during "scarcity" ($F > T$) and "abundant" ($F \leq T$) conditions are given. Section 2.1 defines mathematical symbols and equations. Commonly employed scarcity and stress indices [35] may be expressed using these mathematics (Appendix A). A discussion of the interpretation of impact and cost is provided in Appendix B.

F is the sum of the "free" and "adverse" components, such that $F = F^T + F^f$. By definition, if $c \leq T'$, then $F^T = 0$ and $F = F^f$. The Free Fraction R of the footprint, $R = F^f/F$, is a sort of efficiency metric for the footprint. For example, if a river has a flow S of 1 Million m³/year and an adverse threshold T of 100,000 m³/year, and the total footprint F is 150,000 m³/year, then $F^f = 100,000$ m³/year and $F^T = 50,000$ m³/year; $R = 100,000/150,000$ and thus two thirds of this footprint is "free".

The initial adverse capacity above the threshold is S_0^T, where $S_0^T = S_0 - T_0$, and the initial free capacity below the threshold is S_0^f, where $S_0^f = T_0$ (Figure 1). After an initial footprint F_0 is applied, the remaining adverse capacity is S_1^T (Equation (3), Figure 1), and the remaining free capacity is S_1^f (Equation (4), Figure 1).

$$S_1^T = min \left\{ \begin{array}{c} S_0 - T_0 \\ S_0 - F_0 \end{array} \right\} \tag{3}$$

$$S_1^f = max \left\{ \begin{array}{c} 0 \\ T_0 - F_0 \end{array} \right\} \tag{4}$$

F_x is the footprint of an individual process x. F_x is not bounded by F when there are multiple processes, because the net impacts of different processes may be positive or negative and may offset, such that $F = \sum_x F_x$. As with the aggregated footprint, the process's footprint is the sum of the free and the adverse components, so $F_x = F_x^T + F_x^f$. Processes may possess their own thresholds, T_x; a U.S. Environmental Protection Agency (EPA) regulation placing a Total Maximum Daily Load (TMDL) limit on a factory's emission of water pollution is an example of a process having a threshold.

The relationship between the stock-level footprints and process-level footprints is complicated and is contextualized based on the policies governing this stock's use. In a seniority-based framework such as the U.S. Western States' Prior Appropriation Doctrine the first process to use water ($x = 1$), would have $T_1 = T$. However, each subsequent and junior processes ($x > 1$) would have its threshold set at current free capacity of the stock, $S_1{}^f$ (Equation (4)), after the sum total of all prior footprints F_0 were deducted, such that $F_0 = \sum_{x\_prior} F_x$. Without any seniority one might choose to weight the threshold, free footprint, and adverse footprint of a process by its contribution to the footprint by using the weighting factor $b = F_x/F$, so $F_x{}^T = b\, F^T$, $F_x{}^f = b\, F^f$, and $T_x = b\, T$. Or, one could use a different weighting factor for a more progressive attribution system.

2.2. Application of the Threshold Concept to Water Footprints and Impact Metrics

The aggregated net consumptive fresh water use in the combined surface and ground water resources of a location is the "blue" Water Footprint (BWF). BWF is an implementation of F. The Threshold-based blue surface flow Water Footprint (TWF) is an implementation of F^T. The Free Water Footprint (FWF) is equal to F_f. If BWF is used in LCA, it is an inventory, volumetric, or pressure type LCA metric, whereas TWF is a mid-point LCA metric. However, both use identical volumetric units, for instance cubic meters or gallons. Crucially, for example, for a company that seeks to measure the impact of its water footprint, for $T > 0$ it is not possible to calculate TWF or FWF without full knowledge of the net aggregated BWF of all other processes impacting that water stock, as well as the stock's threshold. TWF and FWF therefore have a fundamentally higher burden of information than BWF.

The typical application of a Water Stress Index (WSI) [51] utilizes research concerning ecological water scarcity and flow variability at river basin and annual scales to characterize the impact of the consumptive freshwater use in the basin. In this paper [51] WSI is always defined as a logistic function of the stock-scale Withdrawal-to-Availability ratio (WTA) (Equation (5)), and is therefore a dimensionless fraction bounded below one. WSI* factors in low-flow season annual flow variability using a Variation Factor (VF), such that $WSI^* = WSI \times VF$. An empirically estimated median value for VF is 1.8 for annual-timescale river basin stock definitions [52,53]. If the river basin has a Strongly Regulated Flow (SRF) [53] due to a large reservoir storage capacity, WSI* becomes $WSI^*{}_{SRF}$ where the square root of the VF is utilized, reflecting lower flow variability and less low-flow seasonal water stress in an SRF basin, such that $WSI^* = WSI \times VF^{1/2}$. Note that this definition of WSI assumes a constant relationship between the WTA ratio and the water stress in the basin.

$$WSI = \frac{1}{1 + e^{-6.4 \cdot WTA^* \left(\frac{1}{0.01} - 1 \right)}} \tag{5}$$

The Relevant for Environmental Deficiency metric (RED, Equation (19)) is a mid-point metric for the impact created by fresh water consumption at annual timescales for river basins [50], such that $RED = BWF \cdot WSI^*$. The main mathematical difference between RED and TWF is that TWF varies the characterization of impact linearly following a discontinuity at the threshold, whereas RED uses a differentiable and smooth logistic characterization from $WTA = 0$. The main applied difference between TWF and RED as mid-point metrics is that regulations are typically written as thresholds, not smooth logistic functions—for better or worse. Implicit in the definition of RED is the existence of a complement to the characterized water impact, analogous to FWF. This is named Free from Environmental Deficiency (FED), such that $FED = BWF - RED$. These metrics, along with BWF, inform the ISO 14046 water LCA draft standard [54–58].

Similar mathematics have been applied in practice in prior case studies [59]. This study utilized flow depletion thresholds established by the State of Michigan's Department of National Resources within that regulator's Michigan Water Withdrawal Assessment Process (MWWAP) [60]. These thresholds are established uniquely for every individual stream segment in U.S. State of Michigan. Figure 2 reproduces that study's map of the Kalamazoo River, where Mubako et al. [59] calculated the surface water flow Depletion, D, and compared it with the flow depletion threshold, T. Note that Mubako et al.'s [59] "D" is equivalent to F in this paper's mathematics; it is the BWF calculated against the "stock" of surface water flow.

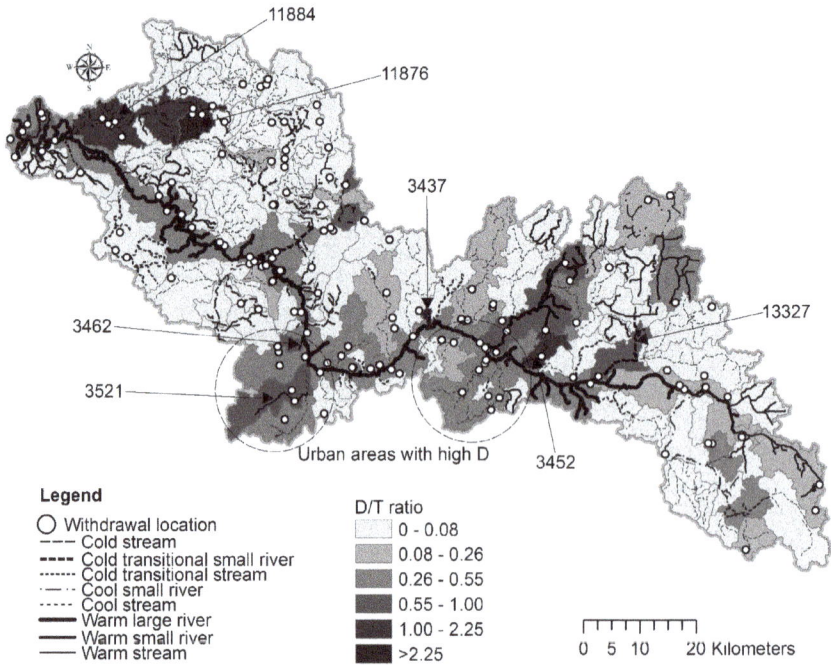

Figure 2. Ratio of the surface water flow (Blue) Water Footprint ($BWF = D = F^T$) to the streamflow depletion threshold, T, for each stream segment of the Kalamazoo River in Michigan, USA. Dark grey colors where $D/T > 1$ indicate the presence of scarcity where $F^T > 0$. This is Figure 5 [59] reproduced with permission of ASCE Press and the Authors.

3. Results

To compare BWF, RED, FED, TWF, and FWF, a synthetic experiment is constructed for a theoretical stream flow. Imagine a stream with a flow of one cubic meter per second, giving initial capacity $S_0 = 1$; hereafter these units will be omitted for simplicity so all results are unitless fractions of a river's total flow. The experiment explores combinations of thresholds T', consumption coefficients c, and Withdrawal-to-Availability (WTA) ratios, so that these five metrics can be compared side by side on a unitless basis. This comparison will make it clear that TWF can give quantitatively similar results to RED depending on the choice of threshold, and that the logistic form chosen by WSI implies an approximate "effective" threshold assumed by RED, a threshold that varies based on the combination of c and WTA.

In Figure 3, the 1:1 line bounds all metrics and approximates FED and FWF for low WTA values that are far below the threshold. RED and TWF are bounded by the c:1 line; RED approaches this line as WTA → 1. $TWF = 0$ and $BWF = FWF$ below the dimensionless value of $WTA = T'/c$. As a result, this dimensionless value defines a critical threshold for water sustainability policy, and it becomes clear that average consumption coefficients and thresholds are essential factors in this policy.

RED's logistic form yields surprising dual peaks and local maxima in FED for higher consumptive use fractions ($c = 0.6$ to 0.9). This high range of consumptive use fractions is common in irrigation-dominated river basins, which are also often water stressed arid or semiarid warm-climate river basins. The smooth logistic function of RED can exceed a slope of one for moderate WTA and high c, meaning that an additional unit of withdrawal creates more than one unit of RED impact under these conditions. FED yields local maxima below the maximum-withdrawal point of *WTA* = 1 in many cases. These local maxima in FED are also local maxima in the Free Fraction R.

For a given c it is possible to calculate a threshold value T' that minimizes the difference between RED and TWF. This minimum-difference threshold can be considered the "effective" flow alteration threshold that is approximately implied by the form of RED and WSI. Best-fit T' is estimated for each c using a linear solver that minimizes the Root of the Mean Squared Error (RMSE) of the error function $e = RED - TWF$. Table 1 gives this best-fit T' for intervals of 0.1 (10%) of c for both the standard and SRF versions of RED. Also in Table 1 is given the RMSE value, the local maxima or "peak" value of FED, and the lowest WTA value at which a FED local maxima occurs.

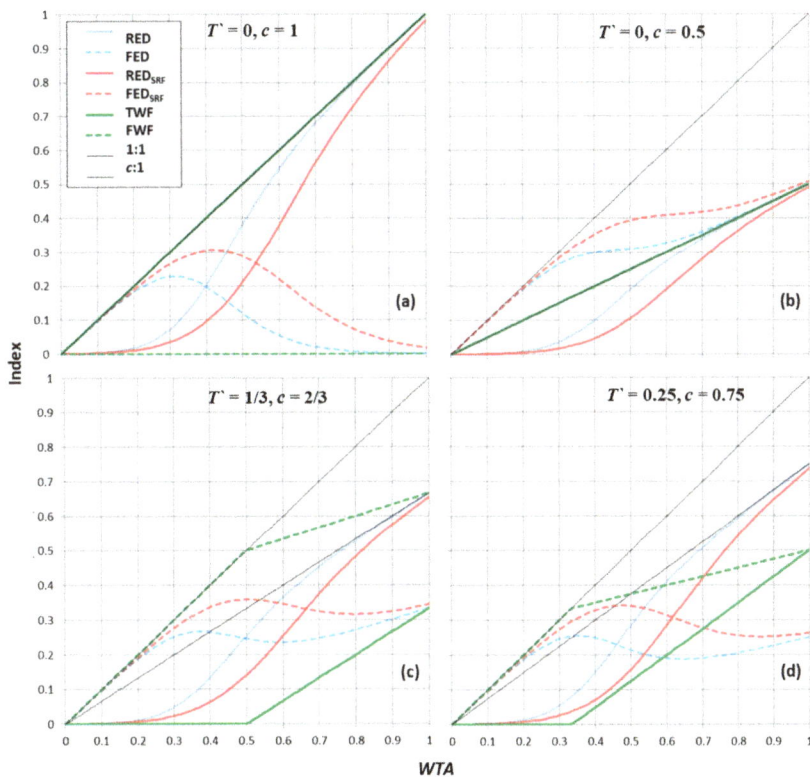

Figure 3. Comparisons between the Relevant for Environmental Deficiency (RED) midpoint impact indicator (in red, with RED Strongly Regulated Flow (SRF) version in blue) and Threshold-based Water Footprint (TWF) (in green), with different choices of threshold T' and mean consumptive use coefficient c, plotted against the Withdrawal-to-Availability index (WTA). BWF is coincident with the c:1 line and bounds RED and TWF. (**a**) T' = 0, c = 1: "Simple Withdrawal" case where all withdrawn water is consumed and counts as adverse impact, resulting in large differences between RED and TWF and additionally Free Water Footprint (*FWF*) = 0. (**b**) T' = 0, c = 0.5: "Symmetry" case resulting in a symmetry of RED and Free from Environmental Deficiency (FED) metrics about TWF and FWF, bounding of TWF and FWF by RED and FED, and *TWF = FWF*. (**c**) T' = 1/3, c = 2/3; "Reversal" case where RED → FWF and TWF → FED as WTA → 1, and *TWF = 0* when *WTA < 0.5*. (**d**) T' = 0.25, c = 0.75; "Convergence" case where TWF converges to FWF at *WTA = 1*, and RED and FED metrics are bounded by TWF and FWF below *WTA = 0.58*.

The effective value of T′ for RED ranges from 1% to 18%, rising with the assumed consumption coefficient. This range of values compares favorably with the river basin freshwater ecosystem flow requirement work [16,39] and specifically with the "presumptive standard" of less than 20% alteration of daily flow due to consumptive use [40], and with the MWWAP's average threshold value of 10% depletion of median summer flows for streams in Michigan [59,60]. When this difference between RED and TWF is expressed relative to the size of the BWF, the relative difference (*RMSE/c*) is constant for all values of c, at 0.08 for the standard RED and 0.09 for the SRF variant of RED. These are relatively small errors that are less than 10% of the total water resource consumption in the system. It is therefore clear that TWF and RED are approximations of each other with a substantial quantitative and qualitative similarity.

To complete the comparison, Figure 4 illustrates the best-fit relationship between RED and WSF. A typical consumptive use coefficient is $c = 0.7$ for heavily dammed, strongly-regulated, and heavily-utilized river basins dominated by irrigated agricultural water uses. This approximates the Lower Colorado River Basin in the United States, or the Nile River Basin below the Aswan dam in Egypt. The best fit between TWF and RED SRF in this case is $T' = 0.125$, or 12.5% of flow (Table 1, row in bold), a threshold that falls in the typical range published for ecological flow requirement thresholds [11,12,16,26,39,40]. Both RED and TWF mid-point impact indices visibly depart from zero at a threshold of approximately $WTA = 0.18$, have a shared value of approximately 0.33 when $WTA = 0.63$, and reach maxima near a value of 0.6 at $WTA = 1$. FED and FWF are also similar, although FWF monotonically increases whereas FED has a local maximum near $WTA = 0.49$. All four indexes are approximately equal when $WTA = 0.63$, and where Free Fraction $R = 0.5$ because $TWF \sim FWF$ here. $WTA = 0.63$ is nearly the inflection point and maximum slope in RED, where the rate of change of impact per unit of water withdrawn switches from positive to negative. Relative to the size of the BWF, the difference between the two metrics is largest when WTA is in the range of 0.3 to 0.4, where TWF is roughly double RED. Figure 4 demonstrates that RED and TWF have quantitative and qualitative similarities, and are both somewhat lower than the c:1 line which delineates the equivalent BWF.

Table 1. The effective flow alteration thresholds T′ that are approximately implied by RED range from 1% to 18% depending on the details. Table gives values of T′ that are the best-fit between RED and TWF, for each consumptive use coefficient (increments of 0.1), achieved by varying T′ and WTA. Also shown is the Root Mean Squared Error (RMSE) of fit and the local-maxima or peak value of FED.

	RED					RED SRF				
c	T′	RMSE	Peak FED	@ WTA	Notes	T′	RMSE	Peak FED	@ WTA	Notes
0	-	0	1	1		-	0	1	1	
0.1	0.010	0.008	0.900	1		0.018	0.009	0.902	1	
0.2	0.019	0.016	0.800	1		0.036	0.018	0.804	1	
0.3	0.029	0.024	0.700	1		0.054	0.027	0.705	1	
0.4	0.038	0.032	0.600	1		0.071	0.036	0.607	1	
0.5	0.048	0.040	0.500	1		0.089	0.045	0.509	1	
0.6	0.058	0.048	0.401	0.4	FED local maxima	0.107	0.054	0.411	0.54	FED local maxima
0.7	0.067	0.056	0.301	0.37	FED local maxima	0.125	0.064	0.351	0.49	FED local maxima
0.8	0.077	0.064	0.248	0.34	FED local maxima	0.143	0.073	0.333	0.46	FED local maxima
0.9	0.086	0.072	0.238	0.33	FED local maxima	0.161	0.082	0.319	0.44	
1	0.096	0.080	0.228	0.31		0.179	0.091	0.306	0.42	
1	0.000	0.122	0.228	0.31		0.000	0.190	0.306	0.42	
	RMSE/c = 0.08					*RMSE/c* = 0.09				

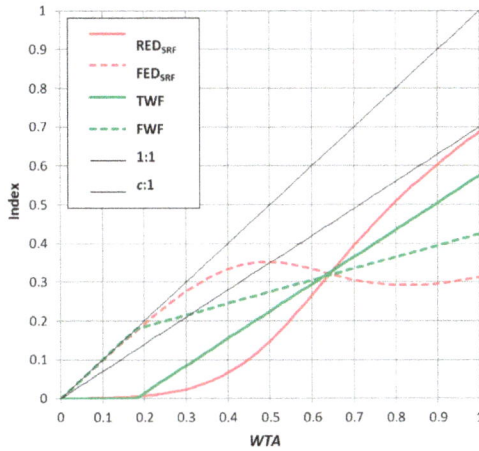

Figure 4. Water impact indices for the SRF irrigation-dominated best-fit special case where $T' = 0.125$ and $c = 0.7$. Note the similarity between RED SRF, in red, and TWF, in green. RED SRF and TWF are approximately equal below $WTA = 0.20$ and at $WTA = 0.63$. BWF is coincident with the c:1 line, and is somewhat higher than RED and TWF.

4. Discussion

Sustainability indexing is more difficult for some parts of the Coupled Natural–Human system than for others. Climate sustainability is relatively simple to assess using Carbon Footprints, because net emissions into a shared global atmosphere have no local context, and there is arguably a single global threshold for carbon dioxide in the atmosphere at around 400 ppm. Water lies at the most difficult extreme on a spectrum of natural resource types, because water resources are physically and ecologically complex, spatio-temporally variable, localized, politicized, regulated, publicly managed, culturally heterogeneous, and values-laden; water impacts are locally "context-based" [61,62]. Water scarcity and water stress is only one dimension of the complex local water context. Work to develop water sustainability indexes, performance indicators, and LCA methods has been difficult because it requires accurate and scale-specific empirically determined scientific knowledge of the system, and implicitly contains socio-political judgments regarding values and thresholds. This will also be true in many other sustainability applications of footprinting and LCA beyond water.

It is interesting to observe what information is required to calculate the quantities in equations one through seven. F and F_x can be calculated with knowledge of withdrawal and the consumption coefficient. F^T, F^f, F_x^T, and F_x^f require knowledge of the footprint and the threshold. The threshold T for the stock is obtained using external information which presumably includes reference to the initially available inventory of the stock, S_0. However, the threshold T_x for an individual process, and by extension its adverse and free footprints, can only be calculated with external knowledge of seniority or weighting between processes, as well as the footprints of all processes. If $F > T$ at the level of a stock, it is impossible to calculate an individual process's Threshold-based Footprint F_x^T (a mid-point metric) without knowing the details of all the other processes' footprints and thresholds, along with their weightings or priorities. If $F < T$ (an abundant stock) it is known that $F_x^T = 0$ and $F_x^f = F_x$, so the Blue Water Footprint of this individual process is sufficient information to calculate its mid-point impact. In other words, a factory or city or power plant needs to know the sum total of all other human and natural agents' water uses and contributions (any significant uses that are upstream or sharing the water stock), along with the water stock's adverse resource impact threshold, in order to calculate the impact of its own water use. In a system with seniority or priority the situation is even worse: each water user needs to know each other user's water use and also each other user's individual threshold and priority.

This information requirement for threshold-based accounting imposes a theoretical limitation with important real-world implications for water users and policymakers. It means that one of two solutions are probably needed to manage a water stock's utilization rate and sustainability; either (1) large water users desiring to account for their water use impacts must take the lead on organizing watershed or aquifer level voluntary data sharing efforts, or (2) comprehensive top-down reporting and detailed public disclosure of (perhaps only large) water uses must be organized by a government or NGO. It is theoretically impossible for a company or city to calculate the impact of its water use without this detailed systemic data. This information requirement motivates costly systemic water data collection so that the more useful context-based mid-point impact metrics (e.g., TWF, WSF, RED) can be calculated. A Blue Water Footprint, by contrast, is easy for a company to calculate without any knowledge of other water users' data- but this footprint is not locally contextualized. This finding highlights an important advantage of the simple inventory/volumetric approach taken by the standard Blue Water Footprint; it is currently much more feasible for an individual company or city to calculate their Blue Water Footprint than a mid-point impact metric like F_x^T, in the absence of systemic and publicly transparent water use and availability data.

Figures 3 and 4 showed that the FED metric can yield local maxima below the maximum-withdrawal point of $WTA = 1$, owing to the nonlinear logistic form of WSI. This is an unexpected result that is either confounding or insightful depending on how much we trust the precise characterization of impact that is contained in WSI. If we trust the WSI characterization, we should perhaps steer policy toward achieving one of these local maxima in FED, because this would locally maximize the Free Fraction R and the (local) efficiency of water use patterns. However, these nonlinear local maxima are not present in the simpler, linear, and discontinuous mathematical form of FWF, which calls into question the robustness of such a maximization. Notably, no local maxima exist for the more important mid-point impact metrics RED or TWF, so policymakers may choose to ignore FED and FWF and rather simply minimize RED or TWF.

Resource thresholds and mid-point and end-point impact measurements depend on the decision maker's point of view; in fact, even simple volumetric inventory metrics depend on point of view because external impacts may be correctly discounted [44,45]. In Figure 1, where would you choose to draw a sharp threshold that distinguishes "acceptable" impacts and costs from those that are unacceptably high? There are many methods and choices for T. TWF makes the choice explicit and visible. Because the use of the WSI characterization factor implies the acceptance of an approximate threshold value, the use of RED requires the acceptance of a specific—but implicit—point of view. This point of view is clearer now that we have an estimate for the thresholds that are implied by RED for each consumption coefficient. Fortunately, these results demonstrate that RED's implicit thresholds range from 1% to 18% of the flow, which is totally compatible with best practices and presumptive standards for thresholds that are commonly held in heavily-managed and irrigation-dominated watersheds.

From the resource stock manager's point of view, the objective is to minimize footprints and/or their social, environmental, and economic impacts [45]. When using Threshold-based Footprints, minimizing F^T is the most urgent objective. This is accomplished by minimizing F, W, and c, but these metrics are equally dependent on S^f and T which can be relaxed through investment in infrastructure or manipulated by changing environmental laws and standards. Resource managers may also seek to maximize the Free Fraction R, or alternatively the ratio of "free" to "adverse" footprints. From the process manager's differing point of view (e.g., as a company with a commitment to sustainability), the objective is to minimize this individual process's adverse footprint F_x^T, which can be done either by reducing the process's own footprint, or by offsetting that footprint by achieving reductions in another process's footprint against the same resource stock.

In summary, this paper derives a mathematics for Threshold-based Footprints and develops a case study that compares the Blue Water Footprint (BWF), Threshold-based surface Water Footprint (TWF), and Relevant for Environmental Deficiency (RED) mid-point LCA impact metric. The findings are general and the Threshold-based Footprint metric is useful as an easily communicated and regulation-compatible "hybrid" between volumetric/inventory footprint metrics and LCA mid-point impact metrics. This new metric is directly applicable for context-based water management. Hopefully this simple threshold-based metric will make it easier to accurately and precisely account for impact in environmental systems that are governed by socio-economically influenced regulations, by harmonizing the impact metrics with the real-world regulations and environmental standards that govern these systems.

Funding: This work was made possible in part by the Great Lakes Protection Fund via Grant #946, by the U.S. Department of Energy's Sandia National Laboratories, and by the U.S. National Science Foundation via the CAP-LTER grant (BCS-1026865), Water Sustainability and Climate grant (EAR-1360509), and INFEWS grant FEWSION (ACI-1639529). The findings are those of the authors, and not necessarily those of the funding agencies and partners.

Acknowledgments: Special thanks are extended to Alex Mayer and Stanley Mubako for their support of this work, and to the water resource professionals of the Great Lakes basin for their pioneering efforts to implement ecological flow thresholds into the scientific and regulatory framework of the region.

Conflicts of Interest: The author declare no conflict of interest.

Appendix A. Various Threshold-Based Water Scarcity and Stress Indices

Commonly employed scarcity and stress indices [35] may be expressed using these mathematics. The ratio of W to S (Equation (A1)) is identical to the commonly employed Withdrawal-to-Availability index (WTA) [47–51]. The ratio I^C (Equation (A2))) of the aggregated footprint F to the Initial Capacity S is identical to the commonly employed Consumption-to-Availability scarcity index (CTA) [47]. Two alternative indices of scarcity are the Threshold-based Adverse Impact Index I^T and the Threshold-based Free Impact Index I^f (EquationS (A3) and (A4)). The Threshold-based Scarcity Index I, is the ratio of the total footprint to the threshold (Equation (A5)), such that the stock is in a "scarce" condition when $F > T$ and the critical value of the dimensionless number is 1. The proportion of the aggregated net footprint that is adverse is P^T, and the free proportion is P^f (Equations (A5) and (A6)). The Free Impact Ratio RI is the ratio of free to adverse impacts, $R = F^f / F^T$ (Equation (A7)). Many other straightforward combinations of these metrics are possible.

$$I^W = \frac{W}{S} \tag{A1}$$

$$I^C = \frac{F}{S} \tag{A2}$$

$$I^T = \frac{F^T}{S} \tag{A3}$$

$$I^f = \frac{F^f}{S} \tag{A4}$$

$$P^T = \frac{F^T}{F} \tag{A5}$$

$$P^f = \frac{F^f}{F} \tag{A6}$$

$$RI = \frac{F^f}{F^T} \tag{A7}$$

Process-specific indices may be constructed. For example, the process's fraction of F^T is the Process-level Scarcity Footprint Fraction, $P_x{}^T$ (Equation (A8)). Similarly, the process's fraction of F^f

is the Process-level Free Footprint Fraction, P_x^f, and the process's fraction of F is the Process-level Footprint Fraction, P_x

$$P_x^T = \frac{F_x^T}{F^T} \qquad (A8)$$

Appendix B. The Economic Interpretation of F^T

Sustainability indexing methods for have been criticized for ignoring economics. In the case of the Water Footprint, for example, the key criticism is that that water has a differing value depending on the place and time of its consumption [63,64]. Therefore, a volumetric Water Footprint cannot be applied uniformly in all locations as a sustainability impact metric, because a unit volume of water consumption could have a large impact in one case and zero impact in another case. We address this criticism by discounting the volumetric impacts below a threshold of "scarcity", and interpreting F^T as an index for the total cost of net aggregated impacts. Water Scarcity Footprints attempt to address this limitation by indexing for the economic condition of scarcity, without attempting to directly address the value of the resource [35,65]. A threshold-based footprint fits this general category.

In what sense is a threshold-based footprint an index of scarcity? In this case "Scarcity" means that there is competition for the stock, and that not all demands can be satisfied at a near-zero (shadow) price. The existence of scarcity implies that some sort of impact is occurring in the system, and that the stock is a "rival" resource [45]. Scarcity is the opposite of "abundance", where no impact or cost is perceived in the system. Scarcity is a normal condition in formal markets, but is a relatively novel condition for most environmental and natural resource management scenarios (especially water), because the responsible human institutions have evolved to prevent water scarcity on the margin (at best) or to imagine that it does not exist (at worst), for example, for water [30,66].

Scarcity footprints exists at specific locations in space and time, and can only be perceived if the spatio-temporal boundaries of the stock r are properly defined [44,67,68]. The Marginal Opportunity Cost of a single additional unit of net impact on a resource stock is illustrated using the Cost Curve in Figure 1. The Opportunity Cost is usually understood as the benefit gained from the most valuable possible alternative application of that resource. In Figure 1 it is clear that the Cost Curve begins with a value near zero and steadily increases as total impacts increase, until at some point the Opportunity Cost is very high or possibly undefined (i.e., the resource is then marginally "priceless") at the point where aggregated net impact F_r on the stock equals the initial capacity S of the stock. The Adverse Impact Threshold for this specific stock T is chosen at the highest value of F where the Opportunity Cost is close enough to zero to be "acceptable" in some socio-environmental-political value judgment. The stock is considered to be "scarce" with adverse marginal impacts and unacceptably high marginal costs when $F > T$, and the stock is "abundant" when $F < T$. Multiple cost curves and thresholds may exist for different types of impacts; in this case, the lowest T should typically be chosen. In the case where water LCA assesses impacts on a sensitive wetland, the acceptable impact threshold might be close to zero [69]. For a nonrenewable resource that is subject to market pricing, this threshold will usually be zero.

We have therefore introduced with the threshold based footprint a sustainability metric that is simultaneously an index for the economic concept of marginal value and scarcity, such that these are zero below the scarcity threshold, and above the scarcity threshold the marginal value is fixed at a single value. This approximation is excellent when $F \leq T$ (an 'abundant' stock where impacts are relatively free of cost), and is generally valid for $F << S$ (a stock not under extreme stress). The total aggregated price (or value) of the resource's utilization is in proportion to the ratio of the threshold to the total cumulative stock impacts, T/F (the ratio is bounded at zero and one). If $T = 0$ there is no discount, and if $F < T$, the discount is 100%. Between these two points, the total aggregated price scales linearly. If $T = S$, all impacts on the stock are 100% discounted. $T = 0$ would occur for a nonrenewable imperishable resource stock such as gold, and $T = S$ might occur for a perishable resource that has only one possible use such as water in an agricultural irrigation system (i.e., a "use-it-or-lose-it" resource).

The shape of the cost curve should have a strong empirical economic basis, but the precise location of the threshold is necessarily informed by subjective and socio-politically contextual judgments as to what is 'acceptable' versus what is 'adverse'. Adverse resource impact thresholds integrate ecology and the subjective socio-environmental-political politics of value, cost, and impact [12,58].

References

1. Liu, J.; Dietz, T.; Carpenter, S.R.; Folke, C.; Alberti, M.; Redman, C.L.; Schneider, S.H.; Ostrom, E.; Pell, A.N.; Lubchenco, J.; et al. Coupled human and natural systems. *Ambio* **2007**, *36*, 639–649. [CrossRef]
2. Cohen, S. We Need to Accelerate the Development of Sustainability Metrics. Available online: https://www.whu.edu/en/faculty-research/entrepreneurship-and-innovation-group/chair-of-entrepreneurship-i/sustainability-blog/measuring-sustainability-the-need-for-precise-metrics/ (accessed on 30 July 2018).
3. Marten, S.; Carpenter, S.R.; Lenton, T.M.; Bascompte, J.; Brock, W.; Dakos, V.; van de Koppel, J.; van de Leemput, I.A.; Levin, S.A.; van Nes, E.H; et al. Anticipating critical transitions. *Science* **2012**, *338*, 344–348.
4. Walker, B.; Holling, C.S.; Carpenter, S.R.; Kinzing, A. Resilience, adaptability, and transformability in socio-ecological systems. *Ecol. Soc.* **2004**, *9*, 5. [CrossRef]
5. RA/SFI. *Thresholds and Alternate States in Ecological and Social–Ecological Systems*; Resilience Alliance report #183; Resilience Alliance and Santa Fe Institute: Santa Fe, NM, USA, 2009.
6. Couttenier, M.; Soubeyran, R. *Drought and Civil War in Sub-Saharan Africa*; Working Paper no.21; Paris School of Economics: Paris, France, 2011.
7. Duarte, C.M.; Agusti, S.; Agawin, N.S.R. Response of a Mediterranean phytoplankton community to increased nutrient inputs: A mesocosm experiment. *Mar. Ecol. Prog. Ser.* **2000**, *195*, 61–70. [CrossRef]
8. Solimini, A.G.; Cardoso, A.C.; Heiskanen, A.-S. *Indicators and Methods for the Ecological Status Assessment under the Water Framework Directive*; #EUR-22314-EN; European Commission Joint Research Centre Institute for Environment and Sustainability: Ispra, Italy, 2006.
9. Lawn, P.A. An assessment of the valuation methods used to calculate the Index of Sustainable Economic Welfare (ISEW), Genuine Progress Indicator (GPI), and Sustainable Net Benefit (SNBI). *Environ. Dev. Sustain.* **2005**, *7*, 185–208. [CrossRef]
10. Wang, L.Z.; Lyons, J.; Kanehl, P. Impacts of urbanization on stream habitat and fish across multiple spatial scales. *Environ. Manag.* **2001**, *28*, 255–266. [CrossRef]
11. Steinman, A.D.; Nicholas, J.R.; Seelbach, P.W.; Allan, J.W.; Ruswick, F. Science as a fundamental framework for shaping policy discussions regarding the use of groundwater in the State of Michigan: A case study. *Water Policy* **2011**, *13*, 69–86. [CrossRef]
12. Reeves, H.W.; Hamilton, D.A.; Seelbach, P.W.; Asher, A.J. *Ground-Water-Withdrawal Component of the Michigan Water-Withdrawal Screening Tool*; United States Geological Survey Scientific Investigations Report 2009–5003; United States Geological Survey: Reston, VA, USA, 2009; p. 36.
13. Swartz, W.; Sala, E.; Tracey, S.; Watson, R.; Pauly, D. The Spatial Expansion and Ecological Footprint of Fisheries (1950 to Present). *PLoS ONE* **2010**, *5*, e15143. [CrossRef] [PubMed]
14. EPA. *Clean Water Act Sec. 303(d), 33 U.S.C. § 1313(d)*; EPA: Washington, DC, USA, 1972.
15. EPA. *Water Quality Planning and Management*; Code of Federal Regulations, 40 CFR 130.7.; EPA: Washington, DC, USA, 1992.
16. Richter, B.D.; Mathews, R.; Harrison, D.L.; Wigington, R. Ecologically sustainable water management: Managing river flows for ecological integrity. *Ecol. Appl.* **2003**, *13*, 206–224. [CrossRef]
17. Sayre, N.F. The Genesis, History, and Limits of Carrying Capacity. *Ann. Assoc. Am. Geogr.* **2008**, *98*, 120–134. [CrossRef]
18. Rees, W.E.; Wackernagel, M. *Ecological Footprints and Appropriated Carrying Capacity: Measuring the Natural Capital Requirements of the Human Economy*; Jansson, A., Folke, C., Hammer, M., Costanza, R., Eds.; Island Press: Washington, DC, USA, 1994.
19. Ehrlich, P.R.; Holdren, J.P. Impact of Population Growth. *Science* **1971**, *171*, 1212–1217. [CrossRef] [PubMed]
20. Yang, H.; Reichert, P.; Abbaspour, K.C.; Zehnder, A.J.B. A water resources threshold and its implications for food security. *Environ. Sci. Technol.* **2003**, *37*, 3048–3054. [CrossRef] [PubMed]

21. Galli, A.; Wiedmann, T.; Ercin, E.; Knoblauch, D.; Ewing, B.; Giljum, S. Integrating Ecological, Carbon and Water footprint into a "Footprint Family" of indicators: Definition and role in tracking human pressure on the planet. *Ecol. Indic.* **2012**, *16*, 100–112. [CrossRef]

22. Perez-Dominguez, R.; Maci, S.; Courrat, A.; Borja, A.; Neto, J.; Elliott, M. *Review of Fish-Based Indices to Assess Ecological Quality Condition in Transitional Waters*; Available online: http://www.wiser.eu/download/D4.4-1.pdf (accessed on 30 July 2018).

23. Zhang, Y.; Singh, S.; Bakshi, B.R. Accounting for ecosystem services in life cycle assessment, Part I: A critical review. *Environ. Sci. Technol.* **2010**, *44*, 2232–2242. [CrossRef] [PubMed]

24. Fiksel, J.; Bruins, R.; Gatchett, A.; Gilliland, A.; Ten Brink, M. The triple value model: A systems approach to sustainable solutions. *Clean Technol. Environ. Policy* **2014**, *16*, 691–702. [CrossRef]

25. Palmer, M.A.; Lettenmaier, D.P.; Poff, N.L.; Postel, S.L.; Richter, B.; Warner, R. Climate change and river ecosystems: Protection and adaptation options. *Environ. Manag.* **2009**, *44*, 1053–1068. [CrossRef] [PubMed]

26. Poff, N.L.; Richter, B.D.; Arthington, A.H.; Bunn, S.E.; Naiman, R.J.; Kendy, E.; Acreman, M.; Apse, C.; Bledsoe, B.P.; Freeman, M.C.; et al. The ecological limits of hydrologic alteration (ELOHA): A new framework for developing regional environmental flow standards. *Freshw. Biol.* **2010**, *55*, 147–170. [CrossRef]

27. Postel, S.; Richter, B.D. *Rivers for Life: Managing Water for People and Nature*; Island Press: Washington, DC, USA, 2003; ISBN 978-1559634441.

28. Witmer, M.C.H.; Cleij, P. *Water Footprint: Useful for Sustainability Policies*; #500007001; PBL Netherlands Environmental Assessment Agency: The Hague, The Netherlands, 2012.

29. Tharme, R.E. A global perspective on environmental flow assessment: Emerging trends in the development and application of environmental flow methodologies for rivers. *River Res. Appl.* **2003**, *19*, 397–442. [CrossRef]

30. Postel, S.L. Entering an Era of Water Scarcity: The Challenges Ahead. *Ecol. Appl.* **2000**, *10*, 941–948. [CrossRef]

31. Hoekstra, A.Y.; Wiedmann, T.O. Humanity's unsustainable environmental footprint. *Science* **2014**, *344*, 1114–1117. [CrossRef] [PubMed]

32. Falkenmark, M. Water and sustainability: A reappraisal. *Environment* **2008**, *50*, 4–16. [CrossRef]

33. Meadows, D.H.; Meadows, D.H.; Randers, J.; Behrens, W.W., III. The Limits to Growth: a report to the Club of Rome. Available online: http://www.donellameadows.org/wp-content/userfiles/Limits-to-Growth-digital-scan-version.pdf (accessed on 30 July 2018).

34. Hussen, A. *Principles of Environmental Economics and Sustainability: An Integrated Economic and Ecological Approach*, 3rd ed.; Routledge: Abingdon, UK, 2012.

35. Brown, A.; Matlock, M.D. A Review of Water Scarcity Indices and Methodologies. *White Pap.* **2011**, *106*, 19.

36. Gini, C. Variabilità e mutabilità. In *Memorie di Metodologica Statistica*; Pizetti, E., Salvemini, T., Eds.; Libreria Eredi Virgilio Veschi: Rome, Italy, 1955.

37. Jackson, L.E.; Kurtz, J.C.; Fisher, W.S. *Evaluation Guidelines for Ecological Indicators*; #EPA/620/R-99/005; U.S. Environmental Protection Agency: Washington, DC, USA, 2000.

38. Samhouri, J.F.; Levin, P.S.; Ainsworth, C.H. Identifying Thresholds for Ecosystem-Based Management. *PLoS ONE* **2010**, *5*, e8907. [CrossRef] [PubMed]

39. Richter, B.D.; Warner, A.T.; Meter, J.L.; Lutz, K. A collaborative and adaptive process for developing environmental flow recommendations. *River Res. Appl.* **2006**, *22*, 297–318. [CrossRef]

40. Richter, B.D.; Davis, M.M.; Apse, C.; Konrad, C. A presumptive standard for environmental flow protection. *River Res. Appl.* **2012**, *28*, 1312–1321. [CrossRef]

41. Wiedmann, T.; Barrett, J. A Review of the Ecological Footprint Indicator—Perceptions and Methods. *Sustainability* **2010**, *2*, 1645–1693. [CrossRef]

42. Solanes, M.; Gonzalez-Villarreal, F. *The Dublin Principles for Water as Reflected in a Comparative Assessment of Institutional and Legal Arrangements for Integrated Water Resources Management*; Global Water Partnership: Stockholm, Sweden, 1999.

43. Glennon, R. *Unquenchable*; Island Press: Washington, DC, USA, 2010.

44. Ruddell, B.L.; Adams, E.A.; Rushforth, R.; Tidwell, V.C. Embedded resource accounting for coupled natural-human systems: An application to water resource impacts of the western US electrical energy trade. *Water Resour. Res.* **2014**, *50*, 7957–7972. [CrossRef]

45. Rushforth, R.R.; Adams, E.A.; Ruddell, B.L. Generalizing ecological, water and carbon footprint methods and their worldview assumptions using Embedded Resource Accounting. *Water Resour. Ind.* **2013**, *1*, 77–90. [CrossRef]

46. Kates, R.W.; Clark, W.C.; Corell, R.; Hall, J.M.; Jaeger, C.C.; Lowe, I.; McCarthy, J.J.; Schellnhuber, H.J.; Bolin, B.; Dickson, N.M.; et al. Sustainability Science. *Science* **2001**, *292*, 641–642. [CrossRef] [PubMed]

47. Berger, M.; Finkbeiner, M. Water footprinting: How to address water use in life cycle assessment? *Sustainability* **2010**, *2*, 919–944. [CrossRef]

48. Berger, M.; Finkbeiner, M. Methodological Challenges in Volumetric and Impact-Oriented Water Footprints. *J. Ind. Ecol.* **2012**, *17*, 79–89. [CrossRef]

49. Hoekstra, A.Y.; Chapagain, A.K.; Aldaya, M.M.; Mekonnen, M.M. *The Water Footprint Assessment Manual: Setting the Global Standard*; Water Footprint Network; Earthscan Publishing: Oxford, UK, 2011.

50. Pfister, S.; Bayer, P.; Koehler, A.; Hellweg, S. Environmental Impacts of Water Use in Global Crop Production: Hotspots and Trade-Offs with Land Use. *Environ. Sci. Technol.* **2011**, *45*, 5761–5768. [CrossRef] [PubMed]

51. Pfister, S.; Koehler, A.; Hellweg, S. Assessing the environmental impacts of freshwater consumption in LCA. *Environ. Sci. Technol.* **2009**, *43*, 4098–4104. [CrossRef] [PubMed]

52. Nilsson, C.; Reidy, C.A.; Dynesius, M.; Revenga, C. Fragmentation and flow regulation of the world's large river systems. *Science* **2005**, *308*, 405–408. [CrossRef] [PubMed]

53. Ridoutt, B.; Pfister, S. A revised approach to water footprinting to make transparent the impacts of consumption and production on global freshwater scarcity. *Glob. Environ. Chang.* **2010**, *20*, 113–120. [CrossRef]

54. Chenoweth, J.; Hadjikakou, M.; Zoumides, C. Review article: Quantifying the human impact on water resources: A critical review of the water footprint concept. *Hydrol. Earth Syst. Sci. Discuss.* **2013**, *10*, 9389–9433. [CrossRef]

55. Kounina, A.; Margni, M.; Bayart, J.; Boulay, A.; Berger, M.; Bulle, C.; Frischknecht, R.; Koehler, A.; Mila i Canals, L.; Motoshita, M.; et al. Review of methods addressing freshwater use in life cycle inventory and impact assessment. *Int. J. Life Cycle Assess.* **2013**, *18*, 707–721. [CrossRef]

56. Mila i Canals, L.; Chenoweth, J.; Chapagain, A.; Orr, S.; Anton, A.; Clift, R. Assessing freshwater use in LCA: Part I—Inventory modelling and characterization factors for the main impact pathways. *Int. J. Life Cycle Assess.* **2008**, *14*, 28–42. [CrossRef]

57. Pfister, S.; Ridoutt, B.G. Water footprint: Pitfalls on common ground. *Environ. Sci. Technol.* **2013**, *48*, 4. [CrossRef] [PubMed]

58. ISO (2014), ISO 14046:2014(en): Environmental Management—Water Footprint—Principles, Requirements and Guidelines. Available online: https://www.iso.org/obp/ui/#iso:std:iso:14046:ed-1:v1:en (accessed on 30 July 2018).

59. Mubako, S.T.; Ruddell, B.L.; Mayer, A.S. Relationship between water withdrawals and freshwater ecosystem water scarcity quantified at multiple scales for a Great Lakes watershed. *J. Water Resour. Plan. Manag.* **2013**, *139*, 671–681. [CrossRef]

60. Zorn, T.G.; Seelbach, P.W.; Rutherford, E.S.; Wills, T.C.; Cheng, S.T.; Wiley, M.J. *A Regional-Scale Habitat Suitability Model to Assess the Effects of Flow Reduction on Fish Assemblages in Michigan Streams*; Fisheries Research Report 2089; Michigan Department of Natural Resources: Ann Arbor, MI, USA, 2008.

61. Pacific Institute. Exploring the Case for Corporate Context-Based Water Targets. 2017. Available online: https://www.ceowatermandate.org/files/context-based-targets.pdf (accessed on 28 June 2018).

62. Alliance for Water Stewardship. The AWS International Water Stewardship Standard. 2014. Available online: http://a4ws.org/wp-content/uploads/2017/04/AWS-Standard-Full-v-1.0-English.pdf (accessed on 28 June 2018).

63. Wichelns, D. Virtual water and water footprints offer limited insight regarding important policy questions. *Int. J. Water Resour. Dev.* **2010**, *26*, 639–651. [CrossRef]

64. Hanemann, W.M. Chapter 4: The economic conception of water. In *Water Crisis: Myth or Reality*; Rogers, P.P., Llamas, M.R., Martinez-Cortina, L., Eds.; Taylor & Francis Group: Oxford, UK, 2006; pp. 61–91.

65. Hoekstra, A.Y.; Mekonnen, M.M.; Chapagain, A.K.; Mathews, R.E.; Richter, B.D. Global monthly water scarcity: Blue water footprints versus blue water availability. *PLoS ONE* **2012**, *7*, e32688. [CrossRef] [PubMed]

66. Zetland, D. *The End of Abundance: Economic Solutions to Water Scarcity*; Aguanomics Press: Mission Viejo, CA, USA, 2011; ISBN 978-0615469737.

67. Fang, K.; Heijungs, R.; de Snoo, G.R. Understanding the complementary linkages between environmental footprints and planetary boundaries in a footprint–boundary environmental sustainability assessment framework. *Ecol. Econ.* **2015**, *114*, 218–226. [CrossRef]
68. Heistermann, M. HESS Opinions: A planetary boundary on freshwater use is misleading. *Hydrol. Earth Syst. Sci.* **2017**, *21*, 3455–3461. [CrossRef]
69. Verones, F.; Pfister, S.; Hellweg, S. Quantifying Area Changes of Internationally Important Wetlands Due to Water Consumption in LCA. *Environ. Sci. Technol.* **2013**, *47*, 9799–9807. [CrossRef] [PubMed]

water

MDPI

Article

Informing National Food and Water Security Policy through Water Footprint Assessment: The Case of Iran

Fatemeh Karandish [1,*] and Arjen. Y. Hoekstra [2,3]

1   Water Engineering Department, University of Zabol, P.O. Box 538-98615, Zabol 9861673831, Iran
2   Twente Water Centre, University of Twente, P.O. Box 217, 7500 AE Enschede, The Netherlands;
    a.y.hoekstra@utwente.nl
3   Institute of Water Policy, Lee Kuan Yew School of Public Policy, National University of Singapore,
    Singapore 259770, Singapore
*   Correspondence: karandish_h@yahoo.com

Received: 9 September 2017; Accepted: 23 October 2017; Published: 29 October 2017

Abstract: Iran's focus on food self-sufficiency has led to an emphasis on increasing water volumes available for irrigation with little attention to water use efficiency, and no attention at all to the role of consumption and trade. To better understand the development of water consumption in relation to food production, consumption, and trade, we carried out the first comprehensive water footprint assessment (WFA) for Iran, for the period 1980–2010, and estimated the water saving per province associated with interprovincial and international crop trade. Based on the AquaCrop model, we estimated the green and blue water footprint (WF) related to both the production and consumption of 26 crops, per year and on a daily basis, for 30 provinces of Iran. We find that, in the period 1980–2010, crop production increased by 175%, the total WF of crop production by 122%, and the blue WF by 20%. The national population grew by 92%, and the crop consumption per capita by 20%, resulting in a 130% increase in total food consumption and a 110% increase in the total WF of national crop consumption. In 2010, 26% of the total water consumption in the semi-arid region served the production of crops for export to other regions within Iran (mainly cereals) or abroad (mainly fruits and nuts). Iran's interprovincial virtual water trade grew by a factor of 1.6, which was mainly due to increased interprovincial trade in cereals, nuts, and fruits. Current Iranian food and water policy could be enriched by reducing the WFs of crop production to certain benchmark levels per crop and climatic region and aligning cropping patterns to spatial differences in water availability and productivities, and by paying due attention to the increasing food consumption per capita in Iran.

Keywords: food security; food self-sufficiency; water footprint; water scarcity; crop trade; virtual water trade; water productivity; water saving

---

1. Introduction

Iran, the second largest country in the Middle East, is facing great water scarcity, which becomes manifest in drying lakes and rivers, dropping groundwater tables, land subsidence, the increasing contamination of water, water supply rationing and disruptions, crop losses, salt and sand storms, the increasing migration of people away from the hardest hit areas, and damage to ecosystems. Iran is mostly arid to semi-arid (Figure 1), with an average annual precipitation of 228 mm (72% less than the global average of 814 mm), and internal renewable water resources of 129×10^9 m$^3 \cdot$y^{-1} (0.32% of the global renewable water resources) [1]. Precipitation ranges from less than 50 mm in central Iran to about 1000 mm at the Caspian coast. Most regions receive less than 100 mm of precipitation per year, and 75% of the country's precipitation falls over only 25% of the country's area. About 75% of the precipitation is offseason, i.e., it falls when not needed by the agricultural sector [2]. Over the last 20 years, the per capita renewable water resources in the country decreased by 29.1% and reached 1732 m$^3 \cdot$y^{-1} in 2014 [1], which is well below the global average of 7000 m$^3 \cdot$cap$^{-1} \cdot$y^{-1}. The population grew from 38.9 billion in 1980 to 74.5 billion in 2010, and is expected to further increase to 88.5 billion in 2030 [3], which will translate into increasing food and water demands.

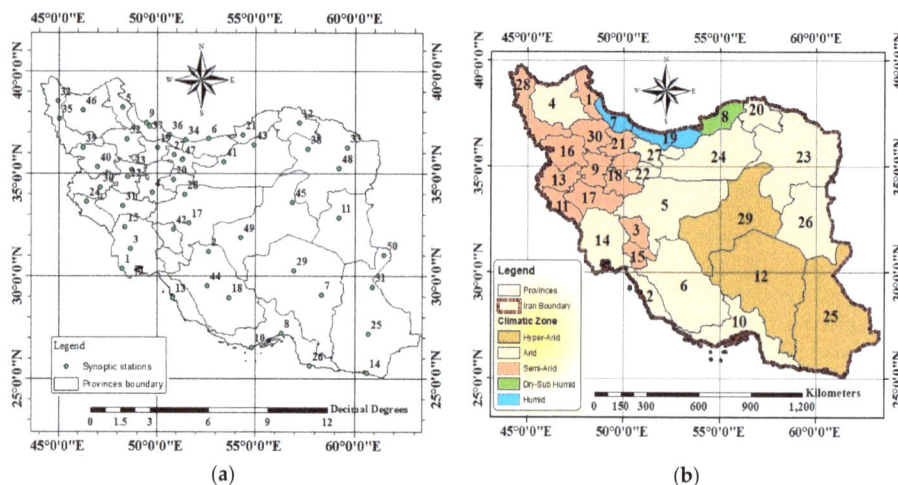

Figure 1. Provinces and the 52 weather stations (**a**) and the climatic regions of Iran (**b**).

In addition to the physical water scarcity, Iran faces a poor management of its water resources. Major infrastructure works are developed without sufficient concern for their long-term impacts, the water governance structure is weak, water management is done based on administrative instead of watershed boundaries, there is insufficient attention to the linkage between development and environment, different government sectors fail to coordinate, and groundwater abstractions are not properly regulated [2]. The mismanagement of water resources has resulted in the shrinking of Urmia Lake in the western part of the country, which is the largest lake in the Middle East and one of the world's largest hypersaline lakes [4]; the disappearance of Hamun Lake in the eastern region [5,6]; and the seasonally drying up of the Zayandeh Rud River, which is the backbone of development in central Iran [7].

Agriculture is the biggest freshwater user in Iran, accounting for 92% of gross blue water abstractions [1], and 97% of net blue water abstractions [8]. Inefficient water management in this sector is thus a main source of the water shortage in the country. In 2004, about 68% of the total renewable water resources was withdrawn [1]. Even though this may look sustainable at first sight, it is far from so, because a substantial percentage of the flow needs to be maintained to protect ecosystems and the livelihoods that depend on them [9–12]. Issues in agricultural policy that require critical attention are the country's aim to achieve food self-sufficiency, the mismatch between the spatial cropping pattern and the geographic spread of water availability, the heavy reliance on irrigation, the low water use efficiency, the low share of rain-fed agriculture in total crop production, the low water and energy prices, the overdraft of aquifers, and the low income level of farmers and their associated inability to adopt better farming practices. The role of the agricultural sector in alleviating the current water scarcity in Iran also gets clear when considering the historical development of the harvested irrigated crop area. The irrigated land area grew by 117% in the period 1980–2010, while the total harvested area, including both rain-fed and irrigated lands, increased only slightly. The growth in irrigation was introduced to meet the increasing food demand of the rapidly increasing population and keep a high food self-sufficiency level. Based on the national statistics, total crop production within the country grew by 175% over the period 1980–2010. With continued population growth as predicted, food demand will keep increasing, as well as the associated water demand when sticking to the food self-sufficiency policy, which again will further aggravate the existing overexploitation of water resources in the country.

As a consequence of Iran's focus on food self-sufficiency, the emphasis has been on increasing the water volume available for irrigation. Little attention has been paid to water use efficiency, and no attention at all has been paid to the role of consumption and trade. In order to better understand the historical development of the relation between food production, consumption, trade, and water consumption, we carried out the first comprehensive water footprint assessment (WFA) for Iran, for the period 1980–2010. In addition, we estimated the water saving per province associated with interprovincial and international crop trade. The water footprint (WF) is a spatially–temporally explicit measure of freshwater used directly or indirectly by a producer or a consumer [13], and could facilitate the analysis of how patterns of consumption, production, and trade relate to patterns of water consumption [14]. The WF of producing a crop comprises a consumptive component, measuring water consumption, and a degradative component, measuring water pollution. In this paper, we focus on the consumptive WF, which again includes two components: the green WF, which refers to the consumption of rainwater, and the blue WF, which refers to the consumption of irrigation water [15]. The WF related to human consumption within a specific region will include an internal and an external component. The former refers to the amount of water consumed within the region for producing products that are consumed within the region; the latter refers to the amount of water consumed in the other regions to produce products that are imported and consumed within the considered region [15]. The trade of food between regions implies a virtual water (VW) flow, which refers to the water consumed in the region of the food origin.

This is the first comprehensive research on the water footprint and virtual water trade for Iran, whereby we also assess the added value of the water footprint assessment for informing Iran's food and water security policy. The main focus in this paper is water use and scarcity, which means that we do not consider other economic, social, and environmental factors that are relevant in policy making, such as labour and land prices, the competitive advantages of different provinces for certain crops, employment, soil degradation, water quality deterioration, and climate change.

2. Results

2.1. Harvested Area and Crop Production

Over the period 1980–2010, the population in Iran grew by 91.5%, but the total harvested area (HA) for the eight crop categories increased by 129%, and total crop production (CP) by 175% (Figure 2). CP grew faster than HA because crop yields increased (by 20% as a weighted average over all crops). Increased crop yields could be attributed to improved field management practices over the period, including better irrigation and soil management practices, and a higher application rates of fertilizers. The percentage of HA irrigated reduced slightly, from 57% in 1980 to 54% in 2010 (with the most pronounced decrease for cereals, but an increase for oil crops). Even though the irrigated percentage in HA decreased, irrigated HA in absolute terms increased by 117%, which aggravated the pressure on the available blue water resources.

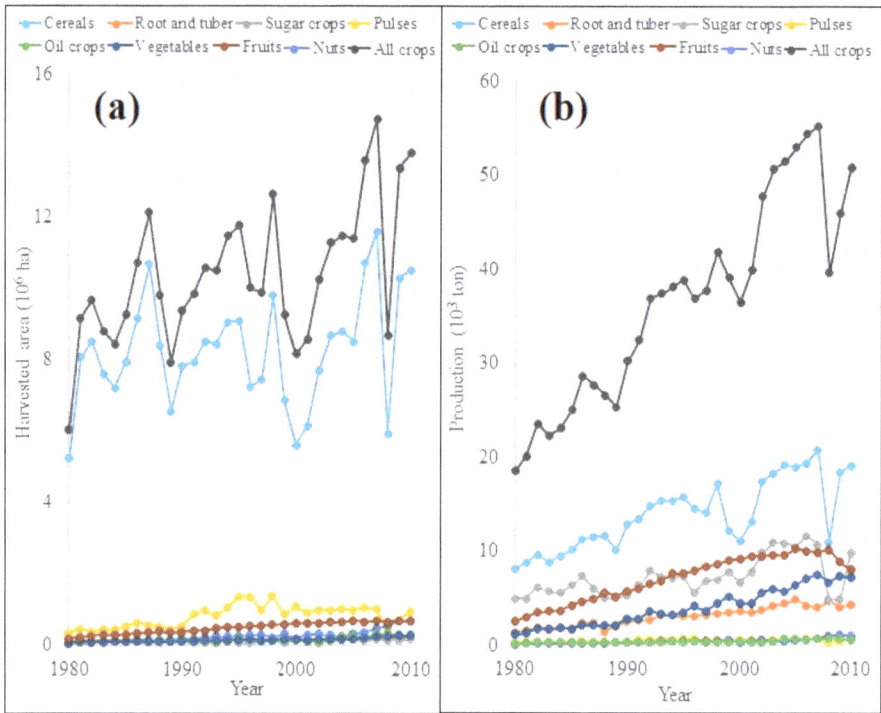

Figure 2. Harvested area (**a**) and production (**b**) in Iran per crop category over the period 1980–2010.

Figure 3 shows the contribution of the different crop categories to HA and CP, per province, as averages over the period 1980–2010. At the national level, cereals were the main crop category over the whole period, but its importance decreased. The contribution of cereals to total HA reduced from 87% in 1980 to 76% in 2010 (with an average of 79% over the period), while the cereal contribution in CP reduced from 44 to 38% (with an average of 39% over the period). Regarding CP at the national level, sugar crops and fruits ranked next to cereals over the whole period, with an average share over the period of 20% and 19%, respectively (but with an overall contribution of 1% and 4.7% to HA, respectively). Regarding HA at the national level, pulses ranked next to cereals over the whole period, with an average share over the period of 7.6% (and an overall contribution of 1.4% to CP). The quickest growth in both HA and CP over the period 1980–2010 was for nuts.

At the national level, the highest crop yields were observed for sugar crops (28 tonne/ha on average), followed by vegetables (27 tonne/ha) and roots and tubers (21 tonne/ha), while the lowest yields were found for nuts (1.8 tonne/ha), cereals (1.7 tonne/ha), and pulses (0.6 tonne/ha). Although cropping patterns are different across provinces, cereals usually dominate HA. Only in the arid province of Hormozgan do fruits take up most of the HA.

a) Contribution of crop category in harvested area

b) Contribution of crop category in total production

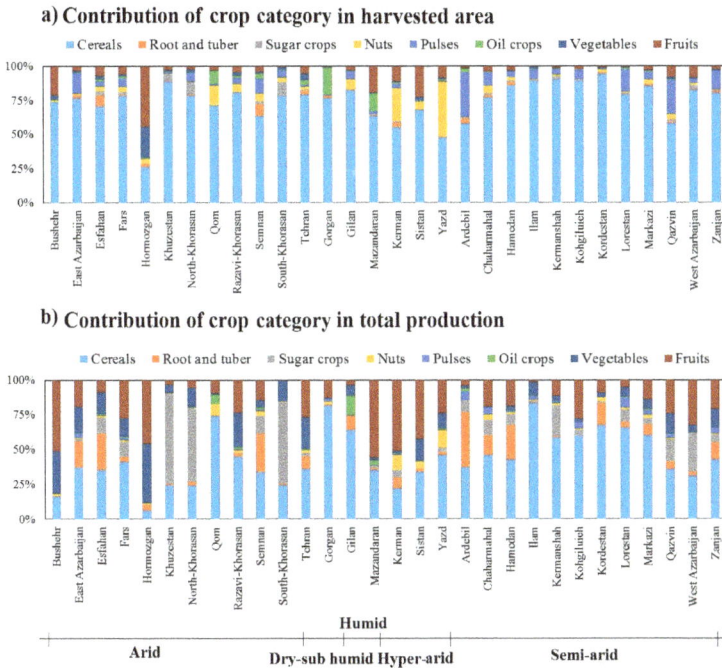

Figure 3. The 30-year average contribution of different crop categories to total harvested area (HA) per province (**a**) and total crop production (CP) per province (**b**). Period 1980–2010.

2.2. WF of Crop Production

The 175% growth in crop production over the period 1980–2010 led to a 122% increase in total WF, from 31.9×10^9 $m^3 \cdot y^{-1}$ in 1980 (42.5% blue) to 70.8 $\times 10^9$ $m^3 \cdot y^{-1}$ in 2010 (62.1% blue) (Figure 4). The growth in total WF at the national level holds for all crop categories. For cereals and sugar crops, the total WF in the country increased, despite the fact that the national average WF per tonne for cereals and sugar crops decreased by 29% and 18%, respectively (Table 1), which was mainly due to the increase in crops yields. The national average WF per tonne for oil crops, pulses, nuts, vegetables, roots and tubers, and fruits increased by 14%, 17%, 18%, 23%, 23% and 50%, respectively. The considerable increase in the WF per tonne for fruits was partly due to a national average reduction of 10% in fruit yield.

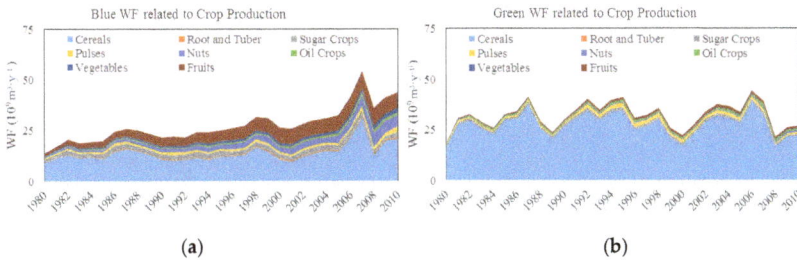

Figure 4. Blue WF (**a**) and green water footprint (WF) (**b**) and per tonne of crop production in Iran over the period 1980–2010.

Table 1 shows that WFs per tonne differ across climatic regions. In general, WFs per tonne are significantly higher in the hyper-arid, arid, and semi-arid regions compared with the dry sub-humid and humid regions. When considering a specific crop category in a specific region, many of the differences between 1980 and 2010 are due to changes in what were the dominant crops per crop category; for instance, rice replaced wheat as the dominant cereal crop in the dry sub-humid region. Differences were also due to changes in the fractions of the land irrigated (for instance, a 420% increase in the irrigated HA in a dry sub-humid region), in changes in yields (for instance, 23%, 12%, and 6.5% reductions in crop yield in semi-arid, arid, and humid regions, respectively), and in changes in climate (as demonstrated by Karandish et al. [16]).

The spatial distribution of the 30-year average total WF of crop production and the blue fraction in the total is shown in Figure 5. The highest WFs, when measured as the total WF in a province divided by the area of the province (expressed in $mm \cdot y^{-1}$), are found in the semi-arid climatic region, because this region has the highest cropland density, while water consumption per unit of cultivated land is also high (at least relative to the humid and dry sub-humid regions). The largest shares of blue WF in the total are found in the provinces where irrigated agriculture dominates over rain-fed agriculture, which is obviously particularly the case in the hyper-arid region, where 93% of the harvested land was irrigated (as an average over 1980–2010). The 30-year total WF ($m^3 \cdot y^{-1}$) of crop production, per province, is summarized in Table 2. The provinces located in the arid and semi-arid regions, the water-scarce regions of the country, are responsible for 87% of the total WF of Iranian crop production of 59.6 billion $m^3 \cdot y^{-1}$. The hyper-arid region ranked next to arid and semi-arid regions, with a contribution of 6.5% to the national WF of crop production over the period.

The 30-year total WF ($m^3 \cdot y^{-1}$) of crop production, per province, is summarized in Table 2. The provinces located in the arid and semi-arid regions, the water-scarce regions of the country, are responsible for 87% of the total WF of Iranian crop production of 59.6 billion $m^3 \cdot y^{-1}$. The hyper-arid region ranked next to arid and semi-arid regions, with a contribution of 6.5% to the national WF of crop production over the period.

Table 1. Regional and national averages of the WF of crop production and the blue share in the total, per crop category, for the years 1980 and 2010.

Climatic Region	Crop Category	1980		2010	
		WF of Crop Production ($m^3 \cdot tonne^{-1}$)	Blue Share (%)	WF of Crop Production ($m^3 \cdot tonne^{-1}$)	Blue Share (%)
Hyper-arid	Cereals	2275	67	2614	85
	Root and tuber	202	59	230	85
	Sugar crops	761	84	953	90
	Pulses	5073	93	5817	96
	Nuts	4948	93	5891	95
	Oil seeds	5728	88	5666	88
	Vegetables	370	80	432	93
	Fruits	1293	94	1541	97
Arid	Cereals	2729	29	2298	54
	Root and tuber	179	69	223	79
	Sugar crops	452	70	343	82
	Pulses	6452	89	7180	84
	Nuts	4397	61	5008	78
	Oil seeds	3595	63	3946	68
	Vegetables	267	84	325	92
	Fruits	885	88	1394	91
Semi-arid	Cereals	4400	39	2600	38
	Root and tuber	167	71	204	75
	Sugar crops	368	63	492	78
	Pulses	4431	78	5842	83
	Nuts	4286	55	5216	70
	Oil seeds	4639	42	4639	70
	Vegetables	323	81	437	88
	Fruits	499	74	835	82
Dry sub-humid	Cereals	570	5	1178	41
	Root and tuber	114	0	132	63
	Sugar crops	1099	21	1973	89
	Pulses	3164	53	5409	91
	Nuts	2438	29	3428	67
	Oil seeds	1785	11	2478	75
	Vegetables	114	46	181	85
	Fruits	352	27	503	88

Table 1. *Cont.*

Climatic Region	Crop Category	1980		2010	
		WF of Crop Production ($m^3 \cdot tonne^{-1}$)	Blue Share (%)	WF of Crop Production ($m^3 \cdot tonne^{-1}$)	Blue Share (%)
Humid	Cereals	1070	36	1182	53
	Root and tuber	192	11	229	20
	Sugar crops	519	8	611	19
	Pulses	4460	38	6299	51
	Nuts	3006	29	3292	45
	Oil seeds	2669	6	2425	13
	Vegetables	287	19	274	39
	Fruits	297	33	364	43
Iran	Cereals	3158	35	2239	48
	Root and tuber	172	67	212	76
	Sugar crops	440	69	362	82
	Pulses	5405	87	6331	87
	Nuts	4289	57	5077	73
	Oil seeds	2663	35	3031	62
	Vegetables	277	82	341	91
	Fruits	732	83	1094	88

Table 2. The 30-year average total water footprint of crop production and the blue share in the total, per province and crop category.

Climatic Region	Province Code *	Total WF of Crop Production ($10^6 \cdot m^3 \cdot y^{-1}$)									Blue Share (%)
		Cereals	Root and Tuber	Sugar crops	Pulses	Nuts	Oil crops	Vegetables	Fruits	All crops	All crops
Hyper-arid	12	758	27	59	38	623	27	15	588	2136	89
	25	505	2	0.0	5	114	1	30	568	1225	87
	29	156	1	4	3	303	6	6	59	538	91
Arid	2	415	0.48	0.0	0.06	22	1	39	460	939	56
	4	3088	88	17	287	111	44	122	319	4076	40
	5	1056	84	110	44	84	24	47	183	1632	74
	6	4221	37	313	213	280	109	135	1090	6398	55
	10	70	4	0.0	1	24	1	93	599	793	89
	14	3029	9	634	67	33	5	95	387	4260	53
	20	1582	14	518	58	9	15	95	86	2377	50
	22	147	0.12	0.00	3	58	33	1	11	253	81
	23	2147	8	0.00	43	299	93	123	271	2984	58
	24	362	32	34	16	58	30	11	39	581	61
	26	1861	6	823	52	158	19	117	40	3076	47
	27	823	19	8	13	30	38	68	94	1094	60
Semi-arid	1	1407	68	21	385	11	156	9	46	2103	28
	3	635	12	24	80	59	4	1	44	860	40
	9	2464	69	66	66	103	13	23	145	2948	35
	11	740	0.30	1	45	6	2	29	10	834	24
	13	2037	4	127	43	52	15	26	66	2370	25
	15	895	0.17	5	38	7	5	5	77	1033	40
	16	2472	27	3	16	87	4	16	46	2671	13
	17	2115	11	59	266	50	53	37	50	2640	31
	18	1645	19	21	115	93	28	18	93	2033	43
	21	651	9	54	98	61	10	30	129	1042	52
	28	1959	14	154	47	34	4	29	307	2549	36
	30	1826	33	26	190	18	6	57	108	2264	30
Dry sub-humid	8	899	25	0.04	11	0.21	557	22	29	1542	27
Humid	7	649	2	0.38	29	77	1	4	32	794	45
	19	860	7	4	18	5	198	9	454	1555	37

Note: * The province codes refer to the provinces shown on the map in Figure 1.

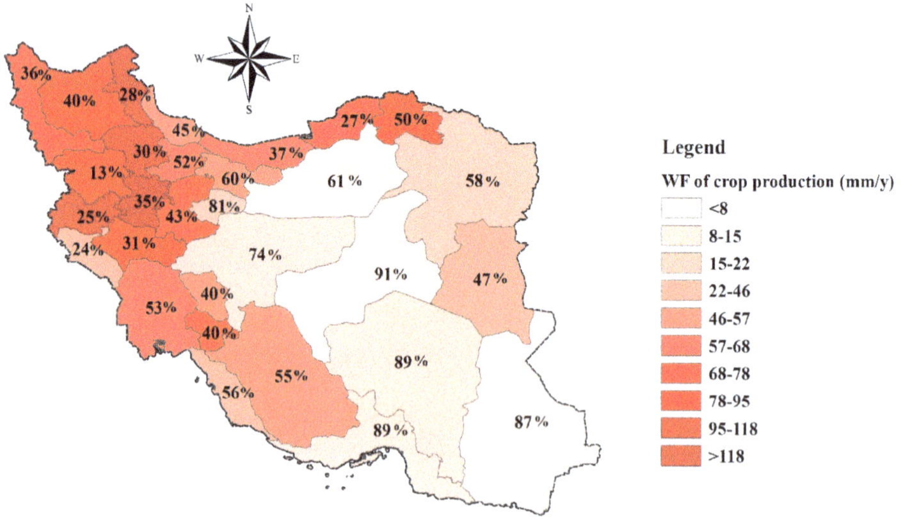

Figure 5. The 30-year average WF of Iranian crop production, per province. The WF in mm·y^{-1} is obtained by dividing the total WF of crop production in the province by the area of the province. Period: 1980–2010. The numbers in the map refer to the blue share in the WF of crop production.

2.3. WF of Crop Consumption

The Iranian crop consumption per capita (considering the 26 crops studied) increased by 20% in the period 1980–2010, from 460 to 552 kg·cap^{-1}·y^{-1}. Given the 92% population growth over this period, total crop consumption (again considering the 26 crops studied) increased by 130%, from 17.9 to 23.7 million t·y^{-1}. The total WF of crop consumption increased by 110%, from 27.7 × 10^9 m^3·y^{-1} in 1980 to 57.3 × 10^9 m^3·y^{-1} in 2010 (Table 3). The blue water fraction increased from 42% to 62% (Figure 6). The increasing WF of consumption per capita is the net result of the growing consumption volume per capita, the changed diet composition, and changes in the WFs per tonne of crop (a decrease for cereals and sugar crops, and an increase for the other crop categories). The contribution of different crops to the WF of consumption considerably changed over the study period. The contribution of cereals to the total WF of crop consumption decreased from 78% in 1980 to 53% in 2010. The contribution of sugar crops decreased as well, from 7.6% to 6.1%. The contributions of all of the other crop categories to the WF of consumption increased. Growing from 1.7% in 1980 to 11% in 2010, the share of the WF related to the consumption of nuts showed the highest increase, followed by oil crops (from 1.7% in 1980 to 7.1% in 2010) and fruits (from 5.8% to 12%), mainly due to the increased proportion of these crops in Iranian consumption and/or increase in WF per tonne of crops in some climatic regions (Table 1).

Table 3. Regional and national averages of the water footprint of crop consumption in Iran per capita and the blue share in the total, per crop category, for the years 1980 and 2010.

Climatic Region	Crop Category	1980		2010	
		WF of Crop Consumption ($m^3 \cdot cap^{-1} \cdot y^{-1}$)	Blue Share (%)	WF of Crop Consumption ($m^3 \cdot cap^{-1} \cdot y^{-1}$)	Blue Share (%)
Hyper-arid	Cereals	494	53	409	62
	Root and tuber	6	63	11	80
	Sugar crops	60	73	51	83
	Pulses	11	93	77	96
	Nuts	25	67	44	77
	Oil seeds	14	56	53	35
	Vegetables	8	83	29	92
	Fruits	41	98	99	93
Arid	Cereals	536	38	397	51
	Root and tuber	6	67	10	76
	Sugar crops	55	69	48	82
	Pulses	12	88	84	85
	Nuts	24	61	46	76
	Oil seeds	12	46	53	35
	Vegetables	7	83	25	91
	Fruits	42	88	96	87
Semi-arid	Cereals	739	27	478	40
	Root and tuber	6	70	11	76
	Sugar crops	47	67	45	81
	Pulses	10	81	75	84
	Nuts	25	57	54	70
	Oil seeds	12	38	61	42
	Vegetables	8	81	31	90
	Fruits	39	81	93	82
Dry sub-humid	Cereals	139	24	225	50
	Root and tuber	4	0	6	63
	Sugar crops	56	69	48	82
	Pulses	11	87	77	86
	Nuts	22	53	40	71
	Oil seeds	7	9	56	78
	Vegetables	5	65	17	84
	Fruits	38	78	85	88
Humid	Cereals	344	22	338	41
	Root and tuber	5	58	10	65
	Sugar crops	56	69	48	81
	Pulses	11	71	77	69
	Nuts	22	48	38	61
	Oil seeds	12	17	52	25
	Vegetables	7	72	25	83
	Fruits	38	78	78	79
Nationwide	Cereals	558	34	408	48
	Root and tuber	5	66	11	76
	Sugar crops	54	69	47	82
	Pulses	12	85	81	85
	Nuts	24	59	47	74
	Oil seeds	12	42	55	37
	Vegetables	7	81	27	90
	Fruits	41	86	94	86

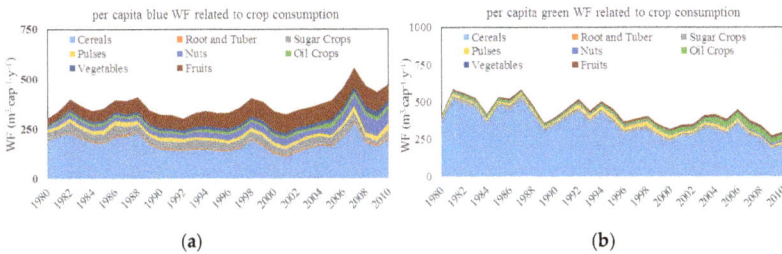

Figure 6. Blue WF (**a**) and green WF (**b**) of Iranian crop consumption per capita, over the period 1980–2010.

The WF of consumption per capita varies across the provinces and climatic regions (Figure 7 and Table 4) as a result of provincial differences in the WF per tonne of crop (Table 1). The 30-year average WF of consumption per capita varies across provinces, in the range of 212–1061 $m^3 \cdot cap^{-1} \cdot y^{-1}$ for cereals (10–54% blue), 7–14 $m^3 \cdot cap^{-1} \cdot y^{-1}$ for roots and tubers (17–80% blue), 31–67 $m^3 \cdot cap^{-1} \cdot y^{-1}$ for sugar crops (61–80% blue), 19–83 $m^3 \cdot cap^{-1} \cdot y^{-1}$ for nuts (51–96% blue), 25–51 $m^3 \cdot cap^{-1} \cdot y^{-1}$ for pulses (31–78% blue), 30–52 $m^3 \cdot cap^{-1} \cdot y^{-1}$ for oil crops (11–54% blue), 14–26 $m^3 \cdot cap^{-1} \cdot y^{-1}$ for vegetables (73–93% blue), and 76–124 $m^3 \cdot cap^{-1} \cdot y^{-1}$ for fruits (77–98% blue). The largest WFs of crop consumption per capita are mainly found in the provinces located in the hyper-arid and semi-arid regions, followed by those located in the arid region.

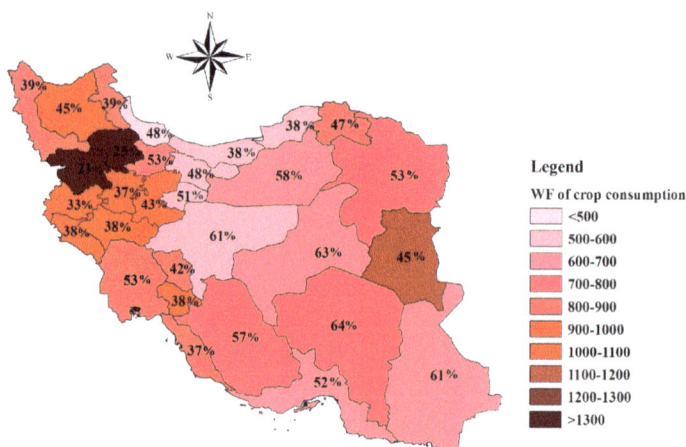

Figure 7. The 30-year average WF of Iranian crop consumption per capita, per province. Period: 1980–2010. The numbers in the map refer to the blue share in the WF of crop consumption.

Table 4. The 30-year average water footprint of crop consumption per capita and the blue share in the total, per province and crop category. Period: 1980–2010.

Climatic Region	Province Code *	WF of Crop Consumption ($m^3 \cdot cap^{-1} \cdot y^{-1}$)									Blue Share (%)
		Cereals	Root and Tuber	Sugar crops	Pulses	Nuts	Oil crops	Vegetables	Fruits	All crops	All crops
Hyper-arid	12	456	9	51	30	35	35	23	113	753	62
	25	377	10	53	83	34	35	19	108	718	64
	29	448	12	63	30	38	40	22	80	733	65
Arid	2	690	10	51	53	37	37	19	90	987	45
	4	598	9	52	22	34	35	21	110	880	39
	5	349	8	49	35	34	30	14	103	624	61
	6	515	11	63	35	34	35	15	95	803	57
	10	387	10	52	43	35	37	19	94	675	54
	14	362	8	51	40	33	35	21	111	661	56
	20	555	9	48	42	38	34	20	100	846	54
	22	650	11	45	33	37	37	19	89	920	47
	23	493	10	52	61	33	37	20	95	801	53
	24	442	11	53	22	25	34	23	96	706	58
	26	815	12	67	69	43	41	25	124	1195	45
	27	367	9	51	34	34	34	18	81	628	50
Semi-arid	1	612	7	31	43	51	52	20	96	912	39
	3	629	10	35	42	38	34	18	80	886	39
	9	611	10	53	26	34	36	19	81	870	42
	11	491	11	46	65	45	35	18	90	801	53
	13	819	9	53	43	38	34	22	78	1098	37
	15	782	10	52	30	41	35	23	103	1076	38
	16	699	11	55	24	32	38	25	82	965	34
	17	773	10	47	19	29	37	23	124	1063	38
	18	1061	11	52	25	39	35	26	88	1336	22
	21	659	11	58	28	30	45	25	97	953	39
	28	679	13	53	33	35	38	17	94	961	43
	30	921	14	49	25	45	37	26	81	1198	32
Dry sub-humid	8	365	10	53	38	30	32	19	80	625	43
Humid	7	342	10	52	33	30	34	17	76	593	42
	19	212	8	52	38	31	34	14	85	474	50

Note: * The province codes refer to the provinces shown on the map in Figure 1.

The hyper-arid region, in which crops usually have the largest WF per tonne, had the highest population growth (2.4-fold over 1980–2010), followed by the arid region (2.0-fold). The humid region, which had the smallest WF per tonne of crops, also had the lowest population growth (1.6-fold).

2.4. Crop and Virtual Water Trade

Crop trade balance per province. While Iran on the whole was a net crop importer over the whole period of 1980–2010, most provinces in the semi-arid and dry-sub humid regions were net crop exporters, due to a large export of cereals to other provinces (Figure 8). Mazandaran province in the humid region was the largest rice-producing province in the country throughout the period. However, upon considering all crops and the whole humid region—which consists of Mazandaran and Gilan provinces—we observe that the region was a crop importer throughout the period. The provinces in the hyper-arid region, which includes Sistan-Baluchestan, Kerman and Yazd provinces, were always the largest net crop-importing provinces, with the crop trade balance (CTB) of the region as a whole increasing from 0.84 million tonnes in 1980 to 2.27 million tonnes in 2010. However, these provinces remained net exporters of fruits and nuts over the period. While most provinces in the arid region were a net crop importer, with an overall regional CTB of 0.84 in 1980 and 2.27 million tonnes in 2010, they had a large contribution in vegetable and fruit exports over the period.

International crop trade. In 1980, the CTB of the country as a whole was 1.91 million tonnes, resulting from a crop import of 1.94 million tonnes and a crop export of 0.03 million tonnes. In 2010, the CTB had not changed, even though both imports and exports increased considerably. CTB was 1.89 million tonnes in 2010, resulting from 3.19 million tonnes of crop import, and 1.30 million tonnes of crop export. Expressed per capita, the national CTB reduced by 49%, from 49.1 to 25.3 kg·cap^{-1}·y^{-1} over the period 1980–2010, which reflects the increased self-sufficiency of the country. Cereals were dominant in the national CTB, both in 1980 (imports of 2.07 million tonnes) and 2010 (imports of 2.31 million tonnes). Oil seeds import grew by 0.86 million tonnes and took second place in the CTB in 2010. The import of pulses increased from 0.004 million tonnes in 1980 to 0.14 million tonnes in 2010. For sugar crops, the CTB changed from zero trade in 1980 to an import of 0.014 million tonnes in 2010. A considerable increase occurred in exporting vegetables and roots and tubers, reaching total exports of 0.002 million tonnes and 0.71 million tonnes in 2010, respectively. The CTB for nuts changed from an export of 0.005 million tonnes in 1980 to an export of 0.16 million tonnes in 2010, respectively.

Interprovincial crop trade. The interprovincial crop trade increased from 5.22 million tonnes in 1980 to 13.6 million tonnes in 2010, which was mainly due to increases in sugar crop and cereal trade (increases of 4.0 and 3.0 million tonnes, respectively). Fruits also experienced a considerable trade increase over the period (of 1.9 million tonnes).

Virtual water (VW) trade balance per province. Net VW import per province for the years 1980 and 2010 is shown in Figure 9. Most of the provinces located in the semi-arid region were VW exporters over the period 1980–2010 (Figure 10). The arid region as a whole was a VW importer over the whole period, although some of the provinces in the arid region had VW exports. In 1980, the largest VW export was from Kohgiluieh-Boyerahmad province in the semi-arid region (4.6 billion m^3·y^{-1} of which 87% was blue water), while in 2010 the largest VW export came from Fars province in the arid region (3.1 billion m^3·y^{-1} of which 68% blue water). In both cases, this was the result of the relatively large VW export related to cereal exports from these provinces. In 2010, 26% of the total water consumption in the semi-arid region served the production of crops for export to other regions (mainly cereals) or internationally (mainly fruits and nuts).

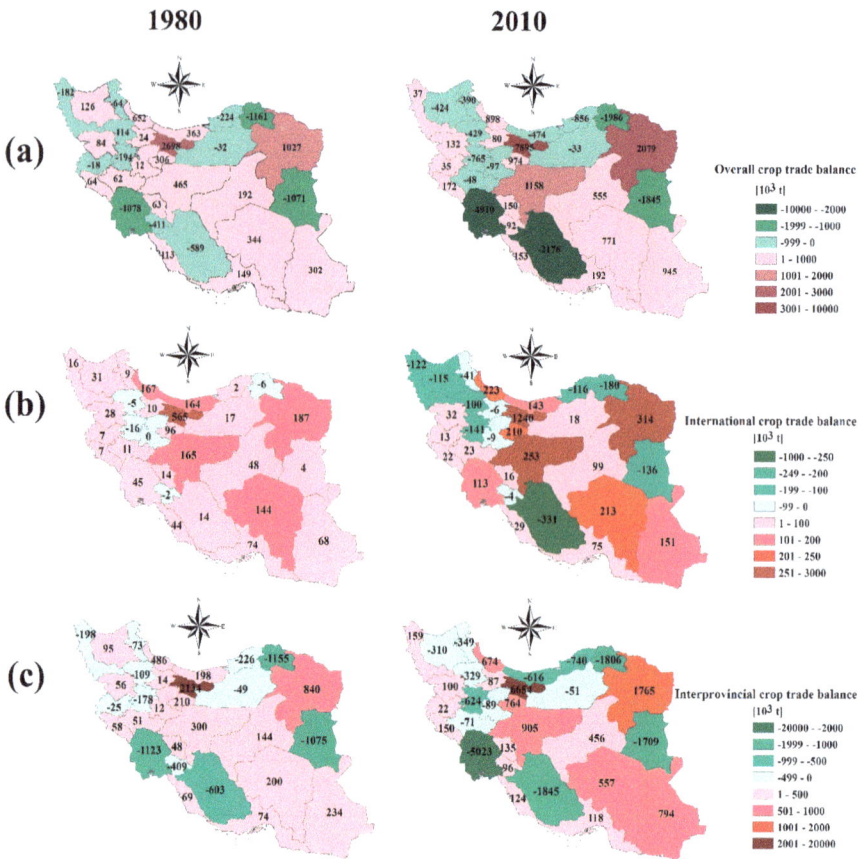

Figure 8. Overall net crop import per province in Iran (**a**), net crop import from abroad (**b**), and net crop import from other provinces (**c**), in the years 1980 (**left**) and 2010 (**right**). Positive signs refer to import; negative signs refer to export.

The changes in crop trade patterns over the period 1980–2010 led to a change in the VW trade pattern as well. Three provinces, namely Golestan (in the dry sub-humid region), Khuzestan (in the arid region) and Kohgiluieh-Boyerahmad (in the semi-arid region), changed from net VW exporters in 1980 to net VW importers in 2010. Vice versa, five provinces in the arid and semi-arid regions, namely Tehran, Qom, Bushehr, East Azarbaijan, and West Azarbaijan, changed from net VW importers in 1980 to net VW exporters in 2010. Besides, Mazandaran province in the humid region changed from a net VW importer to a net VW exporter.

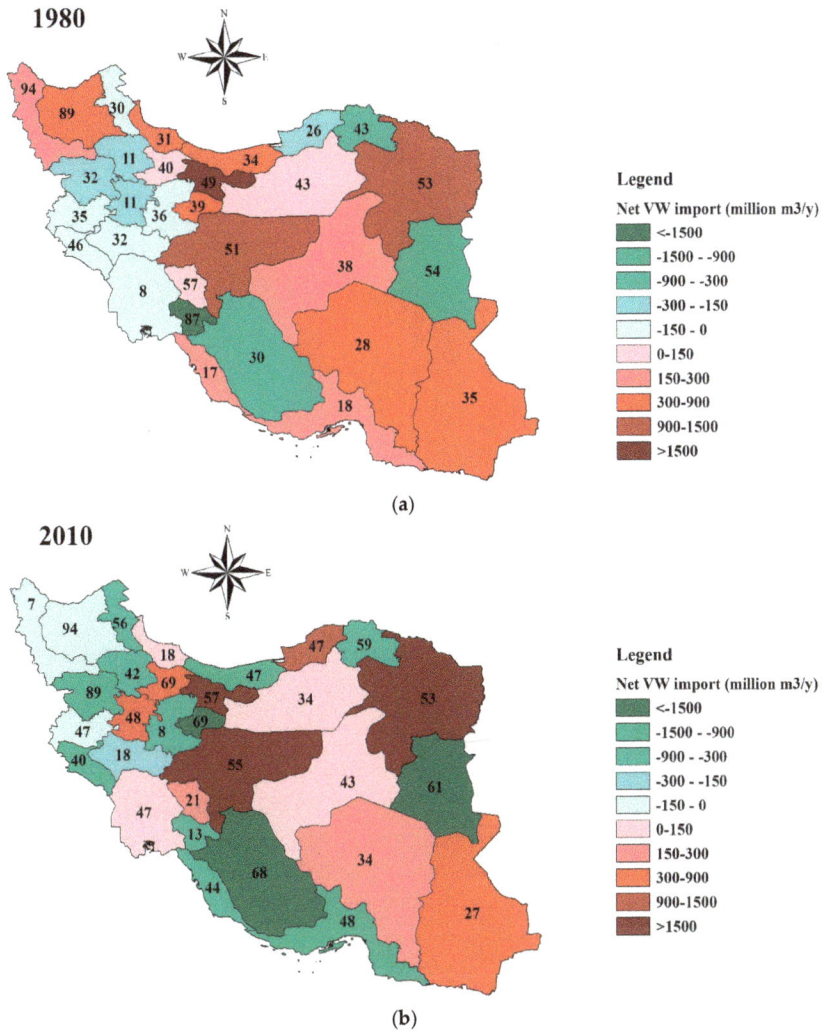

Figure 9. Net total virtual water (VW) import per province in Iran, in the years 1980 (**a**) and 2010 (**b**). Positive signs refer to net virtual water import; negative signs refer to net virtual water export. The figure within each province denotes the percentage of blue water in the VW import of the province.

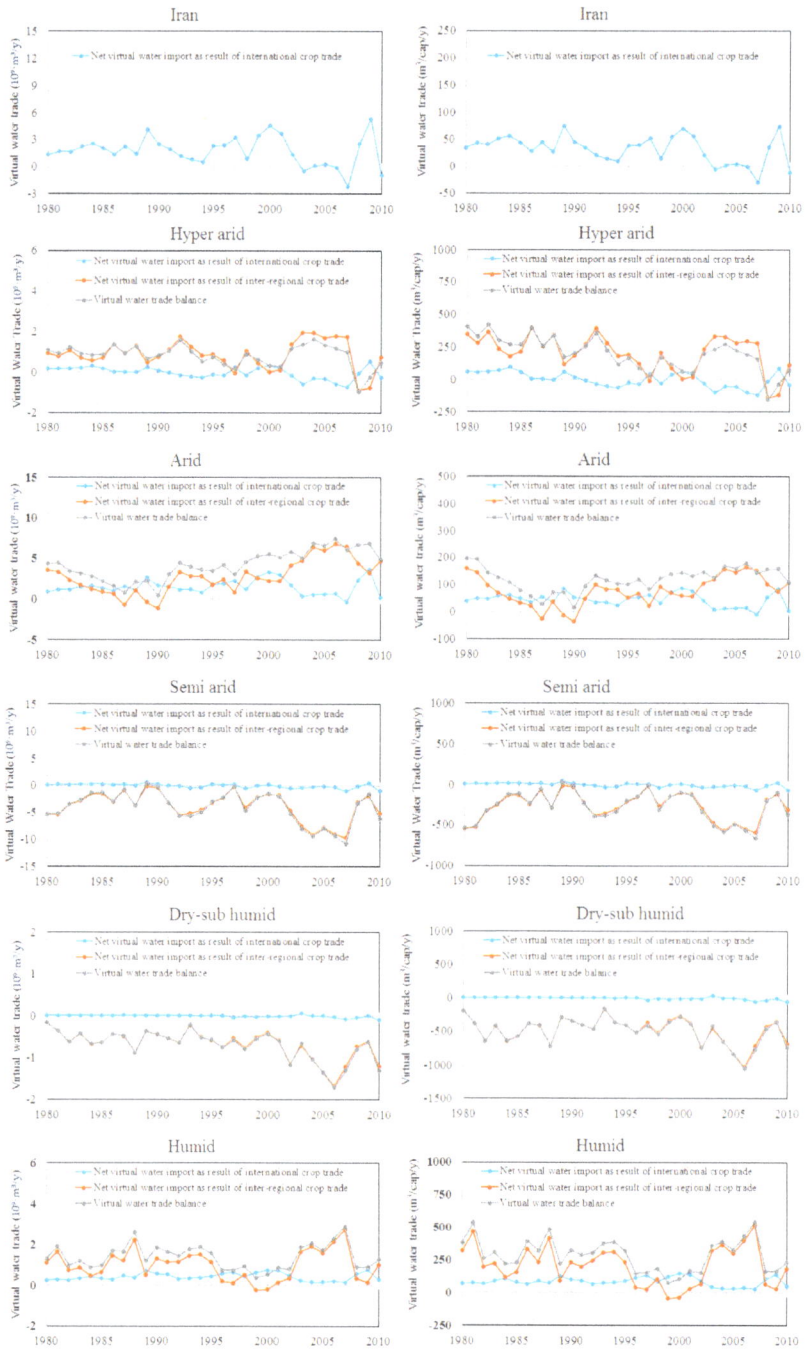

Figure 10. Overall net virtual water trade balance (VWB) and net virtual water import as a result of international and interprovincial crop trade, per climatic region, and for Iran as a whole. Total values in billion $m^3 \cdot y^{-1}$ (**left**) and in $m^3 \cdot cap^{-1} \cdot y^{-1}$ (**right**). Period: 1980–2010.

International virtual water trade. In 1980, the international VW trade of the country as a whole was 1.34 billion $m^3 \cdot y^{-1}$ (with a blue water share of 12.6%), which resulted from a VW import of 1.33 billion $m^3 \cdot y^{-1}$ and a VW export of 0.01 billion $m^3 \cdot y^{-1}$ (Figure 11). In 2010, international VW trade was -0.96 billion $m^3 \cdot y^{-1}$, which resulted from a VW export of 2.68 billion $m^3 \cdot y^{-1}$ and a VW import of 1.72 billion $m^3 \cdot y^{-1}$. While international import in cereals had the largest contribution to the overall VW import in 1980, the import of oil seeds took the first place in 2010. Internationally, Iran exported 0.17 billion $m^3 \cdot y^{-1}$ of blue VW in 1980, and 2.40 billion $m^3 \cdot y^{-1}$ in 2010. The dramatic increase was mainly due to a significant increase in exporting irrigated nuts and fruits in 2010, which are mainly exported from the semi-arid and hyper-arid regions (Figure 12).

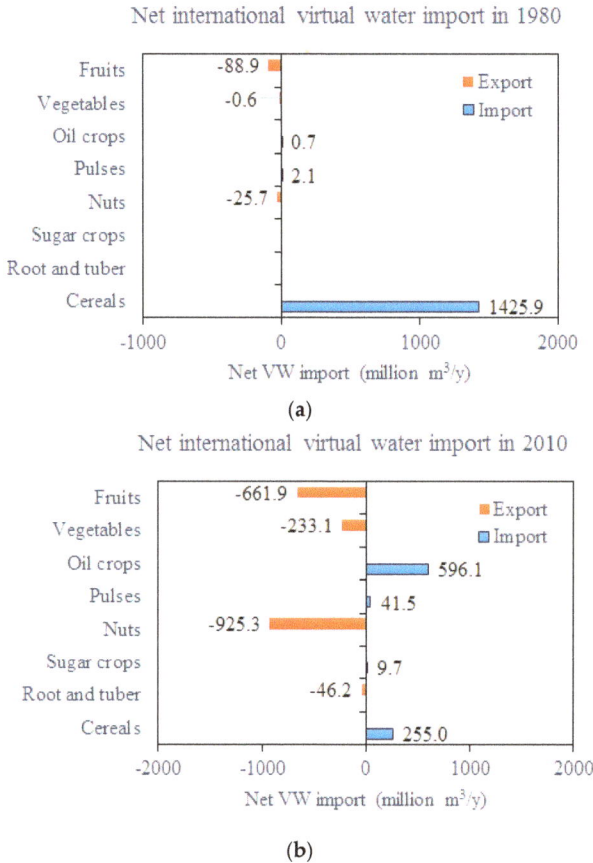

Net international virtual water import in 1980

(a)

Net international virtual water import in 2010

(b)

Figure 11. Net international virtual water import per crop category in 1980 (**a**) and 2010 (**b**).

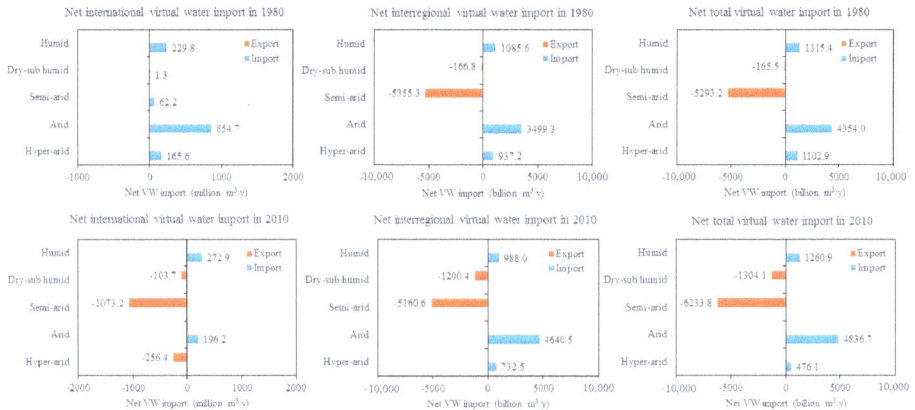

Figure 12. Net international, interregional, and total virtual water import per climatic region in 1980 (**top**) and 2010 (**down**).

Interprovincial virtual water trade. The interprovincial VW trade grew from 9.1 billion $m^3 \cdot y^{-1}$ (59% blue water) in 1980 to 14.8 billion $m^3 \cdot y^{-1}$ (57% blue water) in 2010, which was mainly due to increased interprovincial trade in cereals, nuts, and fruits. The spatial pattern of interprovincial VW trade within the country remained more or less the same over the period, with the semi-arid region responsible for the largest VW export, and the arid region responsible for the largest VW import (Figure 12).

2.5. Water Saving through Crop Trade

Water saving per province. The largest water savings due to trade in the country are found in some provinces in the arid region, most notably Razavi Khorasan and Esfahan (Figure 13). Total water saving in the arid region increased from 5.05 billion $m^3 \cdot y^{-1}$ in 1980 to 13.1 billion $m^3 \cdot y^{-1}$ in 2010. Blue water saving in the arid region increased from 3.71 billion $m^3 \cdot y^{-1}$ in 1980 to 12.0 billion $m^3 \cdot y^{-1}$ in 2010 (Figure 14). However, within the arid region, there are also provinces with water losses due to trade, namely Fars, South Khorasan, and North Khorasan. Most of the provinces in the semi-arid region saved water in relation to international crop trades over the period, but experienced water losses in relation to interprovincial crop trade. The net result of international and interprovincial crop trade for the semi-arid region is an overall water loss of 5.25 billion $m^3 \cdot y^{-1}$ in 1980, and 1.49 billion $m^3 \cdot y^{-1}$ in 2010. For the semi-arid region as a whole, the 3.27 billion $m^3 \cdot y^{-1}$ of blue water loss in 1980 had become 2.35 billion $m^3 \cdot y^{-1}$ in blue water saving in 2010. All three provinces in the hyper-arid region had considerable water saving related to their crop trade, with an increasing trend over time. The two provinces in the humid region, and the one province in the dry sub-humid region, had water savings due to crop trade as well, with again an increasing trend except for Mazandaran province. In 1980, Mazandaran still had a blue water saving due to trade, but in 2010, it had a blue water loss due to the export of irrigated rice.

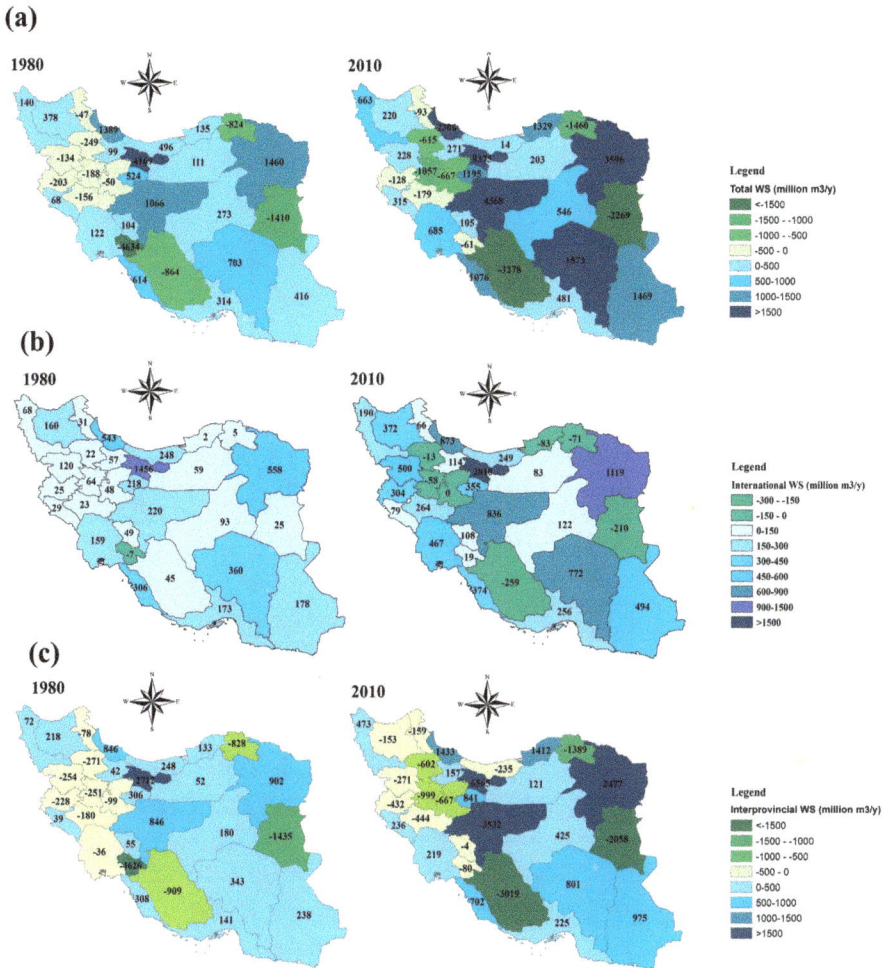

Figure 13. Total (green plus blue) water saving (WS) as a result of total (**a**), international (**b**), and interprovincial (**c**) crop trade, per province, in 1980 (**left**) and 2010 (**right**).

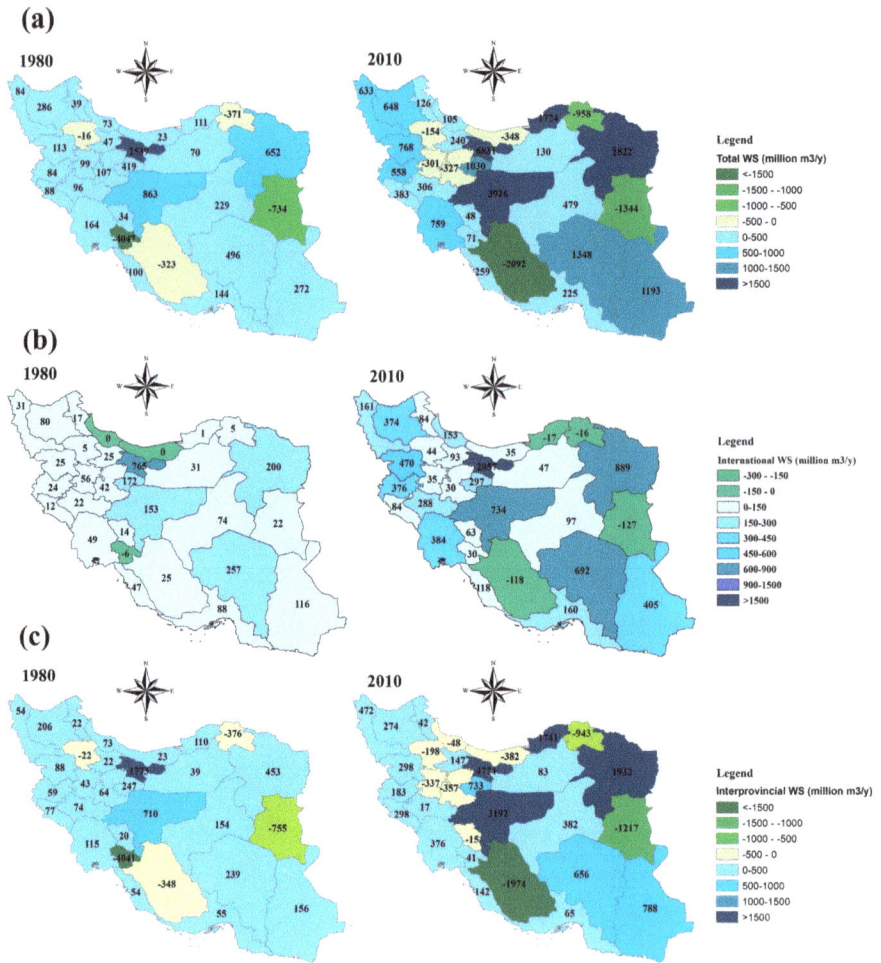

Figure 14. Blue water saving (WS) as a result of total (**a**), international (**b**), and interprovincial (**c**) crop trade, per province, in 1980 (**left**) and 2010 (**right**).

Water saving related to international crop trade. While in 1980, Iran's international crop trade led to a total (green plus blue) water saving of 5.0 billion m^3·y^{-1} (46% blue), this had grown to 9.8 billion m^3·y^{-1} by 2010 (80% blue). However, there was also large variability within this period, which related to the variability in traded crops and their volumes. Cereal imports played the biggest role in the national water saving of Iran through international crop trade, followed by oil crop imports. Overall, the international export of nuts, vegetables, fruits, and root and tubers resulted in water losses through 1980–2010.

Water saving related to interprovincial crop trade. In 1980, interprovincial crop trade was still associated with a total water loss of 1.5 billion m^3·y^{-1}, but this turned into a water saving from 1982 onwards. The water saving related to interprovincial crop trade steadily grew until 2010, when the total water saving amounted to 10.1 billion m^3·y^{-1}. Looking at blue water, we find a blue water loss as a result of interprovincial crop trade of 0.6 billion m^3·y^{-1} in 1980, and a blue water saving of 11.2 billion m^3·y^{-1} in 2010. The water savings due to interprovincial trade refer to most crop categories, but not for cereals and sugar crops, which are traded from provinces with a relatively large WF per tonne (e.g., Fars province in the arid region, with an average WF of 2288 m^3 per tonne of cereals in 2010) to provinces with a smaller WF per tonne (e.g., Tehran, also in the arid region, with a WF of 1731 m^3 per tonne of cereals in 2010).

3. Added Value of WF Assessment for Iran's Food and Water Security Policy

After the 1979 revolution in Iran, the government implemented agricultural policies aimed at achieving food self-sufficiency. The main current policy frameworks governing agriculture and economic development in Iran are [17]: Vision 2025 (adopted in January 2009), Broad Policies for Agriculture (adopted in July 2005), and the fifth Five-Year National Economic, Social and Cultural Development Plan (FYNDP). While the latter plan refers to the period 2011–2016, the sixth Five-Year Plan is still under debate. One of the main objectives of the Iranian government is to achieve national food security through higher agricultural productivity and self-sufficiency in staple crops. In 1999, the government initiated the self-sufficiency strategy for wheat by adopting different policies, which caused Iran to become the 12th largest producer of wheat in the world by 2012 [17]. Thereafter, a guaranteed purchase price was provided for more than 20 crops, with wheat and rice being the most important, which caused a considerable increase in national agricultural production.

Since water availability has direct bearing on food self-sufficiency, the Iranian policy makers implemented ambitious long-term water management plans in the third Five-Year Development Plan of the country (third FYDP) to address the growing gap between demand and supply. Water policies in Iran during this period mainly focused on increasing the amount of water physically available without considering the long-term consequences of this strategy. One of the quantitative goals that have been accomplished after the third FYDP was the increase of total crop production through changing Iran's agricultural system. About 140,000 hectares of irrigation and drainage networks were constructed during the past two decades. To rapidly expand the irrigated lands, the planners and policy makers focused on increasing water availability through constructing dams and the associated infrastructures. During the third FYDP, 12 new dams were constructed, providing an additional water supply of about 3.7 × 10^9 m$^3 \cdot$y^{-1}. Globally, Iran ranks third in dam building, with most dams constructed in the period of 1960–1990 [18]. Currently, about 500 dams are operating in Iran, and 100 more dams are under construction. Moreover, the government considers constructing 400 more dams, which are now in the design or feasibility stages [19]. Based on the reported value in 2016, a volume of 40 billion m$^3 \cdot$y^{-1} of water is currently stored in the Iranian reservoirs [20]. Damming has caused serious environmental problems, such as deteriorating water quality and increasing land desertification and salinization. It has been reported that over two-thirds of Iran's land is rapidly turning into desert as a consequence of environmentally unmanaged damming projects [21].

The expansion of irrigation beyond regional capacity levels caused a dramatic overexploitation of groundwater resources. Now, farmers operate about 500,000 wells in Iran [18], and there is no license or permission for many of them. This has led to the salinization of farmland wells and reduced groundwater access. According to the Institute for Forest and Pasture Research, groundwater levels have dropped by two meters in recent years across 70 plains, affecting as much as 100 million hectares. With little to no metering to ensure that withdrawal limits are not breached, groundwater extraction within Iran has led to a 50% reduction in groundwater availability and significant issues with salinity, as water tables continue to fall [18].

Implementing major interbasin water transfer projects was the other achievement of the third FYDP, mitigating regional water shortages. Transferring desalinated water from the Caspian Sea, and from the Persian Gulf and Sea of Oman to support the dehydrated megacities and parched farmlands within the country are the most recent high-profile projects considered by the Iranian policy makers. While the interbasin water transfer projects are likely to be continued, these plans are unlikely to address water shortages in the long term due to the significant environmental impact these transfers cause. The government also considered the use of unconventional water resources, but as of yet, wastewater use in Iran's agriculture is mostly uncontrolled. There are many local farmers using raw wastewater directly for irrigation without caring about its adverse effects on human health or the environment.

Iran's policy on food self-sufficiency caused a significant increase in total production through increasing water supplies and expanding the irrigated land area, but evaporation losses are large due to the inefficiency of the irrigation systems. Over 70% of the irrigated land is under surface irrigation, with an average irrigation efficiency of 33%, according to the Food and Agriculture Organization of the United Nations (FAO), leaving significant room for water saving through efficiency improvements. Inefficient irrigation can increase the incidence of salinization and waterlogging of agricultural land, and lead to reduced productivity and long-term problems with sustainable land use. In fact, during the past decades, Iran's water problems have mostly been addressed by increasing water availability, while water demand management options have less been considered by the Iranian water authorities.

The water footprint assessment carried out here provides several new insights and management solutions that are currently not considered by the national water strategy of Iran. First, the study shows new insights in how to possibly diminish water consumption in crop production. Our WF assessment demonstrates that the WF per tonne for a specific crop hugely varies within the country, and even within climatic regions. It raises the question of why certain crops are produced in certain provinces, but also why some provinces do better than others. The assessment made here invites the development of benchmarks for the WFs of crops, per crop and per climate region (see for instance Hoekstra [14], Mekonnen and Hoekstra [22], and Zhuo et al. [23]), and for further exploration of what water savings could be achieved when reducing the WF for all crop production in a region to a certain reasonable benchmark level (see for instance Chukalla et al. [24]). WFs can be reduced by diminishing the no beneficial component of evapotranspiration from crop fields, by mulching and better irrigation practices [25]. Adjustments in crop planting dates and selecting appropriate crop varieties that yield more crop per drop are other possible ways to increase water productivity and reduce WFs per tonne of crop [26,27]. In addition, knowledge on the water requirements per unit of crop under certain climatic conditions may result in a reconsideration of the crop production pattern in the country. As we show, for example, oil crops produced in the hyper-arid region have a relatively large WF per tonne of crop, while roots and tubers have a much smaller WF per tonne. Besides, as shown earlier by Karandish et al. [28], roots and tubers also have a smaller WF per hectare, and would give higher economic profit. The question, therefore, is why governmental policies promote planting oil crops in the hyper-arid region.

Second, the study shows how modifying consumption patterns could help to mitigate water scarcity. Iran's water policy makers fully ignore the significant influence of the consumption pattern on exacerbating the water scarcity. Our WF assessment in relation to crop consumption demonstrates the significant influence of diet on water requirements. For example, rice is a common element in the Iranian diet, especially in the northern part of the country, while rice has a much larger WF per tonne compared with alternatives such as wheat or roots and tubers. Besides, even though rice is mostly produced in the humid region, it is mostly irrigated, thus aggravating blue water demands, while wheat and roots and tubers can be produced in the same region under rain-fed conditions.

Third, the analysis in this paper shows that the existing pattern of interprovincial crop trade within the country is counter logical. Although it is the most water-abundant region of the country, the humid region has a net virtual water import through crop trade, due to the relatively small share in total crop production and the decreasing trend in the per capita arable land availability. Economic incentives have encouraged many farmers in northern Iran to change their farms to urban areas. As a consequence, the humid region is a net VW importer, despite being fertile for crop production with a relatively small WF per tonne of crop. On the other hand, the water-scarce semi-arid region, and some provinces in the arid region, produce crops for export to other regions within Iran. Interestingly, the findings here for Iran—of virtual water transfers from water-scarce to more water-abundant regions within a country—is similar to findings for other countries, such as China [29] and India [30]. In recent years, the government has implemented plans for interbasin water transfers, by which water is conveyed from the more water-abundant to the water-scarce regions of Iran. Undoubtedly, this will result in the continued expansion of irrigated agriculture in the arid and semi-arid regions of Iran, where crops have the relatively high WF per tonne. During the few past decades, a strong motivation has been created among local farmers to replace their rain-fed practices with irrigation systems in order to achieve higher annual income through the increased yield. Our findings indeed show that the expansion of the irrigated area has led to a considerable increase in the proportion of the blue WF in the total WF.

Finally, the study demonstrates that Iran's food self-sufficiency policy may be detrimental to maintaining food security in the long run. It has promoted the export of water-intensive products from water-scarce regions, such as cereals from Fars province in the arid region for export to other provinces, which results in groundwater level decline, aquifer depletion, soil salinization, and groundwater quality deterioration. Another example is the promotion of growing cereals, fruits, and sugar crops in West Azarbaijan for export to other provinces, which leads to increased water consumption and contributes to the drying of Urmia Lake. Therefore, knowledge about the virtual water flows entering or leaving a province or climatic region can cast a completely new light on how trade mitigates or aggravates the water scarcity of the province or region.

4. Conclusions

Our analysis shows that food self-sufficiency increased in line with Iranian policy. Besides, the water savings related to international and interprovincial trade increased over time. However, the WF of production substantially increased, particularly within the semi-arid region and some provinces in the arid region that are mostly responsible for feeding the country, which resulted in a strong growth of blue WFs and the overexploitation of water resources in these regions. Besides, our analysis shows that consumption increased because of population growth and an increase in consumption per capita. Current Iranian food and water policy could be enriched by reducing the WFs of crop production to certain benchmark levels per crop and climatic region, aligning cropping patterns to spatial differences in water availability and productivities, and reconsidering interbasin water transfer plans to bring water to water-scarce places with relatively high WFs per unit of crop to produce food for export. Furthermore, Iranian food and water policy could be supplemented by paying due attention to the increasing food consumption per capita in Iran. Finally, the country may have to reconsider its food self-sufficiency and food trade policy. Roots and tubers, nuts, vegetables, and fruits were the most exported crops internationally in 2010. Iran may benefit from the international export of vegetables and roots and tubers due to their relatively low WF per tonne, but exporting nuts and fruits, especially from the drier parts of the country to abroad, leads to a significant national water loss. Furthermore, while importing cereals instead of producing them domestically could save a lot of water, our findings indicate that the per capita international cereals import reduced by 42% over 1980–2010, mainly due to Iran's Wheat Self-sufficiency Project over the past decades.

We acknowledge that adapting Iran's food and water policy is a challenge given the conflicts of interests involved, particularly between the short and long term, and between the goal of food self-sufficiency and the need for sustainable water use. Choices that need to be made will need to consider all of the relevant economic, social, and environmental factors, but will include a political component as well, given the trade-offs to be made. While current Iranian food and water policy narrowly focuses on measures to enhance domestic food production through increased water supply, our research suggests that it could be beneficial to additionally consider the potential of measures to improve water productivity, adapt spatial cropping patterns, shift to diets that are less water intensive, and promote forms of trade that save the scarce domestic water resources. Future research will be necessary to quantify the full potential and implications of these alternative measures.

5. Method and Data

5.1. Study Area

Iran lies between 25°00′ N to 38°39′ N latitude and 44°00′ E to 63°25′ E longitude, and spans an area of 1,640,195 km², which is divided into 30 provinces, as illustrated in Figure 1a. The elevations range from −32 m below sea level to 5428 m above sea level, with a national average of 1200 m. The long-term areal average of minimum (T_{min}) and maximum (T_{max}) temperature and annual precipitation (*P*) are 12.4 °C, 25.2 °C, and 244 mm, respectively. The southeastern provinces of Sistan and Balouhestan and the northern province of Gilan receive the lowest and highest annual P, respectively, viz. 104 mm and 1033 mm. Based on the De Martonne climate classification, there are five climatic regions in Iran: hyper-arid, semi-arid, arid, humid, and dry sub-humid (Figure 1b). The dominant climate is arid and semi-arid (Karandish et al., 2016). Despite having the lowest freshwater availability, the arid and semi-arid regions are responsible for producing more than 70% of the total crop production in the country, with most of the crops being irrigated.

5.2. Method and Data

WF of production. All of the calculations were done per crop per province per year for the study period of 1980–2010. The weighted average WF of each crop category (i.e., cereals, root and tubers, sugar crops, pulses, nuts, oil crops, vegetables, and fruits) was then calculated based on the production of different crops in each category. Thereafter, weighted average values were calculated at the climatic region scale. The WFs of crop production were calculated at a daily time step based on the accounting framework of Hoekstra et al. (2011). For each crop, the green and blue WFs ($m^3 \cdot t^{-1}$) were calculated as the daily green and blue evapotranspiration (*ET*, $m^3 \cdot ha^{-1}$) aggregated over the full growing period, divided by the crop yield (*Y*, $t \cdot ha^{-1}$). ET and Y were simulated using AquaCrop, FAO's water balance and crop growth model [31]. The initial soil moisture content was estimated by running the model for a period of five years, and taking the outcome after the five years as the initial value for our calculation, a procedure followed also for example by Siebert and Döll [32] and Zhuo et al. [33]. Per crop, province, and year, yield data were scaled to fit annual yield statistics at the province level. The model simulates a daily soil water balance for the rooting zone:

$$S_{[t]} = S_{[t-1]} + P_{[t]} + I_{[t]} + CR_{[t]} - ET_{[t]} - RO_{[t]} - DP_{[t]} \tag{1}$$

in which $S_{[t]}$ and $S_{[t-1]}$ are the soil water content at the end of day t and t-1, respectively, $P_{[t]}$ is precipitation on day t, $I_{[t]}$ is irrigation applied on day t, $CR_{[t]}$ is capillary rise, $ET_{[t]}$ is evapotranspiration, $RO_{[t]}$ is surface runoff, and $DP_{[t]}$ is deep percolation. All of the flow terms are in mm/day. Capillary rise is assumed to be zero, since groundwater is assumed to be deeper than one meter below the rooting zone all over Iran. P and I were considered as green and blue water, respectively. The contributions of green (*P*) and blue (*I*) water to RO were calculated based on the ratio of P and I, respectively, to the sum of P and I. The fraction of green and blue water in the total soil water content at the end of the previous day was applied to calculate green and blue DP and ET. Following Chukalla et al. [25] and Zhuo et al. [33], the green soil water content (S_{green}) and blue soil water content (S_{blue}) were calculated as:

$$\begin{cases} S_{green[t]} = S_{green[t-1]} + P_{[t]} + RO_{[t]} \times \frac{P_{[t]}}{P_{[t]}+I_{[t]}} - \left(DP_{[t]} + ET_{[t]}\right) \times \frac{S_{green[t-1]}}{S_{[t-1]}} \\ S_{blue[t]} = S_{blue[t-1]} + I_{[t]} + RO_{[t]} \times \frac{I_{[t]}}{P_{[t]}+I_{[t]}} - \left(DP_{[t]} + ET_{[t]}\right) \times \frac{S_{blue[t-1]}}{S_{[t-1]}} \end{cases} \tag{2}$$

WF of consumption. Following the bottom-up approach [15], per crop and per province, the WF related to consumption of a specific crop ($m^3 \cdot y^{-1}$) was calculated as the crop consumption volume ($t \cdot y^{-1}$) multiplied by the average WF of the crop available in the province ($m^3 \cdot t^{-1}$). As consumption in Iran, we counted all of the components reported under 'utilization' in FAO's food balance sheet.

We calculated the average utilization per crop per capita in Iran and assumed this as the consumption level per capita for each province. Total consumption per province follows from multiplying this with the population in each province. Per province, the average WF of a crop was calculated as a weighted average of the WF of the crop produced in the province, and the WFs of the crops imported from other provinces or abroad:

$$WF_{Prov}[P] = \frac{P_{Prov}[P] \times WF_{prod,Prov}[P] + \sum_e \left(I_e[P] \times WF_{prod,e}[P] \right)}{P_{Prov}[P] + \sum_e I_e[P]} \tag{3}$$

where, $P_{Prov}[P]$ (t·y^{-1}) is the production quantity of crop p, $I_e[P]$ (t·y^{-1}) is the imported quantity of crop p from exporting place e (other provinces in Iran or other countries), $WF_{prod,Prov}[P]$ (m^3·t^{-1}) is the specific WF of crop production in the province, and $WF_{prod,e}[P]$ (m^3·t^{-1}) is the WF of the crop as produced in exporting place e.

International and interprovincial crop trade and virtual water trade. To understand interprovincial trade, we determined, per crop, which provinces had surpluses and which had deficits. The crop origin (abroad or other provinces) for importing into deficit provinces is estimated, per crop, based on the ratio of total Iranian import of that crop to the sum of surpluses in the provinces that have a surplus of that crop. We add, per crop, all provincial exports and calculate the average WF of that sum of provincial surpluses (as a weighted average of the WFs in the surplus provinces). For all of the importing provinces, we assume the WF of the imported crop from other provinces to equal this calculated average. At the province level, the net VW import (m^3·y^{-1}) related to crops is the sum of the interprovincial net VW import plus the international net VW import in the considered province. Data on the WFs related to the crops imported from abroad were obtained from Mekonnen and Hoekstra [34].

Provincial water savings or losses resulting from trade. Water saving (WS) as a result of international or interprovincial crop trade was estimated per province following the method as introduced by Chapagain et al. [35]. WS related to the international crop trade of a province (m^3·y^{-1}) was estimated by multiplying the net import volume of the province from abroad (t·y^{-1}) by the WF per tonne of the crop in the province (m^3·t^{-1}). Similarly, WS related to interprovincial crop trade (m^3·y^{-1}) was computed per province as the net import volume of the province from other provinces (t·y^{-1}) times the WF per tonne of the crop in the importing province (m^3·t^{-1}). We took the national average WF of a crop (m^3·t^{-1}) in instances in which a specific crop was imported to a province, but not grown in that province at all. The provincial WS resulting from trade has a negative sign when there is gross export of a crop rather than gross import. The overall WS related to all interprovincial trade flows within Iran was calculated as the sum of the water savings (or losses) in all of the provinces.

5.3. Data

For the study period of 1980–2010, all of the required data were obtained per crop per province per year. To get the meteorological data, 52 weather stations (Figure 1) located in the five climatic regions were selected [36]. Based on these data, provincial averages of T_{min}, T_{max} and reference evapotranspiration (ET_o) were calculated. ET_o was calculated based on the FAO Penman–Monteith equation [37]. Soil texture data and the total soil water holding capacity were obtained from Batjes [38]. For the hydraulic characteristics for each type of soil, the indicative values provided by AquaCrop were used. The population statistics were obtained from the Statistical Center of Iran [39]. We consider 26 crops common to Iran, which were classified into eight crop categories based on the FAO classification [37]: cereals (wheat, barley, and rice), roots and tubers (potato), sugar crops (sugar beet and sugar cane), pulses (bean, pea, and lentil), nuts (pistachio, walnut, almond, and hazelnut), oil crops (cottonseed, soybean, and canola), vegetables (tomato and onion) and fruits (apple, banana, date, grape, lime, lemon, tangerine, orange, and grapefruit). Agricultural data for the irrigated and rain-fed crops, including crop sowing area (ha), irrigated area (ha), crop planting and harvesting dates, and crop yield (kg·ha^{-1}), were collected per crop per province per year from Iran's Ministry of Agriculture Jihad [40]. Data on Iran's international trade per crop (in t·y^{-1}) were taken from FAO (2016a). Data on national crop consumption per capita, in terms of primary crop equivalents, were obtained from the Supply and Utilization Accounts of FAOSTAT [17].

Acknowledgments: The work of F.K. was funded by the University of Zabol. The work of A.H. was funded by the University of Twente.

Author Contributions: F.K. and A.H. conceived and designed the research; F.K. performed the modeling work; F.K. and A.H. analyzed the data and wrote the paper.

Conflicts of Interest: The authors declare no conflict of interest.

References

1. *AQUASTAT*; Aqua Statistics of Food and Agriculture Organization of the United Nations: Rome, Italy, 2016; Available online: http://www.fao.org/nr/water/aquastat/main/index.stm (accessed on 1 January 2016).
2. Madani, K. Water management in Iran: What is causing the looming crisis? *J. Environ. Stud. Sci.* **2014**, *4*, 315–328. [CrossRef]
3. United Nations. *World Population Prospects: The 2015 Revision*; United Nations: New York, NY, USA, 2015.
4. Fathian, F.; Morid, S.; Kahya, E. Identification of trends in hydrological and climatic variables in Urmia Lake basin, Iran. *Theor. Appl. Climatol.* **2014**, *119*, 443–464. [CrossRef]
5. Sharifikia, M. Environmental challenges and drought hazard assessment of Hamoun Desert Lake in Sistan region, Iran, based on the time series of satellite imagery. *Nat. Hazards* **2013**, *65*, 201–217. [CrossRef]
6. Najafi, A.; Vatanfada, J. Environmental challenges in trans-boundary waters, case study: Hamoon Hirmand Wetland (Iran and Afghanistan). *Int. J. Water Resour. Arid Environ.* **2011**, *1*, 16–24.
7. Madani, K. Iran's water crisis: Inducers, challenges and counter-measures. In Proceedings of the ERSA 45th Congress of the European Regional Science Association, Free University, Amsterdam, The Netherlands, 23–27 August 2005.
8. Hoekstra, A.Y.; Mekonnen, M.M. The water footprint of humanity. *Proc. Natl. Acad. Sci. USA* **2012**, *109*, 3232–3237. [CrossRef] [PubMed]
9. Meijer, K.S.; Van der Krogt, W.N.M.; Van Beek, E. A new approach to incorporating environmental flow requirements in water allocation modeling. *Water Resour. Manag.* **2012**, *26*, 1271–1286. [CrossRef]
10. Tavassoli, H.R.; Tahershamsi, A.; Acreman, M. Classification of natural flow regimes in Iran to support environmental flow management. *Hydrol. Sci. J.* **2014**, *59*, 517–529. [CrossRef]
11. Abdi, R.; Yasi, M. Evaluation of environmental flow requirements using eco-hydrologic–hydraulic methods in perennial rivers. *Water Sci. Technol.* **2015**, *72*, 354–363. [CrossRef] [PubMed]
12. Nia, E.S.; Asadollahfardi, G.; Heidarzadeh, N. Study of the environmental flow of rivers, a case study, Kashkan River, Iran. *J. Water Supply Res. Technol. AQUA* **2016**, *65*, 181–194.
13. Hoekstra, A.Y.; Chapagain, A.K. Water footprints of nations: Water use by people as a function of their consumption pattern. *Water Resour. Manag.* **2007**, *21*, 35–48. [CrossRef]
14. Hoekstra, A.Y. *The Water Footprint of Modern Consumer Society*; Routledge: London, UK, 2013.
15. Hoekstra, A.Y.; Chapagain, A.K.; Aldaya, M.M.; Mekonnen, M.M. *The Water Footprint Assessment Manual: Setting the Global Standard*; Earthscan: London, UK, 2011.
16. Karandish, F.; Mousavi, S.S.; Tabari, H. Climate change impact on precipitation and cardinal temperatures in different climatic zones in Iran: analysing the probable effects on cereal water-use efficiency. *Stoch. Environ. Res. Risk Assess.* **2016**. [CrossRef]
17. *FAOSTAT*; Food and Agriculture Organization of the United Nations Statistics: Rome, Italy, 2014; Available online: http://www.fao.org/faostat/en/ (accessed on 1 January 2014).
18. Lehane, S. *The Iranian Water Crisis*; Future Directions International Pty Ltd.: Perth, Australia, 2014; p. 11.
19. Zafarnejad, F. The contribution of dams to Iran's desertification. *Int. J. Environ. Stud.* **2009**, *66*, 327–341. [CrossRef]
20. IWRM. Iran Water Resources Management Company, 2016. Available online: http://wrm.ir/ (accessed on 1 January 2016).

21. Nikouei, A.; Ward, F. Pricing irrigation water for drought adaptation in Iran. *J. Hydrol.* **2013**, *503*, 29–46. [CrossRef]
22. Mekonnen, M.M.; Hoekstra, A.Y. Water footprint benchmarks for crop production: A first global assessment. *Ecol. Indic.* **2014**, *46*, 214–223. [CrossRef]
23. Zhuo, L.; Mekonnen, M.M.; Hoekstra, A.Y. Benchmark levels for the consumptive water footprint of crop production for different environmental conditions: A case study for winter wheat in China. *Hydrol. Earth Syst. Sci.* **2016**, *20*, 4547–4559. [CrossRef]
24. Chukalla, A.D.; Krol, M.S.; Hoekstra, A.Y. Marginal cost curves for water footprint reduction in irrigated agriculture: Guiding a cost-effective reduction of crop water consumption to a permit or benchmark level. *Hydrol. Earth Syst. Sci.* **2017**, *21*, 3507–3524. [CrossRef]
25. Chukalla, A.D.; Krol, M.S.; Hoekstra, A.Y. Green and blue water footprint reduction in irrigated agriculture: Effect of irrigation techniques, irrigation strategies and mulching. *Hydrol. Earth Syst. Sci.* **2015**, *19*, 4877–4891. [CrossRef]
26. Zhuo, L.; Mekonnen, M.M.; Hoekstra, A.Y. Sensitivity and uncertainty in crop water footprint accounting: A case study for the Yellow River basin. *Hydrol. Earth Syst. Sci.* **2014**, *18*, 2219–2234. [CrossRef]
27. Lopez, L.I.F.; Bautista-Capetillo, C. Green and blue water footprint accounting for dry beans (Phaseolus Vulgaris) in primary region of Mexico. *Sustainability* **2015**, *7*, 3001–3016. [CrossRef]
28. Karandish, F.; Salari, S.; Darzi-Naftchali, A. Application of virtual water trade to evaluate cropping pattern in arid regions. *Water Resour. Manag.* **2015**, *29*, 4061–4074. [CrossRef]
29. Ma, J.; Hoekstra, A.Y.; Wang, H.; Chapagain, A.K.; Wang, D. Virtual versus real water transfers within China. *Philos. Trans. R. Soc. Lond. B* **2006**, *361*, 835–842. [CrossRef] [PubMed]
30. Verma, S.; Kampman, D.A.; Van der Zaag, P.; Hoekstra, A.Y. Going against the flow: A critical analysis of inter-state virtual water trade in the context of India's National River Linking Program. *Phys. Chem. Earth* **2009**, *34*, 261–269. [CrossRef]
31. Steduto, P.; Hsiao, T.C.; Raes, D.; Fereres, E. AquaCrop-The FAO crop model to simulate yield response to water: I. Concepts and underlying principles. *Agron. J.* **2009**, *101*, 426–437. [CrossRef]
32. Siebert, S.; Döll, P. Quantifying blue and green virtual water contents in global crop production as well as potential production losses without irrigation. *J. Hydrol.* **2010**, *384*, 198–217. [CrossRef]
33. Zhuo, L.; Mekonnen, M.M.; Hoekstra, A.Y.; Wada, Y. Inter- and intra-annual variation of water footprint of crops and blue water scarcity in the Yellow River Basin (1961–2009). *Adv. Water Resour.* **2016**, *87*, 21–41. [CrossRef]
34. Mekonnen, M.M.; Hoekstra, A.Y. The green, blue and grey water footprint of crops and derived crop products. *Hydrol. Earth Syst. Sci.* **2011**, *15*, 1577–1600. [CrossRef]
35. Chapagain, A.K.; Hoekstra, A.Y.; Savenije, H.H.G. Water saving through international trade of agricultural products. *Hydrol. Earth Syst. Sci.* **2006**, *10*, 455–468. [CrossRef]
36. IMO. Iran Meteorological Organization: Tehran, Iran, 2016. Available online: www.irimo.ir/far (accessed on 1 January 2016).
37. Allen, R.G.; Pereira, L.S.; Raes, D.; Smith, M. *Crop Evapotranspiration Guidelines for Computing Crop Water Requirements*; FAO Irrigation and Drainage Paper 56; Food and Agriculture Organization of the United Nations: Rome, Italy, 1998.
38. Batjes, N.H. *ISRIC-WISE Global Data Set of Derived Soil Properties on a 5 by 5 Arc-Minutes Grid (Version 1.2)*; Report 2012/01; ISRIC World Soil Information: Wageningen, The Netherlands, 2012.
39. SCI. Statistical Center of Iran: Tehran, Iran, 2016. Available online: http://www.amar.org.ir (accessed on 1 January 2016).
40. IMAJ. Iran's Ministry of Agriculture Jihad: Tehran, Iran, 2016. Available online: www.maj.ir (accessed on 1 January 2016).

Article

Water Footprint Accounting Along the Wheat-Bread Value Chain: Implications for Sustainable and Productive Water Use Benchmarks

Pascalina Matohlang Mohlotsane [1], Enoch Owusu-Sekyere [1,*], Henry Jordaan [1], Jonannes Hendrikus Barnard [2] and Leon Daniel van Rensburg [2]

[1] Department of Agricultural Economics, University of the Free State, Posbus 339, Bloemfontein 9300, South Africa; MohlotsaneMP@ufs.ac.za (P.M.M.); JordaanH@ufs.ac.za (H.J.)
[2] Department of Soil- and Crop- and Climate Sciences, University of the Free State, Posbus 339, Bloemfontein 9300, South Africa; BarnardJH@ufs.ac.za (J.H.B.); VRensBL@ufs.ac.za (L.D.v.R.)
* Correspondence: kofiwusu23@gmail.com

Received: 2 July 2018; Accepted: 28 August 2018; Published: 31 August 2018

Abstract: Efficient and wise management of freshwater resources in South Africa has become critical because of the alarming freshwater scarceness. The situation requires a thorough examination of how water is utilized across various departments that use water. This paper reports on an examination of the water footprint and economic water productivities of the wheat-bread value chain. The assessment methodology of the Water Footprint Network was employed. The findings reveal that 954.07 m^3 and 1026.07 m^3 of water are utilized in the production of a ton of wheat flour in Bainsvlei and Clovelly in South Africa. The average water footprint for wheat bread was 954.53 m^3 per ton in Bainsvlei and 1026.53 m^3 per ton in Clovelly. More than 99% of the water is used in producing the grain at the farm level. The processing stage of the value chain uses less than 1% of the total water footprint. About 80% of all the water utilised along the wheat bread value chain is attributed to blue water. The findings revealed a significant shift from green water consumption to higher blue water use, and this is a major concern for water users and stakeholders along the wheat-bread value chain, given that blue water is becoming scarce in South Africa. The groundwater contributes about 34% and 42% of the average total water footprint of wheat at the farm level in Clovelly and Bainsvlei, respectively, suggesting the need to have an idea of the contribution of groundwater in water footprint evaluation and water management decision of farmers. This insight will aid in minimizing irrigation water use and pressure on groundwater resources. A total of ZAR 4.27 is obtained for every m^3 of water utilized along the wheat-bread value chain. Water footprint assessment has moved away from sole indicator assessment, as a deeper awareness of and insight into the productive use of water at different stages has become vital for policy. To make a correct judgment and to assess the efficient and wise use of water, there is a need for catchment- or region-specific water footprint benchmarks, given that water footprint estimates and economic water productivities vary from one geographical area to another.

Keywords: economic water productivities; groundwater; wheat-bread; water footprint accounting; South Africa; value addition

---

1. Introduction

Freshwater is a renewable resource, but, when considering its availability regarding unit per time per region, the limitations of this resource cannot be ignored [1,2]. In global terms, agriculture accounts for 99% of freshwater consumption [3] and is therefore considered as the single largest freshwater user globally. Hoekstra and Chapagain [4] show that visualizing the amount of water used in producing products can further increase our understanding of the global picture of freshwater utilization—a concept that is explored by the Global Water Footprint Network Standard approach (GWFNS). The GWFNS approach has become apparent as an important sustainability indicator in the agricultural sector, as well as in the agri-food-processing industry [5–7]. This assessment includes both the indirect and direct use of freshwater by a consumer or a product along with its value chain [8].

South Africa is deemed as water scarce and water limited country [9]. Irrigated agriculture uses about 60% of South Africa's available surface and freshwater resources [10]. Nonetheless, 30% to 40% of this water is lost through leaks and evaporation, which gives the impression that water use in this sector is inefficient [10]. According to the Department of Agriculture Forestry and Fisheries (DAFF) [11], South Africa's agricultural sector is the least direct contributor to the gross domestic product (GDP) measured in per million cubic meters of freshwater use and is also the least direct employer per million cubic meters of freshwater [11,12]. This is in contrast with the commitment of the National Water Research aim of achieving sustainable and efficient use of freshwater by all South Africans, especially among producers of key food crops.

Wheat is the largest cultivated commercial crop globally [13]. In South Africa, wheat is the largest winter cereal grain with a total requirement of 2.7 million metric tons per year [11]. Most of the wheat used for bread production is produced locally. Wheat production is spread among 32 of the 36 crop-production regions, with an estimated 3200 to 4000 producers. South Africa's wheat production is estimated at 1.88 million metric tons for the 2016/17 production year [11]. About 69.63% of South Africa's total wheat demand is produced locally, and 30.37% is imported. About 60% of the wheat flour is used to produce bread. In South Africa, existing statistics indicate that 2.8 billion loaves of bread are consumed per year. This indicates that, in a year, sixty-two loaves of bread, with an average weight of 700 g are consumed per person per year, with a noticeable difference in preference and consumption amongst the provinces [11]. Given the relative importance of the crop and the water scarcity situation in the country, potential strategies that will reduce and identify large water uses along the value chain water use will be deduced.

Two well-known concepts are applied in the assessment of water footprint. These are the Life Cycle Assessment (LCA) approach and the Water Footprint Assessment Manual (WFAM). Recently, some developments in the Water Footprint framework have taken place within the framework of Life Cycle Assessment [14]. The LCA approach proposed to weight the original volumetric water footprint by the water scarcity in the catchment where the water footprint is situated (ISO, 2014), with the aim of attaining a water-scarcity weighted water footprint that portrays the possible local environmental impact of water usage [14]. This proposal has received some critique in recent years [15]. The critique as elaborated by Hoekstra [15] is that there will be confusion about water scarcity if volumes of water use are counted differently based on the level of local water scarcity [15]. This relates to allocation of water resources to opposing uses and reduction at a global scale. Secondly, the LCA approach ignores green water usage, and this neglect suggests that the LCA does not accept the fact that green water is scarce amidst changing climates. The third critique is that since water scarcity in a given geographical area increases with increasing total water consumption in the area, multiplying the consumptive water use of a given process with water scarcity suggests that the subsequent weighted water footprint of a process will be impacted by the water footprints of other processes [15]. The fourth critique is that the manner in which the LCA approach treats the water footprint is inconsistent with definitions of other environmental footprints. Finally, the Water Stress Index as described by the LCA approach lacks relevant physical understanding [15].

In terms of the water footprint of wheat, the latter approach (WFAM) has been employed by some authors in recent years. For instance, Mekonnen and Hoekstra [16] gave an overview of the green, blue and grey water footprint of several crops and derived crop products worldwide, including for South Africa. Mekonnen and Hoekstra [13] estimated the water footprint of wheat. Aldaya and Hoekstra [5] calculated the water footprint of pasta and pizza margarita in Italy. Ahmed and Ribbe [17] explored the green and blue water footprints of rain-fed and irrigated wheat in Sudan. Neubauer [18] calculated the water footprint required to produce 1 kg of bread in Hungary. Sundberg [19] conducted a water footprint assessment of winter wheat and derived wheat products in Sweden. Ababaei and Etedali [20] calculated the water footprint of wheat produced without irrigation in Iran.

None of these studies considered an assessment of the water footprint along the entire wheat value chain in South Africa. For instance, Le Roux et al. [21] evaluated the water footprint of wheat in South Africa, but they only focused on quantifying the water footprint at the farm level, without considering water utilization along the entire wheat-bread value chain. Nonetheless, Mekonnen and Hoekstra [3] quantified the water footprint of wheat for several countries not excluding South Africa. Additionally, Mekonnen and Hoekstra [16] evaluated the water footprint of several crops and derived crop products worldwide, including South Africa. Their estimates were reported at the national and provincial levels, and, as such, there is no current information on the water footprint of wheat at the catchment- specific level in South Africa. Catchment- or regional-specific estimates are needed to better inform water managers and policy makers about water management policies across different regions. Also, it has been found that catchment- or region-specific water footprints vary from national footprint estimates [22,23].

Furthermore, no current studies examined the economic water productivity of bread along its respective value chain, which include farm level, milling, and bakery stages, in South Africa. Aldaya et al. [5] estimated the economic water productivity of wheat in Central Asia. Similarly, Chouchane et al. [24] and Zoumides et al. [25] added water productivities evaluation when assessing the water footprint of crops in Tunisia and Cyprus, respectively. The main objective of this study was to account for the water footprint and economic productivity of water along the wheat-bread value chain. The present study contributes to the existing literature on water footprints and economic water productivities of crops. The water footprint estimates calculated from this study can act as benchmarks for the catchment area considered in this study. The findings of this study can potentially advise policymakers and water users on economically efficient and sustainable water management strategies.

2. Methodology

2.1. Choice of Theoretical Framework and Models

This study followed the water footprint concept of Hoekstra et al. [8]. The definition of blue, green and grey water footprints followed that of Hoekstra et al. [8] in the Water Footprint Assessment Manual. The study employed this method because it involves several dimensions, showing the sources of water utilization in quantities [26]. The conceptualization procedure of the study is presented in Figure 1.

According to the water footprint concept adopted in this study, the water footprint can be calculated in four phases, namely, goal setting, water footprint accounting, sustainability assessment and formulation of response [8]. However, in this study, our third phase focuses on water productivity assessment. In the first phase of this study, the step-wise accumulation approach was followed because, along the wheat value chain, each output product serves as an input for the next product. The total water footprint will include proportional water footprints of the various inputs within the value chain [8]. The analysis was for a single production year.

The step-wise accumulation approach is expressed empirically in Equation (1). By this approach, the water footprint of wheat bread (W), which the main output product, is stated to be made from z inputs (e.g., wheat, flour, etc.). We denote the z inputs to range from j = 1 ... z. Given that z inputs are utilized to produce w wheat products, we denote the different wheat output products as W = 1 ... w. The wheat products' (W) water footprints are specified as:

$$WF_{prod}[W] = \left(WF_{proc}[W] + \sum_{i=1}^{z} \frac{WF_{prod}[j]}{f_w[W,j]} \right) * f_v[W], \tag{1}$$

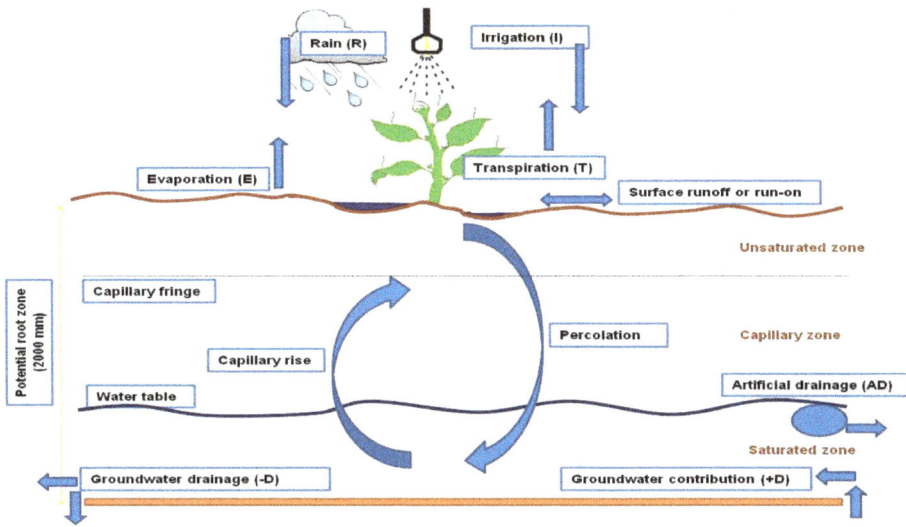

Figure 1. Procedure conceptualization of the field experiment.

$WF_{prod}[W]$ represents the total water used in producing W. $WF_{prod}[j]$ denotes the water footprint of input j. The water utilized in the processing z inputs to W outputs is represented by $WF_{proc}[W]$ (Hoekstra et al., 2011). $f_w[W, j]$ and $f_v[W]$ are the product and value fractions, respectively [8]. Thus, the water footprint of wheat along the product cycle at the farm level is the sum of a process water footprint of the different sources of water used in production according to Aldaya and Hoekstra [6] and Ababaei and Etedalie [20]. The process water footprint is specified as:

$$WF_{proc,blue,green,grey}[W] = \left(\frac{CWU_{blue}}{Y_t}\right) + \left(\frac{CWU_{green}}{Y_t}\right) + \left(\frac{(\alpha \times AR)/(c_{max} - c_{nat})}{Y_t}\right), \qquad (2)$$

The blue, green, and grey water footprints of wheat at the farm level are denoted as $WF_{proc,blue,green,grey}[W]$. The blue component of the water footprint is represented by $\frac{CWU_{blue}}{Y_t}$, where CWU_{blue} represent the blue water used in producing wheat and Y_t is wheat yield [8]. In this paper, blue water use was categorized into surface and groundwater sources. This will give an idea about the proportion of water extracted from the ground and surface according to Hoekstra et al. [8].

$$WF_{blue} = \frac{CWU_{surface}}{Y_t} + \frac{CWU_{ground}}{Y_t}, \qquad (3)$$

The green component of the water footprint is represented as $\frac{CWU_{green}}{Y_t}$ and CWU_{green} indicates the green water used in producing wheat [8]. The crop water use components in Equation (2) summed the daily evapotranspiration over the complete growing period of the wheat crop [8] and are stated empirically as:

$$CWU_{blue,green} = 10 \times \sum_{d=1}^{lgp} ET_{blue,green}, \qquad (4)$$

$ET_{blue,green}$ characterizes the blue and green water evapotranspiration. The water depths are changed from millimetres to volume per area by using the factor 10 [8]. The last part of Equation (2) is the grey water footprint component. This is calculated by taking the chemical application rate for the field per hectare (AR, kg/ha) and multiplying by the leaching-run-off fraction (α). The product is divided by the difference between the maximum acceptable concentration (c_{max}, kg/m^3) and the natural concentration of the pollutant considered (c_{nat}, kg/m^3) [8]. It is worth mentioning that grey water was not used at the processing stage. Grey water was only used at the farm level and the pollutant considered is nitrogen.

At the milling stage, the water footprint of flour is specified as in Equation (5):

$$WF_{milling}[flour] = \frac{TWU_{mill}}{Q_{flour}}, \tag{5}$$

where TWU_{mill} is the total water used to produce a given quantity of flour (Q_{flour}) and at the water utilized at the bakery (TWU_{bakery}) for a given quantity of bread (Q_{bread}) is specified as in Equation (6).

$$WF_{baking}[bread] = \frac{TWU_{bakery}}{Q_{bread}}, \tag{6}$$

The total water footprint of bread along the wheat- bred value chain is a combination of all the footprints in this value chain. After calculating the water footprint ($WF_{prod}[W]$), we estimated physical water productivity (PWP) of the output products (W) in kilograms per cubic meter, expressed as:

$$PWP(kg/m^3) - \frac{1}{WF_{prod}[W](m^3/tonne)} \times 1000, \tag{7}$$

Subsequently, we estimated the economic water productivities for the different outputs at different stages by multiplying the physical water productivity by the monetary value added to each w output per kilogram. The value added to the output products along the value chain is calculated by subtracting the cost per kilogram of w from the sales revenue obtained from selling one kilogram of w at each stage of the value chain [27,28]. Consequently, the value added to the output product (w) becomes the total revenue of the output product minus the cost of all intermediate inputs (z) used to produce it. Let the value added to w at a specific stage of the value chain be represented by $VAD_{jvc}[W]$ and specified as:

$$VAD_{jvc}[W] = Rev_{Jvc}(W) - Cost_{jvc}(W), \tag{8}$$

where $Rev_{jvc}(W)$ represents sales revenue attained from one kilogram of w and $Cost_{jvc}(W)$ denotes all intermediate inputs costs, including the cost of water usage, capital, land, labour, feed, taxes, conveyance, packing, fuel, repairs and maintenance, etc. The sum of the value added at each stage of the product cycle became the total value added ($TVAD[W]_{vc}$), stated as:

$$TVAD_{vc}[W] = \sum_{j=1}^{3} VAD_{jvc}, \tag{9}$$

Value added to water along the wheat-bread value chain is quantified as the ratio of the value added to the output product (w) at a given stage over the volume of water used at that stage [27,28]. From this, we calculated the marginal value of water $MVAD[water]$ as the partial derivative of total value added ($TVAD_{jvc}$) with respect to water use (WU_{jvc}):

$$MVAD[water]_{vc} = \partial \frac{TVAD_{jvc}}{WU_{jvc}}, \tag{10}$$

The marginal value added to water is then multiplied by physical water productivity to attain the economic water productivity according to Chouchane et al. [24] and Owusu-Sekyere et al. [28]:

$$EWP\left(ZAR/m^3\right) = PWP\left(kg/m^3\right) \times VAD(ZAR/kg), \tag{11}$$

2.2. Data Description

This paper employed primary data that cover the wheat-bread value chain. Data on water usage for wheat production were sourced from Van Rensburg et al. [29], who conducted a lysimeter experiment to solicit spatiotemporal data from the Vaalharts and Orange-Riet regions. This study made use of actual measurements through a lysimeter trial to avoid any assumptions that come with water use models. The experiment consisted of five treatments replicated three times, and an average was taken to represent each sample. The cultivars used were selected by their wide use in all the central parts of South Africa. Aboveground biomass was harvested when the crops were dry by cutting it just above the soil surface. The lysimeter trial evaluation procedure for the different treatments employed in the two study areas captured data on groundwater levels, irrigation, drainage and changes in soil moisture content.

The lysimeter procedure consisted of five treatments for groundwater levels, namely, no groundwater considered (control), one meter to constant, 1.5 m to constant, one meter to falling, and 1.5 m to falling. The results are presented in Table 1. Table 1 presents the recorded data used in the estimation of the blue and green water footprints.

Table 1. Collective data for wheat production.

Treatments	Cum. ET	R	WUE	I + R	I	G	DM	Yield
BAINSVLEI								
Control	880	183	11.23	864	681	0	15,999	9881
1 m—Constant	954	183	11.00	371	188	605	16,123	10,475
1.5 m—Constant	914	183	10.87	481	298	467	16,319	9921
1 m—Falling	906	183	11.57	400	217	532	16,776	10,458
1.5 m—Falling	881	183	11.63	460	277	443	15,578	10,230
CLOVELLY								
Control	825	183	9.83	834	651	0	14,708	8375
1 m—Constant	869	183	10.40	469	286	424	13,995	9010
1.5 m—Constant	860	183	10.63	540	357	330	15,185	9161
1 m—Falling	830	183	10.77	426	243	408	15,230	8937
1.5 m—Falling	824	183	10.47	472	289	360	14,898	8620

Cum. ET = cumulative evapotranspiration; R = effective rain; I + R = irrigation and rain; I = irrigation; G = groundwater; DM = dry matter. Source: Authors' calculations.

The data used in the estimation of grey water footprints are presented in Table 2. The nitrogen (Kg N/ha) and phosphorus fertilization (Kg P/ha) were based on targeted yields under irrigation. The wheat farmers usually apply nitrogen and phosphorus fertilizers. Some farmers apply potassium fertilizers. However, in this trial, only nitrogen and phosphorus fertilizers were considered. The leaching runoff coefficients of nitrogen for the two areas are 0.074 and 0.138 whereas those of phosphorus were 0.080 and 0.280. The nitrogen application rates were presented for different potential yields. The target yields range from 2–5 tons per hectare to above 8 tons per hectare, with corresponding nitrogen application rates ranging from 80-130 kg N/ha to 200+ kg N/ha. Prior to the phosphorus application, the soil phosphorus status was examined to know the quantity of phosphorus to apply. The quantities of phosphorus already in the soil were categorized into less than 5 mg/kg, 5–18 mg/kg, 19–30 mg/kg and above 30 mg/kg. The application rates of phosphorus varied depending on the amount that is already available in the soil. Soils with less than 5 mg/kg received a higher amount of phosphorus, followed by 5–18 mg/kg with soils containing above 30 mg/kg receiving the least amount of phosphorus applied.

Table 2. Nitrogen (Kg N/ha) and phosphorus fertilization (Kg P/ha) based on targeted yield under irrigation.

Nitrogen Application Rates	
Target Yield (ton/ha)	**Nitrogen (kg N/ha)**
4–5	80–130
5–6	130–160
6–7	160–180
7–8	180–200
8+	200+

Phosphorus Application Rates **(Kg P/ha)**				
Target Yield (ton/ha)	**Soil Phosphorus Status (mg/kg)**			
	>5 *	**5–18**	**19–30**	**>30**
4–5	36	28	18	12
5–6	44	34	22	15
6–7	52	40	26	18
7+	>56	>42	>28	21

* Minimum quantity that should be applied at the low soil phosphorus level. Source: DAFF [30].

The field trial captured data on evapotranspiration (ET), rainfall, irrigation, ground and surface water consumed by the crop, as well as yield in Bainsvlei and Clovelly. From Table 1, it can be seen that the yield of wheat from the different trials varied depending on the scale of measurement, ranging from 9881 to 10,458 per hectare in Bainsvlei, and between 8375 to 9161 kg per hectare in Clovelly. The cumulative ET, crop total evapotranspiration, indicates the crop water requirement. Bluewater is further distinguished as either surface or groundwater. The average crop water requirement (Cum. ET) is between 880 mm and 954 mm in Bainsvlei and between 824 mm and 869 mm in Clovelly. Effective rainfall in this period was only 183 mm per annum. Figure 2 shows the map of the catchment area where the study took place.

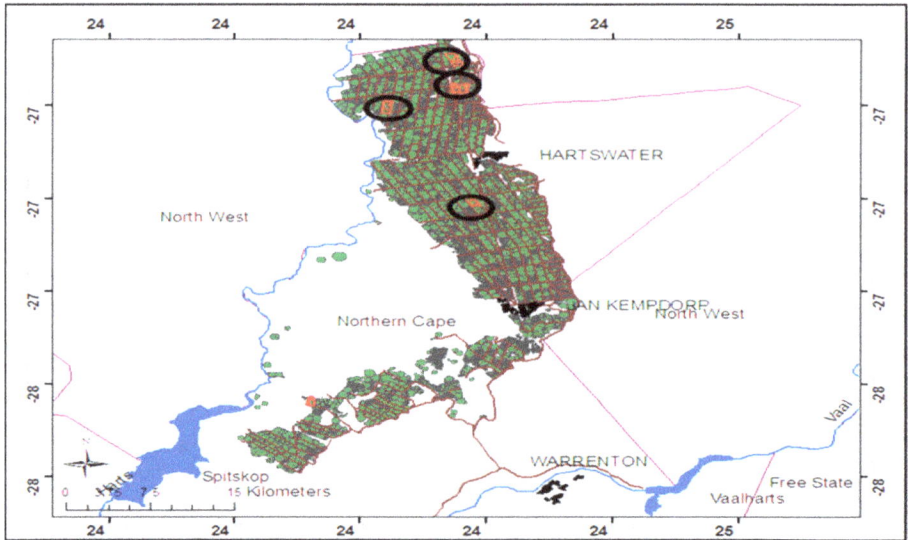

Figure 2. Map of the study area. Source: Van Rensburg et al. [29].

At the processing level, primary data were acquired through a questionnaire from a bread milling company that has a total of five mills and 15 commercial bakeries across South Africa. Data collected from this source included the quantities of wheat milled, quantities of flour used, volumes of water used to produce a specified quantity of flour and bread, as well as total water used at the mill and bakery. Also, the cost of water and the prices of wheat, flour and bread were obtained. Thus, production costs and income received along the flour-bread supply chain were known. In the case of wheat, the producer prices were obtained from GrainSA [31].

3. Results and Discussion

3.1. Water Footprint of Wheat Production at the Farm Level in Bainsvlei and Clovelly

The estimated water footprint of the two areas is presented in Table 3. The results show that, for all the trials and the control, wheat production uses more blue water relative to green water. In Bainsvlei, blue water ranged from 7200 m^3 to 7930 m^3, whereas that of Clovelly ranged from 6490 m^3 to 7100 m^3. This shows that the crop water use in Bainsvlei is higher than that of Clovelly. For both study areas, the blue water use for the control group was lower than the trial estimates. In addition, the blue water use for the control group in Bainsvlei was 6810 m^3 per hectare, and that of Clovelly was 6510 m^3 per hectare.

Regarding the water footprint, the results show that the blue water footprint was higher than the green water footprint. This implies that wheat farmers in the two areas rely mostly on blue water resources. The green water footprint in Bainsvlei ranged from 174 m^3 per ton to 185 m^3 per ton. The blue water footprint from the surface fluctuated from 179 m^3 per ton to 272 m^3 per ton in Bainsvlei for the treatment group, while the blue water footprint from the groundwater ranged from 432 m^3 per ton to 576 m^3 per ton for the treated group in Bainsvlei. This suggests that much of the blue water footprint arises from groundwater resources. Regarding percentage usage, the results show that the proportion of groundwater used is about 37% to 69% of the total blue water footprint in Bainsvlei. The grey water footprint in Bainsvlei ranged from 52 to 55 cubic meters per ton, suggesting that about 52 to 55 cubic meters are required to reduce nitrogen and phosphorus pollutants to ambient levels. The total water footprint in Bainsvlei ranges from 928 to 983 m^3 per ton. The total water footprint for the control is lower than that of the treated groups. These water footprint estimates are lower relative to the global average water footprint of 1827 m^3 per ton reported for wheat by Mekonnen and Hoekstra [16]. At the national level, the water footprint of wheat in South Africa was found to be 1363 m^3 per ton for the period of 1996–2005 [16], whereas our findings revealed a range of 928 to 983 m^3 per ton.

In Clovelly, the green water footprint ranged from 199 to 218 m^3 per ton. In terms of blue water footprint, the results indicate that the blue water footprint from the surface ranges from 273 to 388 m^3 per ton for the treated group, while the blue water footprint from the ground ranges from 370 to 471 m^3 per ton for the treated group. The volume of water utilized to reduce the nitrogen and phosphorus pollutants to ambient levels ranged from 54 to 60 cubic meters. The total water footprint in Clovelly ranges from 993 to 1053 m^3 per ton. For the surface blue water footprint, we observed that the water footprint for the control group was higher than for the treated group. Furthermore, it is clear from the results that the total water footprints vary in the two areas and for the different treatments. Further, the total water footprint estimates for wheat in Clovelly are lower than the global and South African averages reported by Mekonnen and Hoekstra [16].

From the results, it was found that the water footprint estimates for the control and treatment groups in Clovelly were higher than those of Bainsvlei. The high water footprint in Clovelly may be attributed to the low wheat yield compared with the Bainsvlei yield per hectare. The high water footprint can also be attributed to the high surface water (irrigation) utilization in Clovelly relative to the irrigation water usage in Bainsvlei. Also, more groundwater was used in Bainsvlei relative to Clovelly. This may be attributed to the high water-holding capacity of the soil in Bainsvlei. The average water footprint of wheat in Bainsvlei was 954 m^3 per ton, and that of Clovelly was 1026 m^3 per ton.

Table 3. Summary of the blue and green water footprint of wheat at the Vaalharts and Orange-Riet sites.

SAMPLE	ET Crop (mm)	ET Green (mm)	ET Blue S (mm)	ET Blue G (mm)	CWU (m³)	CWU Green (m³ ha)	CWU Blue (m³ ha)	Yield (ton ha)	WF Green (m³ ton)	WF$_{blue}$ Surface (m³ ton)	WF$_{blue}$ Ground (m³ ton)	WF Grey (m³ ton)	Total WF (m³ ton)
							BAINSVLEI						
Control	880	183	681	0	8800	1830	6810	9.9	185	688	0	55	928
1 m—Constant	954	183	188	605	9540	1830	7930	10.5	174	179	576	52	981
1.5 m—Constant	914	183	268	467	9140	1830	7350	9.9	185	271	472	55	983
1 m—Falling	906	183	217	532	9060	1830	7490	10.5	174	207	507	52	940
1.5 m—Falling	881	183	277	443	8810	1830	7200	10.2	179	272	434	54	939
							CLOVELLY						
Control	825	183	651	0	8250	1830	6510	8.4	218	775	0	60	1053
1 m—Constant	869	183	286	424	8690	1830	7100	9.0	203	318	471	56	1048
1.5 m—Constant	860	183	357	340	8600	1830	6970	9.2	199	388	370	54	1011
1 m—Falling	830	183	243	408	8300	1830	6510	8.9	206	273	458	56	993
1.5 m—Falling	824	183	289	360	8240	1830	6490	8.6	213	336	419	58	1026

S = Surface water; G = Groundwater.

In Bainsvlei, the average green water footprint was found to be 180 m^3 ton^{-1}, and this accounted for only 19% of the total average water footprint in this region. The average blue water from the surface (323 m^3 per ton) accounted for about 34%, and that from the groundwater (398 m^3 per ton) accounted for about 42% of the average total water footprint in Bainsvlei. In Clovelly, the average green water footprint was found to be 208 m^3 per ton, and this accounted for about 20% of the average total water footprint. The average blue water footprint from the surface (irrigation) and ground was 418 m^3 per ton and 344 m^3 per ton, respectively. These estimates accounted for 41% and 34% of surface and groundwater, respectively. Generally, the average blue water footprints in Bainvlei and Clovelly are 721 m^3 per ton and 762 m^3 per ton, respectively.

3.2. Blue Water Footprint Benchmarks and Economic Water Productivities at the Farm Level

Table 4 presents the blue water footprint benchmarks for wheat production at different groundwater levels in the Vaalharts and Orange-Riet regions. In this section, we estimated water footprints for a control group where irrigation was done without considering the water from the ground. Secondly, four treatments for different groundwater levels were considered. Figure 3 presents the different ground water levels considered in this study.

Table 4. Blue water footprint benchmarks for different groundwater levels at the Vaalharts and Orange-Riet regions.

SAMPLE	Yield (ton ha)	WF$_{blue}$ Surface (m^3 ton)	WF$_{blue}$ Ground (m^3 ton)	Total Blue WF (m^3 ton)	PWP Surface (kg m^3)	PWP Ground (kg m^3)	Total PWP (kg m^3)	EWP Surface (ZAR m^3)	EWP Ground (ZAR m^3)
				BAINSVLEI					
Control	9.9	688	0	688	1.45	-	1.45	5.81	-
1 m—Constant	10.5	179	576	755	5.59	1.74	7.32	22.35	12.71
1.5 m—Constant	9.9	271	472	743	3.69	2.12	5.81	14.76	12.31
1 m—Falling	10.5	207	507	714	4.83	1.97	6.80	19.32	13.42
1.5 m—Falling	10.2	272	434	706	3.68	2.30	5.98	14.71	13.78
				CLOVELLY					
Control	8.4	775	0	775	1.29	-	1.29	5.16	-
1 m—Constant	9.0	318	471	789	3.14	2.12	5.27	12.58	11.18
1.5 m—Constant	9.2	388	370	758	2.58	2.70	5.28	10.31	14.27
1 m—Falling	8.9	273	458	731	3.66	2.18	5.85	14.65	12.77
1.5 m—Falling	8.6	336	419	755	2.98	2.39	5.36	11.90	12.80

S = Surface water; G = Groundwater; PWP = Physical water productivity; EWP = Economic water productivity.

The results indicate that the yield of wheat varies depending on the level of groundwater available to the crop and this impacts on the water footprint estimates. The blue water footprints calculated for the two areas can act as a benchmark for water utilization in wheat production in Bainsvlei and Clovelly soils. In Bainsvlei, the results indicate that without considering the groundwater, 688 m^3 per ton of blue water from the surface is required. However, with the consideration of blue water from the ground, the results indicate that farmers will require between 179 to 272 cubic meters of water from the surface (irrigation) to produce a ton of wheat in the study area. This is because about 434 to 576 m^3 per ton is contributed by groundwater. In Bainsvlei, the optimal blue water footprints for 1 m—Constant, 1.5 m—Constant, 1 m—Falling and 1.5 m—Falling groundwater levels are 755 m^3 per ton, 743 m^3 per ton, 714 m^3 per ton and 706 m^3 per ton, respectively. About 61% to 76% of the total blue water footprint is from groundwater. This provides the rationale for the consideration of available groundwater contribution to crop water requirement. This gives an understanding of how the groundwater is depleted.

Similarly, in Clovelly, the results indicate that 775 cubic meters of blue water from the surface (irrigation) are required to produce a ton of wheat, without accounting for water from the ground. When the groundwater levels were considered, it was revealed that the total blue water footprint for the different groundwater levels ranges from 731 to 789 m^3 per ton. Nonetheless, about 370 to 471 m^3 per ton of the total blue water footprint originated from the groundwater source, emphasizing the significant contribution of water from the ground to total water footprint. In Bainsvlei, the optimal blue water footprints for 1 m—Constant, 1.5 m—Constant, 1 m—Falling and 1.5 m—Falling groundwater levels are 789 m^3 per ton, 758 m^3 per ton, 731 m^3 per ton and 755 m^3 per ton, respectively. The results from the two areas imply that without considering the groundwater and the volume of water it provides to the root zones of crops, water will be utilized inefficiently.

We calculated economic water productivities for both surface and groundwater utilization to understand the how much can be saved in monetary terms of if the contribution of water from the ground is taking into consideration. The results indicate that in Bainsvlei, only 5.81 ZAR is attained per cubic meter of water used without considering contribution from the ground (controlled). When the contribution of the ground was accounted for, about 14.71 to 22.35 ZAR m^3 can be attained due to the reduced surface irrigation requirement and cost of irrigation. The water from the ground can contribute about 12.31 to 13.78 ZAR m^3 as indicated in Table 4. In Clovelly, an amount of 5.16 ZAR is attained per cubic meter of water used for the control group. When the contribution from the ground was considered, about 10.31 to 14.65 ZAR was attained per cubic meter of blue water (surface) used. The increase in economic water productivities was as a result of reduced irrigation cost due to water contribution from the ground. Economic water productivities from the ground range from 11.18 to 12.80 ZAR. The results imply that it is economical to account for water contribution from the ground when taking water management decisions at the farm level.

Given that blue water from the surface (irrigation) contributes to the production cost, it can be said that adoption of objective irrigation which takes into account volume of water available to the crop from the ground before irrigating is more efficient and economical. Thus, objective irrigation scheduling conserves water (better utilization of rainfall and shallow groundwater as water sources) relative to subjective irrigation scheduling.

Figure 3. Lysimeter trial for evaluation of different groundwater levels.

3.3. Water Footprint at the Processing Stage of the Wheat-Bread Value Chain

In this section, water footprint estimates are calculated for wheat flour and bread. The results are presented in Table 5. Water utilization at the processing level of the value chain consisted of the volume of water utilized at the milling and bakery units. Given the volume of water used in the milling process and the mass of flour produced, the water footprint of wheat flour at the milling stage was found to be 0.07 m^3 per ton. At the bakery stage, 0.46 m^3 of water was utilized to produce a ton of bread. Summing the water footprint of the milling and bakery stages resulted in 0.53 m^3 per ton.

Table 5. Water use at the processing stage of the value chain (milling and bakery).

Parameter	Unit	Quantity
Milling stage		
Quantity of wheat	ton	767,545
Volume of water used	m^3	46,053
Quantity of flour	ton	632,348
Water footprint (flour)	m^3 ton	0.07
Bakery stage		
Quantity of bread produced	ton	379,803
Volume of water used	m^3	174,452
Water footprint (bread)	m^3 ton	0.46
Total water footprint processing	m^3 ton	0.53

Source: Authors' calculations.

The physical and economic water productivity of the individual products involved in this value chain is presented in Table 6. We found that wheat is considerably high in terms of physical and economic productivities. Therefore, more value is created per m^3 of water utilized to produce the grain than for other products, such as wheat flour and bread, along the wheat-bread value chain. The physical water productivity estimates show that 1.037 kg of wheat is gained per cubic meter of water utilized.

Table 6. Physical and economic water productivity of wheat, flour and bread along the wheat-bread value chain.

Parameters	Wheat	Flour	Bread
Physical and Economic Water Productivities			
Yield	9.010 ton ha	632,348 ton	379,803 ton
Total water use	8690 m^3 ha	46,053 m^3	17,447 m^3
Physical water productivity	1.037 kg m^3	0.014 kg m^3	0.022 kg m^3
Value added	4.0 ZAR kg	5.7 ZAR kg	1.7 ZAR kg
Economic water productivities	4.15 ZAR m^3	0.08 ZAR m^3	0.04 ZAR m^3

Source: Authors' calculations.

Also, 0.014 kg of flour and 0.022 kg of bread are gained per cubic meter of water utilized at the milling and bakery stages, respectively. In the case of value addition, results indicated that the total value added to wheat along the wheat-bread value chain is ZAR11.43 per kilogram. Of this amount, the highest value was added in the milling stage, followed by the farm-level and bakery stages. Regarding percentage contribution to the total value added, the results indicate that about 65% of the value is from the processing level and only 35% is from the farm level (see Table 7). Economically, more value is obtained per cubic meter of water used at the farm gate, followed by the milling stage and bakery stage.

Table 7. Summary of the value added to wheat along the wheat-bread value chain.

Production Stage	Value Added	% Share of Value Added
Farm level	4.0 ZAR kg	35.1
Processing level		
Milling	5.7 ZAR kg	50.0
Bakery	1.7 ZAR kg	14.9
Sub-total	7.4 ZAR kg	64.9
Total value added	11.4 ZAR kg	100

Average exchange rate for December 2016: US$1 = 14.62ZAR.

Summing the water footprint of the different stages resulted in an average total water footprint of 954.07 m^3 per ton and 954.53 m^3 per ton for wheat flour and bread, respectively, in Bainsvlei. In Clovelly, the average water footprint for wheat flour and bread are found to be 1026.07 and 1026.53 m^3 per ton, respectively.

4. Conclusions and Implications

The efficient and sustainable management of freshwater resources in South Africa has become a critical policy issue in recent years because water scarcity in the country is becoming alarming. The situation requires a thorough examination of water utilization. One of the sectors that is gaining particular attention is the agricultural sector because it is known to utilize more freshwater, globally. This paper examined the water footprint of the wheat-bread value chain, with a particular emphasis on the contribution of groundwater.

From the findings of the study, it is concluded that it takes 991 m^3 of water to produce one ton of bread in the Vaalharts and Orange-Riet regions of South Africa. The water footprint estimates obtained for wheat flour and bread in this study are lower than the global and national averages reported by Mekonnen and Hoekstra [16]. In Bainvlei and Clovelly, the total water footprint estimates for wheat flour are 31% and 26% lower than the South African average reported from 1996 to 2005 [16]. For bread, the total water footprint estimates for Bainvlei and Clovelly are 21% and 15% lower than the national average reported for South Africa. The water footprint of wheat in the study areas is lower than the global average. This may be attributable to the high yields. Higher yields result in low water footprint estimates. Blue water footprint accounted for about 80% of the total water footprint of bread.

Although the total water footprints in these areas are significantly lower, what is crucial for policy concerns is the share of the blue WF which is much larger than in the study of Mekonnen and Hoekstra [16] from 1996 to 2005. For instance, the current blue water footprint estimates for wheat in Bainvlei and Clovelly are about 68% and 69% higher than the blue water footprint for estimated for the period of 1996–2005. From 1996 to 2005, much of the water used in wheat production was green water, suggesting that there has been a significant shift from green water usage to higher blue water consumption over the years. This might be as a result of changes in climate and rainfall patterns over the years. The significant differences support the rationale for area-specific estimates and seasonal evaluation of water footprints to understand the dynamics of water consumption.

The shift to higher blue water consumption is a major concern for water users and stakeholders along the wheat-bread value chain, given that blue water is becoming scarcer in South Africa. Therefore, it is important that wheat farmers adopt good farm management practices that will continue to improve wheat yields. Such practices can include the adoption and breeding of high-yielding wheat cultivars which are drought resistant.

The water utilized in the processing stage is insignificant, as it accounts for less than 1% of the total water footprint and as such, much attention should be paid to water consumption at the farm or production level. We conclude that the water footprint of wheat varies from one production area to another and from season to season.

Of further importance is the conclusion that groundwater contributes about 34% and 42% of the average total water footprint in Clovelly and Bainsvlei, respectively. This provides the rationale for the consideration of the contribution of water from the ground to total water footprint. Previous studies aggregated blue water footprint without an indication of the proportion contributed by the ground water source [6,17,19]. Meanwhile, an understanding of this contribution to ET can help minimize irrigation water usage and also reduces the cost of production since blue water is a constituent of production cost. Our findings support the idea that the adoption of objective irrigation scheduling conserves water through the better utilization of rainfall and shallow groundwater available to the root zone of crops. This approach is also proven to be economically efficient regarding water usage. The depth of the groundwater has a significant influence on the contribution of groundwater to the total blue water footprint and, as such, the depth of the groundwater should be examined. Furthermore, it is revealed that the total water footprint varies in the two areas and for different groundwater levels. It is worth concluding that, by not accounting for the water available to the crop (controlled) from the ground, more blue water will be applied and this leads to an upsurge in the blue water footprint (surface).

More value is gained at the farm gate, followed by the milling stage and the bakery stage for every m^3 of water utilized. Also, we conclude that more value is added to wheat at the milling stage, followed by the farm gate and bakery stages. The study recommends that to minimize blue water utilization, wheat farmers should investigate the groundwater levels and to know the water available to the crop before irrigation. In other words, accounting for the water contribution of groundwater to the total water footprint will provide a better understanding of water utilization in crop production and how it influences the surface water needed. Secondly, objective irrigation scheduling can be adopted to reduce irrigation water usage. Wheat farmers and breeders can rely on drought-resistant wheat varieties or cultivars that can depend on the available rainfall and available water from the ground. Generally, water footprint assessment has moved away from sole indicator assessment, and a deeper awareness of the productive usage of different sources of water has become vital for policy.

Given the absence of benchmarks or metrics for different catchment areas in South Africa, our findings can potentially act as blue water footprint benchmarks for wheat production in Bainsvlei and Clovelly, particularly for the same ground water levels in Bainvlei and Clovelly. A similar assessment should be conducted in other regions or catchment areas to make a correct judgment and to assess the efficiency and wise use of water, given that water footprint estimates and economic water, productivities vary from one geographical area to another. This will help in achieving the objective of the National Water Research bodies, which seeks to achieve sustainable and efficient water use for the benefit of all users. Finally, we recommend the inclusion of economic water productivities as well as value addition to a water footprint assessment along a given production chain.

Author Contributions: P.M.M. and E.O.-S. performed the analysis and wrote the paper, H.J. conceptualized the project idea and secured the funding as well as project administration; J.H.B. and L.D.v.R. performed the field experiment. E.O.-S. and H.J. supervised the study.

Funding: This research and the APC were funded by the Water Research Commission (WRC) of South Africa as part of a research project, "Determining the water footprints of selected field and forage crops towards sustainable use of fresh water, grant number "K5/2397//4".

Acknowledgments: Financial and other assistance by the WRC is gratefully acknowledged. Financial assistance by the University of the Free State through the Interdisciplinary Research Grant is gratefully acknowledged.

Conflicts of Interest: The authors declare no conflict of interest.

References

1. Jefferies, D.; Munoz, I.; Hodges, J.; King, V.J.; Aldaya, M.; Eric, A.E.; Canals, L.M.; Hoekstra, A.Y. Water footprint and the life cycle assessment as approaches to assess potential impacts of products on water consumption: Key learning points from pilot studies on tea and margarine. *J. Clean. Prod.* **2012**, *33*, 155–166. [CrossRef]

2. Agudelo-Vera, C.M.; Mels, A.R.; Keesman, K.J.; Rijnaarts, H.H.M. Resource management as a key factor for sustainable urban planning. *J. Environ. Manag.* **2011**, *92*, 2295–2303. [CrossRef] [PubMed]

3. Mekonnen, M.; Hoekstra, A.Y. *National Water Footprint Accounts: The Green, Blue and Grey Water Footprint of Production and Consumption*; Value of Water Research Report No. 50; UNESCO-IHE Institute for Water Education: Delft, The Netherlands, 2011.

4. Hoekstra, A.Y.; Chapagain, A.K. The water footprint of Morocco and the Netherlands: Global water use as a result of domestic consumption of agricultural commodities. *Ecol. Econ.* **2007**, *65*, 143–151. [CrossRef]

5. Aldaya, M.M.; Munoz, G.; Hoekstra, A.Y. *Water Footprint of Cotton, Wheat and Rice Production in Central Asia*; Value of Water Research Report Series No. 56; UNESCO-IHE Institute for Water Education: Delft, The Netherlands, 2010. Available online: http://doc.utwente.nl/77193/1/Report41-CentralAsia.pdf (accessed on 20 September 2015).

6. Aldaya, M.M.; Hoekstra, A.Y. The water needed for Italians to eat pasta and pizza. *Agric. Syst.* **2010**, *103*, 401–415. [CrossRef]

7. Bulsink, F.; Hoekstra, A.Y.; Booij, M. *The Water Footprint of Indonesia, Provinces Related to the Consumption of Crop Products*; Value of Water Research Report Series No. 37; UNESCO-IHE Institute for Water Education: Delft, The Netherlands, 2009. Available online: http://doc.utwente.nl/77199/1/Report37-WaterFootprint-Indonesia.pdf (accessed on 9 October 2015).

8. Hoekstra, A.Y.; Chapagain, A.K.; Aldaya, M.M.; Mekonnen, M.M. *The Water Footprint Assessment Manual*, 1st ed.; Earthscan: London, UK, 2011. Available online: http://www.waterfootprint.org/downloads/TheWaterFootprintAssessmentManual.pdf (accessed on 30 January 2015).

9. Mukheibir, P. Local water resource management strategies for adapting to climate induced impact in South Africa. In *Rural Development and the Role of Food, Water and Biomass*; University of Cape Town: Cape Town, South Africa, 2005.

10. Department of Water Affairs (DWA). *Proposed National Water Resource Strategy 2: Summary, Managing Water for an Equitable and Sustainable Future*; DWA: Pretoria, South Africa, 2013. Available online: http://www.gov.za/sites/www.gov.za/files/Final_Water.pdf (accessed on 20 August 2015).

11. Department of Agriculture Forestry and Fisheries (DAFF). *Market Value Chain Profile*; DAFF: Pretoria, South Africa, 2012. Available online: http://www.nda.agric.za/docs/AMCP/Wheat2012.pdf (accessed on 15 August 2015).

12. World Wide Fund (WWF). *Innovation in the South African Water Sector; Danish Investment into Water Management in South Africa*; WWF: Pretoria, South Africa, 2015. Available online: http://www.wwf.org.za/media_room/publications/?15461/Innovations-in-the-SA-water-sector (accessed on 17 October 2016).

13. Mekonnen, M.M.; Hoekstra, A.Y. *A Global and High Resolution Assessment of the Green, Blue and Grey Water Footprint of Wheat*; Value of Water Research Report Series, No. 42; UNESCO-IHE Institute for Water Education: Delft, The Netherlands, 2010. Available online: http://doc.utwente.nl/76916/1/Report42-WaterFootprintWheat.pdf (accessed on 15 August 2015).

14. ISO. *ISO 14046: Environmental Management—Water Footprint—Principles, Requirements and Guidelines*; International Organization for Standardization: Geneva, Switzerland, 2014.

15. Hoekstra, A.Y. A critique on the water-scarcity weighted water footprint in LCA. *Ecol. Indic.* **2016**, *66*, 564–573. [CrossRef]

16. Mekonnen, M.M.; Hoekstra, A.Y. *The Green, Blue and Grey Water Footprint of Crops and Derived Crop Products Volume 1: Main Report*; Value of Water Research Report Series; UNESCO-IHE: Delft, The Netherlands, 2010. Available online: http://www.waterfootprint.org/Reports/Report47-WaterFootprintCrops-Vol1.pdf (accessed on 16 February 2015).

17. Ahmed, S.M.; Ribbe, L. Analysis of water footprints of rainfed and irrigated crops in Sudan. *J. Natl. Resour. Dev.* **2011**, *1*, 1–9.

18. Neubauer, E. Water footprint in Hungary. *APSTRACT* **2012**, *6*, 83–91. [CrossRef] [PubMed]

19. Sundberg, H. The Water Footprint of Winter Wheat in Sweden. In *Human Development Report: The Rise of the South Human Progress in a Diverse World*; Technical Notes; University of Lund: Lund, Sweden, 2012.

20. Ababaei, B.; Etedali, H.R. Estimation of water footprint components of Iran's wheat production: Comparison of global and national scale estimates. *Environ. Sci. Process.* **2014**, *1*, 193–205. [CrossRef]

21. Le Roux, B.; van de Laan, M.; Vahrmeijer, T.; Annamdale, J.G.; Bristow, K.L. Estimating Water Footprint of Vegetable Crops: Influence of Growing Season, Solar Radiation Data and Functional Unit. *Water* **2016**, *10*, 473. [CrossRef]

22. Owusu-Sekyere, E.; Scheepers, M.E.; Jordaan, H. Water footprint of milk produced and processed in South Africa: Implications for policy-makers and stakeholders along the dairy value chain. *Water* **2016**, *8*, 322. [CrossRef]

23. Dong, H.; Yong, G.J.S.; Tsuyoshi, F.; Tomohiro, O.; Bing, X. Regional water footprint evaluation in China: A case of Liaoning. *Sci. Total Environ.* **2013**, *442*, 215–224. [CrossRef] [PubMed]

24. Chouchane, H.; Hoekstra, A.Y.; Krol, M.S.; Mekonnen, M.M. The water footprint of Tunisia from an economic perspective. *Ecol. Indic.* **2015**, *55*, 311–319. [CrossRef]

25. Zoumides, C.; Bruggeman, A.; Hadjikakou, M.; Theodoros, Z. Policy-related indicators for semi-arid nations: The water footprint of crop production and supply utilization of Cyprus. *Ecol. Indic.* **2014**, *43*, 205–214. [CrossRef]

26. Berger, M.; Finkbeiner, M. Water footprinting: How to address water use in life cycle assessment. *Sustainability* **2010**, *2*, 919–944. [CrossRef]

27. Crafford, J.; Hassan, R.M.; King, N.A.; Damon, M.C.; de Wit, M.P.; Bekker, S.; Rapholo, B.M.; Olbrich, B.W. *An Analysis of the Social, Economic, and Environmental Direct and Indirect Costs and Benefits of Water Use in Irrigated Agriculture and Forestry: A Case Study of the Crocodile River Catchment, Mpumalanga Province*; Water Research Commission (WRC): Pretoria, South Africa, 2004.

28. Owusu-Sekyere, E.; Scheepers, M.E.; Jordaan, H. Economic water productivities along the dairy value chain in South Africa: Implications for sustainable and economically efficient water-use policies in the dairy industry. *Ecol. Econ.* **2017**, *134*, 22–28. [CrossRef]

29. Van Rensburg, L.; Barnard, J.H.; Bennie, A.T.P.; Sparrow, J.B.; du Preez, C.C. *Managing Salinity Associated with Irrigation at Orange-Riet at Vaalharts Irrigation Schemes*; Water Research Commission (WRC): Pretoria, South Africa, 2012.

30. Department of Agriculture, Forestry and Fisheries (DAFF). *Production Guideline for Wheat*; DAFF: Pretoria, 2016. Available online: https://www.google.com/url?sa=t&rct=j&q=&esrc=s&source= web&cd=1&cad=rja&uact=8&ved=2ahUKEwiMleuIjYjdAhXMFsAKHWZICSsQFjAAegQIAxAC& url=http%3A%2F%2Fwww.daff.gov.za%2FDaffweb3%2FPortals%2F0%2FBrochures%2520and% 2520Production%2520guidelines%2FWheat%2520-%2520Production%2520Guideline.pdf&usg= AOvVaw1o5TQs9FRfAMttcN9HeQMn (accessed on 20 August 2017).

31. Grain SA. Producer Price Framework for Irrigation Wheat. 2016. Available online: http://www.grainsa.co. za/pages/industry-reports/production-reports (accessed on 17 October 2016).

water

MDPI

Article

Water Footprints of Vegetable Crop Wastage along the Supply Chain in Gauteng, South Africa

**Betsie le Roux [1], Michael van der Laan [1,*], Teunis Vahrmeijer [1], John G. Annandale [1]
and Keith L. Bristow [2]**

[1] Department of Plant and Soil Sciences, University of Pretoria, Private Bag X20, Hatfield 0028, South Africa;
    betsielr@gmail.com (B.l.R.); jtv@villacrop.co.za (T.V.); John.Annandale@up.ac.za (J.G.A.)
[2] CSIRO Agriculture & Food, PMB Aitkenvale, Townsville, QLD 4814, Australia; Keith.Bristow@csiro.au
*   Correspondence: Michael.vanderLaan@up.ac.za; Tel.: +27-12-420-3665

Received: 29 November 2017; Accepted: 10 April 2018; Published: 24 April 2018

Abstract: Food production in water-scarce countries like South Africa will become more challenging in the future because of the growing population and intensifying water shortages. Reducing food wastage is one way of addressing this challenge. The wastage of carrots, cabbage, beetroot, broccoli and lettuce, produced on the Steenkoppies Aquifer in Gauteng, South Africa, was estimated for each step along the supply chain from the farm to the consumer. Water footprints for these vegetables were used to determine the volume of water lost indirectly as a result of this wastage. Highest percentage wastage occurs at the packhouse level, which is consistent with published literature. Some crops like lettuce have higher average wastage percentages (38%) compared to other crops like broccoli (13%) and cabbage (14%), and wastage varied between seasons. Care should therefore be taken when applying general wastage values reported for vegetables. The classification of "waste" presented a challenge, because "wasted" vegetables are often used for other beneficial purposes, including livestock feed and composting. It was estimated that blue water lost on the Steenkoppies Aquifer due to vegetable crop wastage (4 Mm^3 year^{-1}) represented 25% of the estimated blue water volume that exceeded sustainable limits (17 Mm^3 year^{-1}).

Keywords: Steenkoppies Aquifer; carrots; cabbage; beetroot; broccoli; lettuce; packhouse; retail; consumers

1. Introduction

Ending hunger, achieving food security, improving nutrition and promoting sustainable agriculture is the second of the 17 Sustainable Development Goals (SDGs) [1]. South Africa has high levels of malnutrition, resulting in 27% of children under the age of five being stunted (low height for age); in addition, 12% are underweight (low weight for age), 5% are wasted (low weight for height), and 15% of infants are born with a low birth weight [2]. Food production is highly dependent on the availability of freshwater and is responsible for an estimated 86% of total freshwater used globally [3]. Freshwater is becoming an increasingly scarce resource, posing direct risks to food production, and already relies on the unsustainable use of groundwater. A study done by Wada, et al. [4] indicated that the global use of non-renewable groundwater abstractions increased by more than three times between the years 1960 and 2000. It was noted that for the year 2000, unsustainable use of groundwater supplied approximately 234 km^3 year^{-1}, which is 20% of the gross irrigation water demand. Climate change is expected to exacerbate the risks of water scarcity, while population growth puts further pressure on the agricultural sector to produce more food.

Water footprint (WF) assessments have been proposed as a way of improving water resource management. Allan (1998) introduced the term "virtual water", indicating that economically and logistically it is more reasonable to import, for example, one tonne of grain instead of the 1000 tonnes of water required to produce the one tonne of grain. Hoekstra [5] further developed this concept of virtual water by saying that a nation's WF, for example, does not only consist of locally sourced water used, but also includes the water used to produce the imported products that are consumed. In an agricultural context, a WF according to the Water Footprint Network (WFN) can be defined by the volume of water required to produce a certain mass of crop yield. Hoekstra, et al. [6] distinguish between blue, green and grey WFs. Surface and underground water resources, which are available to multiple users, are defined as blue water. In a crop production context, the blue WF therefore consists predominantly of the irrigation water applied. Green water is water originating from rainfall that is stored in the soil and is available for vegetation growth only. In order to account for water quality impacts, Hoekstra, Aldaya, Chapagain and Mekonnen [6] proposed the concept of a grey WF, which is the volume of water required to dilute emitted pollutants to ambient water quality levels. The WFs, according to the WFN, quantify water consumption along the entire production chain of products, processes, businesses and within nations or catchments [6].

Water management in South Africa is particularly challenging because of severe water shortages in most parts of the country and a highly variable climate [7]. In many catchments throughout South Africa, water supply no longer meets demand, and as surface water resources in South Africa are already almost fully developed, exploiting alternative sources will be done at significantly higher costs than previously [8]. Irrigated agriculture uses approximately 40% of South Africa's exploitable runoff on around 1.7 million hectares of land [9]. Nieuwoudt, et al. [10] estimated that 90% of vegetable and fruit products are grown under irrigation in South Africa because of low and erratic rainfall and the high value of these crops. The vulnerability of food production in South Africa was emphasized by the drought of 2015 which was, according to the South African Weather Bureau, the driest calendar year since nationwide recordings started in 1904 [11].

There is therefore a need to find ways of producing more food with available water resources, and getting more of what is produced to consumers. Reducing food wastage is one way to increase food supplies without increasing the volume of freshwater required for production Lundqvist, et al. [12] reported that up to 50% of production can be lost between the field and consumer, or from "field to fork". From a pilot study conducted by the World Wildlife Fund (WWF) and Woolworths in South Africa, it was estimated that only 26% of spinach that was produced was consumed. The remainder was wasted along the supply chain, mostly at the farm level (29%) and during packaging and processing (38%) [13].

There is very limited information published on wastage of specific vegetable crops. Nahman, et al. [14] studied the cost of household waste in South Africa, which arises from the actual value of the food that is wasted as well as from cost of disposing the food to landfill sites. They estimated the cost of food waste to be R32.5 billion, which represents 1.22% of South Africa's annual GDP [14]. Oelofse and Nahman [15] estimated food wastage along the supply chain relative to production. Oelofse and Nahman [15] estimated that over 9 million tonnes of food waste are generated in South Africa per annum and that 177 kg of food was wasted per capita per year in 2007. Literature sources often report food wastage of different commodity groups, such as cereals, fruit and vegetables, and meat, and there is a lack of data on wastage of particular crops or crop types. Reasons for this may include a lack of record keeping by different stakeholders along the supply chain or an unwillingness to make these data easily available due to perceived sensitivities, for example, in terms of institutional reputation and competitiveness. In South Africa, food waste is regulated as part of waste management in general, and there is no legislation that obligates the recording of food wastage per se [13].

The aim of this study was to quantify indirect water losses through the wastage of crops produced in a major production region on the Steenkoppies Aquifer (Lat: 26.03° S to 26.19° S, Long: 27.65° E to 27.48° E; Altitude 1560 to 1650 m) located west of Tarlton in Gauteng, South Africa. Water footprints of the crops were used to determine the volumes of water associated with current food wastage. The Steenkoppies Aquifer is a dolomitic karst aquifer under stress due to increased competition from various stakeholders. As estimated by le Roux, et al. [16], current annual production on the aquifer requires 25 Mm^3 of blue water, which exceeds sustainable limits of blue water abstractions by 17 Mm^3. We determine whether reductions in food wastage, assuming this could lead to reductions in production, at the farm level, could provide a way to achieve sustainable water use on this aquifer. It is hypothesized that there will be an important difference between the wastage fractions for different vegetable crops due to differences in growth forms (above or below ground crops), harvesting, handling and processing techniques, marketable properties for each crop type, differences in shelf-life between crops, and exposure to various pests and diseases, which will translate into differences in indirect water losses through wastage.

2. Materials and Methods

2.1. Obtaining Data on Percentage Wastage along the Supply Chain

Measured or estimated data were obtained on the wastage of the main vegetable crops produced on the Steenkoppies Aquifer, namely, carrots (*Daucus carota*), beetroot (*Beta vulgaris*), cabbage and broccoli (*Brassica oleracea*) and lettuce (*Lactuca sativa*) at different stages along the supply chain. The stages along the supply chain included the packhouse level (which is on-farm), the market or distribution point, retail and consumer levels.

For each stage, the percentage wastage was normalised according to the volume of vegetables delivered to that particular stage. Therefore, the percentages did not represent total wastage along the supply chain, but for that stage only. Total production, which was derived from cropped areas on the Steenkoppies Aquifer [16,17], was then used to determine total wastage from field to fork. For each stage along the supply chain, wastage was determined by subtracting wastage at all preceding stages from total production and multiplying the remainder with the percentage wasted in that particular stage. These calculations are given in the equations below and are illustrated in Figure 1.

$$TW_1 = TP \times WP_1$$

$$TW_2 = TP(1 - WP_1)WP_2$$

$$TW_3 = TP(1 - WP_1)(1 - WP_2)WP_3$$

$$TW_4 = TP(1 - WP_1)(1 - WP_2)(1 - WP_3)WP_4$$

$$TC = TP(1 - WP_1)(1 - WP_2)(1 - WP_3)(1 - WP_4)$$

where TP is total production, TW is the total wastage (mass) and WP is the percentage wastage at a particular level along the supply chain. Subscripts 1, 2, 3 and 4 represent the packhouse, market, retail and consumer levels, respectively, and TC is the total amount (mass) consumed. Offcuts were not counted as wastage because they are not considered fit for human consumption and were not included in total production figures. Calculations were carried out for each crop separately in each of the four seasons (Summer: November to February; Autumn: March and April; Winter: May to August and Spring: September and October).

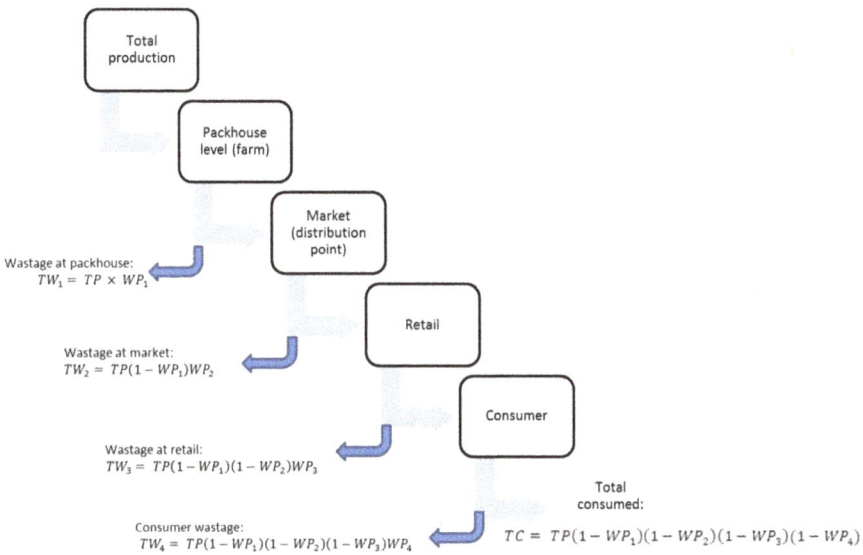

Figure 1. Calculating wastage for the stages along the supply chain of vegetables (carrots, cabbage, beetroot, broccoli and lettuce) from the farmer to consumer.

2.1.1. Wastage at the Packhouse Level

At the farm, there are three stages during which crop material can be discarded, namely:

- Discards at the planting stage, which represent seedlings that do not grow,
- Discards during growth stages, which represent crops that do not develop into a harvestable product,
- Discards at the packhouse, which represent vegetables that are not marketable.

Discards during planting and growing are not considered wastage for our purposes because these plants never develop into an edible product and are also not recorded as production. Seedlings use relatively little water and therefore do not have a significant impact on water resources [18]. Vegetables wasted at harvest represent an edible product and are therefore considered as food wastage.

Daily production reports for the year 2015 for a packhouse on a major farm on the Steenkoppies Aquifer provided the input and output volumes of carrots, cabbage and lettuce. The difference between input and output weights or volumes was assumed to equal wastage. Beetroot and broccoli are not packed on the Steenkoppies Aquifer, so data on wastage in the packhouse were not available for these two crops. Wastage of beetroot in the packhouse was assumed to be the same as that of carrots because both are root crops, and treatment in the packhouse will be similar. Wastage of broccoli in the packhouse was assumed to be the same as that of cabbage because these two crops are similar. Although cabbage and lettuce data were given as number of heads, they were used to calculate the percentage wastage at the packhouse, which was multiplied by total yield to provide total wastage in terms of weight. Therefore, all calculations on wastage for the remainder of the supply chain were performed in terms of weight.

2.1.2. Wastage at the Fresh Produce Market or Distribution Point

The Tshwane Fresh Produce Market (Lat: 25.74° S, Long: 28.17° E) provided data on all produce received daily from the Steenkoppies Aquifer as well as sold and discarded from July 2011 to July 2014 [19]. This market serves one of two key metropolitan centres in the region, namely, Pretoria, the capital city of South Africa and the Joburg Market, which services Johannesburg, South Africa's economic capital. The data from the Tshwane Fresh Produce Market were highly detailed and reflected the specific masses of each vegetable received, sold and discarded for each farm on the Steenkoppies Aquifer.

2.1.3. Wastage at the Retailer Level

Despite efforts to obtain wastage data from multiple retailers in the region, quantitative data on wastage at the retail level were not available because retailers do not normally record produce losses, and those who do are unwilling to disclose the data. Theoretically, it can be assumed that the difference between products bought and sold by the retailer will be equal to the wastage. In reality it is more complicated because although the processing of vegetables into combined, "ready-to-use" packets reduces the percentage of food losses, it also complicates estimations of food losses. The quantities of specific vegetables in pre-packed products are not usually recorded. It was, therefore, not possible to record exactly how much of a particular vegetable was sold. Even if wasted products were weighed, there is the problem that the vegetables that are wasted often have much lower water contents than the fresh products as they begin to decompose, potentially underestimating the wastage in terms of the original mass of fresh product. Estimations of wastage at the retail level were therefore based on information obtained during several semi-structured interviews with experienced retailers.

2.1.4. Wastage by Consumers

Crop-specific data were not available on wastage at the consumer level. Estimated wastage by consumers was acquired from relevant literature, namely, Gustavsson, et al. [20], cited by Oelofse and Nahman [15].

2.2. Estimating the Water Footprints of the Wastage of Selected Vegetables

This study focusses on blue water footprints of wastage because it was determined that agriculture on the Steenkoppies Aquifer uses blue water unsustainably [16]. The volume of blue water lost due to the wastage of selected vegetables produced on the Steenkoppies Aquifer in a year was estimated using the WFs calculated by Le Roux, et al. [21]. These WFs followed the WFN methodology [6], represented the mean over ten years (2004–2013), and include only water used for production. The methods used to calculate these WFs are described by Le Roux, van der Laan, Vahrmeijer, Bristow and Annandale [21]. The WF aims to be a robust, simplified metric, and for that purpose it was assumed that the WFs are relevant for all years. Water footprints were determined for wastage, for each season specifically, at each step of the supply chain by multiplying the total wastage at each step with the blue crop WFs. Globally, according to Hoekstra and Chapagain [3], agriculture is responsible for 86% of water used. Therefore, additional water used at each stage along the supply chain, for example, water used for washing, was considered to be relatively low compared to water used for production and was excluded from this study. Potential savings in green water used through reductions in food wastage were assumed to be negligible because these crops replaced natural vegetation that would use green water anyway. Grey water was also excluded in this assessment.

3. Results

3.1. Estimation of Total Wastage

3.1.1. Wastage at the Packhouse Level

At the packhouse level, wastage is mostly due to pests and diseases or because crops have unmarketable characteristics. The farm that was assessed was the sole provider for a large supermarket group, and no cases have been reported where vegetables were wasted because of low demand or flooded markets. The percentages of carrots, cabbage and lettuce wasted at the packhouse level in each season are given in Figure 2. Compared to carrots and lettuce, the percentage wastage of cabbage in the packhouse is very low. Wastage during this stage is not closely correlated with seasons because the wastage is due more to unmarketable traits as opposed to rotting during this first stage.

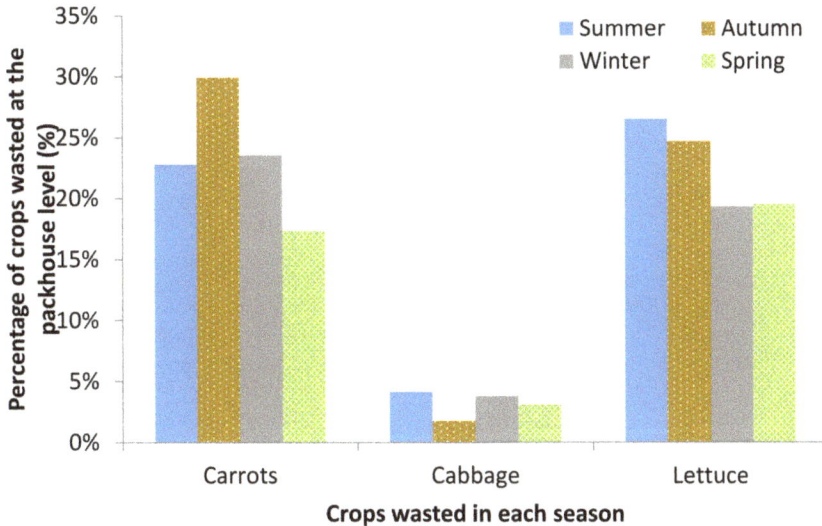

Figure 2. Wastage of carrots, cabbage and lettuce in each season in a packhouse on the Steenkoppies Aquifer. Carrot values are assumed to represent beetroot values, and cabbage values are assumed to represent broccoli values.

The carrot production report was in kilograms of yield, while cabbage and lettuce were reported as "heads". Carrots that are not marketable or sold include broken pieces that are too short to be marketed in a low value pack, as well as grossly malformed, cracked, extremely thick or thin carrots. In the case of cabbage and lettuce, most waste heads are edible except those with serious insect infestation and those that are rotten or decayed. Cabbage heads that are not marketable include those that have decay, worm damage, black rings, discolouration, dehydration, *Anthropoda* infestation and those with incorrect head sizes. Lettuce heads that are not marketable include those that have browning, decay, worms, sun scorch, deep cuts, incorrect sizes, malformation and bruising. The trimmed leaves and non-marketable vegetables are fed to the cattle on the farm.

3.1.2. Wastage at the Market/Distribution Point

The percentage discarded in terms of what the market received from the Steenkoppies Aquifer for each crop in each season is presented in Figure 3. At this stage of the supply chain, wastage is due to rotting of the crops, which is why waste percentages are higher in summer and higher for more perishable crops, like lettuce. Wastage of beetroot is particularly low in all seasons, except for summer. Wasted products at the market are now used to make compost in a digester on site, which is a more recent development that was launched in 2014.

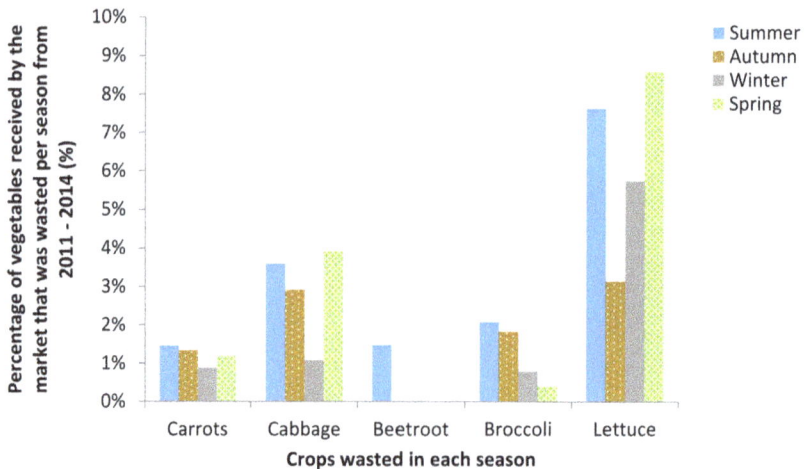

Figure 3. Percentage of crops received by the Tshwane Fresh Produce Market from 2011–2014 that was wasted.

3.1.3. Wastage at the Retailer Level

Wastage at the retailers mostly occurs when vegetables reach the end of their sell-by date or shelf-life. Weather conditions impact food wastage at the retailer level, but management decisions also play an important role in terms of percentage food losses. Retailers that order too many vegetables once or twice a week generally have more losses than retailers who order fewer vegetables more often, even daily. Most green grocers cut and combine vegetables that approach the end of their shelf life into pre-packed products for salads, soups or stir-fry vegetables. In supermarkets, ageing vegetables are used to make salads and sandwiches in the supermarket delis. This greatly reduces food losses at the retail level, but in the case of lettuce, for example, there is a limit to how much salad or sandwiches can be sold in the deli, and wastage cannot be completely avoided. Wastage from the retailer is often given to soup kitchens, livestock farms or used for composting.

Carrots, cabbage, beetroot and broccoli have a relatively long shelf-life and wastage is generally low. According to experienced retailers [22–24], wastage of these vegetables at the retail level is between 1% and 5%. It was therefore assumed that wastage of these vegetables at the retailer is 5% in summer, 3% in autumn and spring and 1% in winter. Lettuce is more perishable and according to experienced retailers, average wastage of lettuce at the retail level is between 7% and 10%. It was therefore assumed that wastage of lettuce at the retailer is 10% in summer, 9% in autumn and spring and 7% in winter.

3.1.4. Wastage by Consumers

According to Gustavsson, Cederberg, Sonesson, Van Otterdijk and Meybeck [19], cited by Oelofse and Nahman [14], average wastage in South African households in terms of total production is 2% for roots and tubers and 5% for fruit and vegetables. Thus, the wastage of carrots and beetroot was assumed to be 2% of total production, and the wastage of cabbage, broccoli and lettuce was assumed to be 5% of total production at the household level. According to Nahman, De Lange, Oelofse and Godfrey [13], most wastage in South Africa occurs in low-income communities. This is because of the number of low-income households in South Africa, which is much higher than high-income households and does not reflect higher wastage per household in low income communities.

3.1.5. Total Wastage of Vegetables from the Steenkoppies Aquifer along the Supply Chain to the Consumer

Table 1 summarises wastage at each stage of the supply chain in terms of the annual production of each vegetable on the Steenkoppies Aquifer. Wastage of cabbage and broccoli is relatively low because of the low percentage wastage in the packhouse and the generally longer shelf lives of these crops. Lettuce has the highest percentage wastage for all seasons (ranging from 33% in winter to 42% in summer) because of the high percentage wastage in the packhouse and the short shelf life of the crop. An estimated 29% of the annual production of carrots and beetroot (root vegetables) and 32% of the annual production of cabbage, broccoli and lettuce is lost due to wastage. This is much lower than that indicated by Oelofse and Nahman [15], who estimated annual wastage of 44% for roots and tubers and 51.5% for other vegetables in terms of average annual food production. The percentage wastage estimated by Oelofse and Nahman [15] was based on percentage wastage given by Gustavsson, Cederberg, Sonesson, Van Otterdijk and Meybeck [20] for sub-Saharan Africa. The percentage contribution to total wastage (including all five vegetables) at each step along the supply chain, as calculated in this study, is given in Figure 4A and is compared to the findings of food wastage along the supply chain in South Africa as published by Oelofse and Nahman [15] and given in Figure 4B. Oelofse and Nahman [15] estimated that 79% of total wastage occurs before distribution during agricultural production, post-harvest handling and storage, and processing and packaging. Our packhouse level data include all three of these losses combined. The average percentage wastage in the packhouse on the Steenkoppies Aquifer was 70% of total food wastage along the supply chain, which correlates well with estimates from Oelofse and Nahman [15]. Oelofse and Nahman [15] also reported wastage during distribution, which included our market and retail stages. Our percentage wastage for the market and retail stages was 9% and 12%, respectively, in terms of total wastage along the supply chain; the sum of these values correlates well with the 17% wastage during distribution reported by Oelofse and Nahman [15]. We estimate 8% wastage at the household level in terms of total wastage along the supply chain, compared to 4% estimated by Oelofse and Nahman [15]. There is, however, variation in average annual wastage between different crops, which varied from 13% for broccoli to 38% for lettuce, as illustrated in Figure 5.

High inter-seasonal variation in vegetable wastage was observed. For carrots and beetroot, there was a 12% difference between highest food wastage in autumn and lowest food wastage in spring. Maximum wastage of lettuce in summer was 10% more than the minimum wastage of lettuce in winter. Large differences in total production may affect the percentage wastage, where lower production levels may be easier to manage, resulting in less wastage. For all crops, percentage wastage was higher in summer compared to winter, partly because of shorter shelf lives when temperatures are higher.

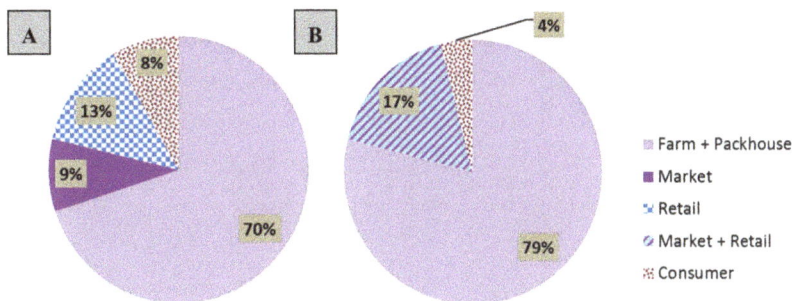

Figure 4. (**A**): Average percentages of total annual wastage of carrots, cabbage, beetroot, broccoli and lettuce produced on the Steenkoppies Aquifer at different stages along the supply chain from the field to the consumer; (**B**): Wastage of food along the supply chain in South Africa estimated by Oelofse and Nahman [15].

Table 1. Summary of wastage of carrots, cabbage, beetroot, broccoli and lettuce along the supply chain from the farm to the consumer in terms of total production on the Steenkoppies Aquifer in 2005.

Crop	Season	Total Annual Production (tonnes)	Percentage Wastage in Terms of Mass Received by Each Stage (%)				Total Wastage at Each Stage (tonnes)					Total Percentage Wastage (%)	Total Consumed
			Farm	Market	Retail	Consumer	Farm	Market	Retail	Consumer	Total		
Carrots	Summer	13,487.3	22.8%	1.445%	5.0%	2.0%	3075.5	150.4	513.1	195.0	3934.0	29.2%	9553.1
	Autumn	8454.6	29.9%	1.331%	3.0%	2.0%	2527.4	78.9	175.5	113.5	2895.3	34.2%	5559.4
	Winter	9193.7	23.6%	0.882%	1.0%	2.0%	2167.3	61.9	69.6	137.9	2436.8	26.5%	6756.9
	Spring	3221.7	17.3%	1.192%	3.0%	2.0%	557.7	31.8	79.0	51.1	719.5	22.3%	2502.2
Beetroot	Summer	3093.8	22.8%	1.470%	5.0%	2.0%	705.5	35.1	117.7	44.7	903.0	29.2%	2190.8
	Autumn	4768.3	29.9%	0.000%	3.0%	2.0%	1425.4	0.0	100.3	64.9	1590.6	33.4%	3177.7
	Winter	4218.3	23.6%	0.020%	1.0%	2.0%	994.4	0.7	32.2	63.8	1091.1	25.9%	3127.2
	Spring	2585.5	17.3%	0.012%	3.0%	2.0%	447.6	0.3	64.1	41.5	553.4	21.4%	2032.1
Subtotal 1 *		49,022.9					11,900.8	359.1	1151.4	712.2	14,123.5	28.8%	
Cabbage	Summer	3699.9	3.4%	3.590%	5.0%	5.0%	125.1	128.3	172.3	163.7	589.4	15.9%	3110.5
	Autumn	1369.1	1.6%	2.915%	3.0%	5.0%	22.2	39.3	39.2	63.4	164.1	12.0%	1205.1
	Winter	2705.1	3.7%	1.079%	1.0%	5.0%	100.0	28.1	25.8	127.6	281.5	10.4%	2423.6
	Spring	2373.1	3.4%	3.925%	3.0%	5.0%	81.2	90.0	66.1	106.8	344.0	14.5%	2029.1
Broccoli	Summer	1015.5	3.4%	2.069%	5.0%	5.0%	34.3	20.3	48.1	45.6	148.3	14.6%	867.2
	Autumn	62.2	1.6%	1.818%	3.0%	5.0%	1.0	1.1	1.8	2.9	6.8	11.0%	55.4
	Winter	482.4	3.7%	0.796%	1.0%	5.0%	17.8	3.7	4.6	22.8	49.0	10.2%	433.5
	Spring	672.4	3.4%	0.399%	3.0%	5.0%	23.0	2.6	19.4	31.4	76.4	11.4%	596.0
Lettuce	Summer	15,855.3	26.5%	7.627%	10.0%	5.0%	4205.1	888.6	1076.2	484.3	6654.1	42.0%	9201.2
	Autumn	2965.1	24.7%	3.129%	9.0%	5.0%	732.2	69.9	194.7	98.4	1095.2	36.9%	1869.9
	Winter	9918.0	19.3%	5.733%	7.0%	5.0%	1917.5	458.6	527.9	350.7	3254.8	32.8%	6663.2
	Spring	6858.5	19.5%	8.589%	9.0%	5.0%	1337.1	474.2	454.2	229.6	2495.2	36.4%	4363.2
Subtotal 2 **		47,976.6					8596.6	2204.7	2630.3	1727.3	15,158.7	31.6%	

* Subtotal 1 for carrots and beetroot (root vegetables), ** Subtotal 2 for cabbage, broccoli and lettuce.

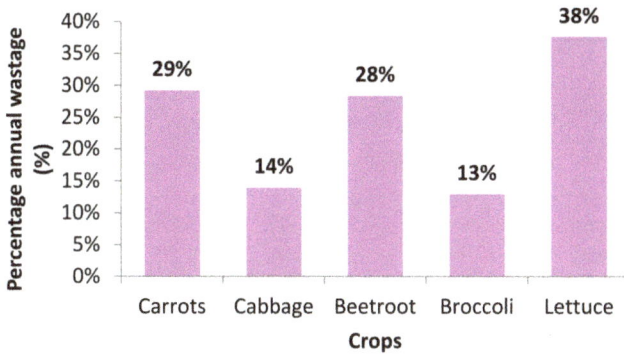

Figure 5. Percentage annual wastage along the supply chain of the five selected vegetable crops in terms of total production on the Steenkoppies Aquifer in 2005.

3.2. Water Footprint of Wastage of Selected Vegetables

The blue WFs of seasonal discards along the supply chain to the consumer of the selected vegetable crops produced on the Steenkoppies Aquifer are given in Table 2 and Figure 6. As shown in Table 2, wastage of carrots in summer, autumn and winter, beetroot in autumn and winter and lettuce in summer, winter and spring used relatively high volumes of blue water during production on the aquifer, which can be considered as lost through food wastage. For carrots and lettuce, this is partly because of the high volumes produced (Table 1) and high percentages losses (29% and 38% for carrots and lettuce, respectively, Figure 5). This resulted in more wastage of the crops and translated into higher water losses. Higher volumes of wastage for lettuce were countered by low crop WFs during production. Beetroot production was lower, but percentage wastage was high (28%, Figure 5). Lower volumes of blue water lost indirectly through wastage of cabbages and broccoli was due to lower production (Table 1) and lower percentage losses (14% and 13% for cabbage and broccoli, respectively, Figure 5). Most of the total wastage of vegetables produced on the Steenkoppies Aquifer occurred in the packhouse and was due to the wastage of lettuce along the whole supply chain.

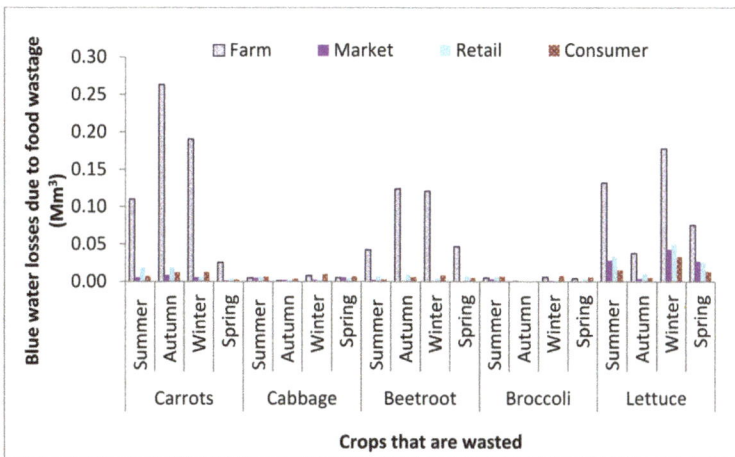

Figure 6. Blue water losses through food wastage along the supply chain from the field to the consumer due to vegetables produced on the Steenkoppies Aquifer in 2005.

Table 2. Blue water footprints [21] and blue water lost due to wastage of vegetables produced on the Steenkoppies Aquifer in 2005.

Crop	Season	Blue Crop Water Footprint (m³ tonne⁻¹)	Blue Water Lost Due to Wastage (Mm³)				Total (Mm³)
			Farm	Market	Retail	Consumer	
Carrots	Summer	35.70	0.110	0.005	0.018	0.007	0.14
	Autumn	104.12	0.263	0.008	0.018	0.012	0.30
	Winter	87.84	0.190	0.005	0.006	0.012	0.21
	Spring	45.21	0.025	0.001	0.004	0.002	0.03
Cabbage	Summer	37.76	0.005	0.005	0.007	0.006	0.02
	Autumn	53.22	0.001	0.002	0.002	0.003	0.01
	Winter	77.46	0.008	0.002	0.002	0.010	0.02
	Spring	63.20	0.005	0.006	0.004	0.007	0.02
Beetroot	Summer	59.83	0.042	0.002	0.007	0.003	0.05
	Autumn	86.62	0.123	0.000	0.009	0.006	0.14
	Winter	121.12	0.120	0.000	0.004	0.008	0.13
	Spring	103.57	0.046	0.000	0.007	0.004	0.06
Broccoli	Summer	142.38	0.005	0.003	0.007	0.006	0.02
	Autumn	225.23	0.000	0.000	0.000	0.001	0.00
	Winter	321.98	0.006	0.001	0.001	0.007	0.02
	Spring	169.61	0.004	0.000	0.003	0.005	0.01
Lettuce	Summer	31.26	0.131	0.028	0.034	0.015	0.21
	Autumn	51.20	0.037	0.004	0.010	0.005	0.06
	Winter	92.63	0.178	0.042	0.049	0.032	0.30
	Spring	56.18	0.075	0.027	0.026	0.013	0.14
Total			1.376	0.143	0.217	0.165	1.901

As shown in Table 2, an estimated 1.9 Mm³ blue water was lost due to wastage of the selected vegetable crops. It was estimated that 12 Mm³ blue water was used to grow maize and wheat on the Steenkoppies Aquifer [18]. The wastage of maize and wheat has not been determined, but Gustavsson, Cederberg, Sonesson, Van Otterdijk and Meybeck [19] reported 19% wastage of cereals in sub-Saharan Africa; therefore, it was estimated that wastage of maize and wheat used 2.3 Mm³ of blue water [18]. Total wastage on the Steenkoppies Aquifer would therefore use approximately 4.2 Mm³ of blue water, which is 24.5% of the estimated volume of the 17 Mm³ year⁻¹ blue water that exceeded sustainable limits [16].

4. Discussion

4.1. Wastage of Vegetable Crops

Our results indicated that 70% of food wastage occurs on-farm at the packhouse level, and similar results were found in other studies [13,15,20]. The packhouse level is therefore the most important link in the supply chain where food wastage should be addressed. WWF [13] observed that different players along the food supply chain all believed that greater wastage occurred at other points along the chain and are therefore not motivated to make any improvements to reduce food wastage within their own operations. Farmers from the Steenkoppies Aquifer expressed similar concerns that high food wastage is occurring at the market and retail levels, so it is possible that potential reductions in food wastage at the packhouse level have not yet been fully explored. Research is required to understand the reasons for the high percentage wastage at the farm or packhouse level to determine if farmers are making the best use of existing technologies to prevent food losses and whether there are other simple ways of reducing current losses. For example, carrots fit for human consumption that are discarded during packaging because of unmarketable properties could instead be cut or grated and sold, rather than being used for animal feed. In this study, data were obtained from a large commercial farm that uses advanced technologies, and the results could be very different for smaller scale farms.

Thus, future research should also be done to determine the impact of technology, farm size and farm infrastructure on crop wastage.

Average wastage for carrots, cabbage, beetroot, broccoli and lettuce along the supply chain as calculated in this study was lower than previous estimates for sub-Saharan Africa [15]. For example, literature sources indicate that 51.5% of fruits and vegetables are wasted along the supply chain [12]. This was an overestimate in the wastage of cabbage, which ranges between 10.4% in winter to 15.9% in summer, but was similar to the estimated 42% lettuce wastage in summer. The large variation in wastage between different crops also translates into significant differences in the wastage-related WFs of these crops.

A key challenge in quantifying food wastage is to classify waste. Most of the wastage reported here was not simply discarded. Wastage at the packhouse level is fed to livestock, wastage at the Tshwane Market is used for composting, and wastage at many of the green grocers that were contacted is given to charity organisations or livestock farmers. The beneficial use of these vegetables could disqualify them from being classified as waste, especially if they substitute other foods which would have been used for livestock feed. However, in the face of food insecurities and water scarcity, it is still worth considering these losses from the food supply chain and figuring out ways to minimize such wastage. Another challenge in quantifying vegetable wastage is the natural loss of the water content of the produce following harvest, which may result in lower measures of wastage weight relative to what was transferred from the previous step of the supply chain. If products are measured in terms of vegetable counts, like cabbage heads with more or less standard sizes, this problem could potentially be overcome.

A major challenge during this research was to obtain comprehensive data along the entire supply chain, especially at the retail level. WWF [13] also experienced a lack of data or unwillingness of companies to share their data during a survey that involved food retailers in South Africa. Policies are needed that require transparency on food wastage at all stages along the supply chain to improve data collection, as well as the active participation of agencies that take responsibility for collecting the necessary data and a central database where data can be recorded. The low wastage estimated for broccoli was mainly because of the assumption that wastage at the packhouse level will be similar to cabbage. This, however, could be an underestimation, and recording data for beetroot and broccoli at the packhouse level is recommended for future research. Recording losses of maize and wheat is more challenging compared to vegetable crops because these crops are processed into and sold as different products, which will have different wastage potential. Future research should also consider the losses of these grain crops in more detail.

It can be argued that reductions in food wastage may be one of the simpler ways to address food insecurities and water scarcities. In a global study on food losses, the minimum wastage recorded for fruits and vegetables was 37%, which was recorded in industrialised Asia, and a minimum of 33% wastage of root and tubers was recorded in northern Africa, and western and central Asia [20]. Losses recorded for some crops in this study are therefore below those minimum values reported previously, and further reductions may require fundamental changes to the current food supply chain system. The fraction of food wastage that cannot be prevented should always be diverted from landfill sites and used for other beneficial purposes such as composting, animal feed and biogas digesters, which is in line with the goals of the National Waste Management Strategy of South Africa [25].

Reducing wastage and the associated ecological impacts is not only difficult but can also be highly complex. For example, there may be a trade-off in reducing the blue WF due to less wastage as a result of increased pesticide use, which also leads to an increase in the grey WF as more of these pesticides are leached into groundwater. Reducing wastage through better refrigeration will most likely lead to a larger carbon footprint. Comprehensive assessments such as those used in Life Cycle Assessments (LCAs) will be valuable to improve understanding of the different impacts of potential mitigation measures.

4.2. Implication of Food Wastage on Water Resources

Under natural conditions, the only known outlet for the Steenkoppies Aquifer, Maloney's Eye, discharges an average of 14.7 Mm^3 per year [16]. This volume has decreased to an average 8.3 Mm^3 per year since 2006, which was the time when irrigated agriculture started to increase [26]. This reduction in outflow together with measured declines in borehole levels emphasize the unsustainable water use of the aquifer by agriculture [16]. If an ecological flow requirement (EFR) of 46%, as stipulated by the Department of Water Affairs [27], is taken into account, the total water available for abstractions is 8 Mm^3 per year. As current agricultural water use was estimated at 25 Mm^3 per year, irrigation has pushed the aquifer 17 Mm^3 per year beyond sustainable blue water use limits.

However, even eliminating food wastage completely will still not bring current blue water use to within sustainable limits, as blue water used to produce crop material that becomes wastage on the Steenkoppies Aquifer is estimated to be 25% of the blue water use that exceeds sustainable limits. Ridoutt, et al. [28] conducted a study on the blue, green and grey WFs of wastage of fresh mangoes in Australia and found that by reducing wastage, the impact of current production could be reduced to within sustainable limits. On the Steenkoppies Aquifer, however, where we only considered the sustainability of blue water, reducing wastage alone is simply not sufficient to achieve sustainable blue water use at current levels of production.

Blue water used to produce crop material that becomes wastage is also much higher than the total blue water used in packhouses to clean and pack these crops [17]. For example, in terms of total production on the Steenkoppies Aquifer for 2005, it was estimated that 0.04 Mm^3 for carrots, 0.003 Mm^3 for cabbage and 0.03 Mm^3 for lettuce were used in the packhouse, compared to 0.7, 0.1 and 0.7 Mm^3 of blue water linked to wastage, respectively. However, according to Le Roux [18], relatively high savings of blue water are possible through reductions of water use during cultivation. For example, by substituting the more common "Iceberg" lettuce, with "Cos" lettuce, which has a lower WF, and delaying harvest to achieve higher yields, blue water use in 2005 could have been reduced by 1 Mm^3. Lettuce only represents 14% of production by area, so similar switching to more water efficient and higher yielding crops could further such savings.

Reductions in food wastage, with concomitant reductions in the need for vegetable production, should therefore be considered as only one of multiple measures to address unsustainable blue water use on the Steenkoppies Aquifer. Mitigation should undoubtedly focus on the production phase, as this is where the highest gains can be made. Other measures may include the reduction of total production to within sustainable blue water limits, selecting crops and cultivars with low water footprints and using water more efficiently, for example, switching from sprinkler to drip irrigation, and using soil water conservation practices such as mulching. Increased use of waste waters may also be part of the solution. These responses are focussed on blue water savings and do not take into account the water quality implications. Such changes in agricultural management practices may be difficult to enforce because the system is primarily driven by economic factors. For example, farmers select crops based on market prices and demands. Irrigation on the aquifer also plays an important role in economic development in South Africa. Change will need to be driven through incentives or changing consumer choices. Consumers should also be encouraged to cultivate crops that have high percentage wastage along the supply chain, such as lettuce, in their own homestead gardens to reduce the decay that happens along the supply chain. People are also more likely to eat crops with unmarketable properties that have been grown in their own gardens.

5. Conclusions

The main objectives of this study were to quantify indirect water losses through the wastage of crops produced on the Steenkoppies Aquifer and to determine whether reductions in food wastage, assuming that this would lead to lower production at the farm level, could provide a way to achieve more sustainable water use. The hypothesis that there will be an important difference between the wastage fractions for different vegetable crops was also tested.

The highest percentage of wastage occurred during the packhouse stage, for reasons including damage by pests and diseases and unmarketable properties, so efforts to reduce wastage should focus on this stage. Quantifying food wastage is complicated by the fact that produce classified as wastage is often used for other purposes such as animal feed and compost. The amount of wastage for different types of vegetables can be highly variable, with small fractions for less perishable crops like cabbage, and high fractions for other crops such as lettuce. Care should therefore be taken when doing calculations with average or generalised published data on the wastage of vegetables.

For the Steenkoppies Aquifer, eliminating wastage completely (which is probably impossible) will accomplish around a quarter of the savings needed to achieve sustainable blue water use. Reductions may also come with other unwanted ecological impacts, for example, through the increased use of pesticides and refrigeration. Increasing water use efficiency through practices such as mulching could be effective but may have cost implications for the producer. Household cultivation of perishable crops may be effective in reducing wastage too, but ultimately more drastic policy changes and system interventions, for example, limiting the extent of production, or restricting what can be produced when, may be the only ways to achieve ambitious sustainability targets. Ideally, such changes will be achieved through incentives or changing consumer choices, involving multiple stakeholders in a more harmonious way.

Author Contributions: The first author, B.l.R., conducted this research and wrote the paper as part of her PhD thesis. M.v.d.L. supervised the research. T.V. contributed to the research in gathering data and developing the crop parameters for modelling. J.G.A. and K.L.B. are specialists in the field and acted as reviewers and mentors.

Acknowledgments: This research originated from a project initiated, managed and funded by the Water Research Commission (WRC project No. K5/2273//4: Water footprint of selected vegetable and fruit crops produced in South Africa), that is now published as WRC Report No. TT 722/17 (see Van der Laan (2017)). The first author, Betsie le Roux, received financial support for research from the WRC and a bursary from the National Research Foundation (NRF) of South Africa (NRF Grant number: 88572).

Conflicts of Interest: The authors declare no conflict of interest.

References

1. United Nations. Sustainable Development Goals. Seventeen Goals to Achieve Sustainable Development. Available online: http://www.un.org/sustainabledevelopment/sustainable-development-goals/ (accessed on 23 June 2017).
2. World Bank. *South Africa-Nutrition at a Glance*; World Bank: Washington, DC, USA, 2011; Available online: http://documents.worldbank.org/curated/en/413801468334804797/South-Africa-Nutrition-at-a-glance (accessed on 7 August 2017).
3. Hoekstra, A.Y.; Chapagain, A.K. *Globalization of Water: SHARING the Planet's Freshwater Resources*; Blackwell Publishing: Oxford, UK, 2011.
4. Wada, Y.; Beek, L.P.H.; Bierkens, M.F.P. Nonsustainable Groundwater Sustaining Irrigation: A Global Assessment. *Water Resour. Res.* **2012**, *48*. [CrossRef]
5. Hoekstra, A.Y. *Virtual Water Trade: Proceedings of the International Expert Meeting on Virtual Water Trade. Delft, The Netherlands, 12–13 December 2002*; Value of Water Research Report Series No. 12; UNESCO—IHE: Delft, The Netherlands, 2003.
6. Hoekstra, A.Y.; Aldaya, M.M.; Chapagain, A.K.; Mekonnen, M.M. *The Water Footprint Assessment Manual: Setting the Global Standard*; Routledge: London, UK, 2011.
7. Smakhtin, V.; Ashton, P.; Batchelor, A.; Meyer, R.; Murray, E.; Barta, B.; Bauer, N.; Naidoo, D.; Olivier, J.; Terblanche, D. Unconventional water supply options in south africa. *Water Int.* **2001**, *26*, 314–334. [CrossRef]
8. Department of Water Affairs. *National Water Resource Strategy 2*; Department of Water Affairs: Pretoria, South Africa, 2013.
9. Backeberg, G.R.; Reinders, F.B. Institutional reform and modernisation of irrigation systems in South Africa. In Proceedings of the 5th Asian Regional Conference of ICID, New Delhi, India, 6–11 December 2009.

10. Nieuwoudt, W.L.; Backeberg, G.R.; Du Plessis, H.M. The value of water in the South African economy: Some implications. *Agrekon* **2004**, *43*, 162–183. [CrossRef]

11. De Jager, E. *South Africa Annual Total Rainfall General Information*; South African Weather Services: Pretoria, South Africa, 2016.

12. Lundqvist, J.; de Fraiture, C.; Molden, D. *Saving Water: From Field to Fork: Curbing Losses and Wastage in the Food Chain*; SIWI Policy Brief; Stockholm International Water Institute (SIWI): Stockholm, Sweden, 2008.

13. World Wide Fund for Nature (WWF). *Food Loss and Waste: Facts and Futures*; WWF: Cape Town, South Africa, 2017. Available online: http://awsassets.wwf.org.za/downloads/WWF_Food_Loss_and_Waste_WEB.pdf (accessed on 4 September 2017).

14. Nahman, A.; De Lange, W.; Oelofse, S.; Godfrey, L. The costs of household food waste in South Africa. *Waste Manag.* **2012**, *32*, 2147–2153. [CrossRef] [PubMed]

15. Oelofse, S.H.; Nahman, A. Estimating the magnitude of food waste generated in South Africa. *Waste Manag. Res.* **2013**, *31*, 80–86. [CrossRef] [PubMed]

16. Le Roux, B.; van der Laan, M.; Vahrmeijer, T.; Bristow, K.L.; Annandale, J.G. Establishing and testing a catchment water footprint framework to inform sustainable irrigation water use for an aquifer under stress. *Sci. Total Environ.* **2017**, *599–600*, 1119–1129. [CrossRef] [PubMed]

17. Vahrmeijer, J. Improved Management of Groundwater in Irrigated Catchments under Stress. Ph.D. Thesis, University of Pretoria, Pretoria, South Africa, 2016. Unpublished.

18. Le Roux, B. A Framework to Apply Water Footprinting for Sustainable Agricultural Water Management: A Case Study on the Steenkoppies Aquifer. Ph.D. Thesis, University of Pretoria, Pretoria, South Africa, 2017.

19. Tshwane Fresh Produce Market. *Data on Vegetables from the Steenkoppies Aquifer*; Tshwane Fresh Produce Market: Pretoria, South Africa, 2014.

20. Gustavsson, J.; Cederberg, C.; Sonesson, U.; Van Otterdijk, R.; Meybeck, A. *Global Food Losses and Food Waste*; Food and Agriculture Organization of the United Nations: Rome, Italy, 2011.

21. Le Roux, B.; van der Laan, M.; Vahrmeijer, J.; Bristow, K.; Annandale, J. Estimating water footprints of vegetable crops: Influence of growing season, solar radiation data and functional unit. *Water* **2016**, *8*, 473.

22. Dos Santos, L. (Fruit and vegetable retailer, Pretoria, South Africa). Personal communication, 15 July 2014.

23. Mentis, D. (Fruit and vegetable retailer, Pretoria, South Africa). Personal communication, 30 June 2016.

24. Gathino, S. (Fruit and vegetable retailer, Pretoria, South Africa). Personal communication, 16 May 2016.

25. Department of Environmental Affairs. *National Waste Management Strategy*; Department of Environmental Affairs: Pretoria, South Africa, 2011.

26. Vahrmeijer, J.; Annandale, J.; Bristow, K.; Steyn, J.; Holland, M. Drought as a catalyst for change: A case study of the steenkoppies dolomitic aquifer. In *Drought in Arid and Semi-Arid Regions*; Schwabe, K., Albiac, J., Connor, J.D., Hassan, R.M., Gonzalez, L.M., Eds.; Springer: Dordrecht, The Netherlands, 2013; Chapter 14, pp. 251–268.

27. Department of Water Affairs. *Classification of Significant Water Resources in the Mokolo and Matlabas Catchment: Limpopo Water Management Area (WMA) and Crocodile (West) and Marico WMA: Ecological Water Requirements Report*; Report no: Rdm/wma 1,3/00/con/cla/0312; Directorate Water Resource Classification: Pretoria, South Africa, 2011.

28. Ridoutt, B.G.; Juliano, P.; Sanguansri, P.; Sellahewa, J. The water footprint of food waste: Case study of fresh mango in australia. *J. Clean. Prod.* **2010**, *18*, 1714–1721. [CrossRef]

water

MDPI

Article

A First Estimation of County-Based Green Water Availability and Its Implications for Agriculture and Bioenergy Production in the United States

Hui Xu and May Wu *

Energy Systems Division, Argonne National Laboratory, 9700 S. Cass Ave, Lemont, IL 60439, USA;
hui.xu@anl.gov
* Correspondence: mwu@anl.gov; Tel.: +1-630-252-6658

Received: 21 November 2017; Accepted: 25 January 2018; Published: 2 February 2018

Abstract: Green water is vital for the terrestrial ecosystem, but water resource assessment often focuses on blue water. In this study, we estimated green water availability for major crops (i.e., corn, soybean, and wheat) and all other users (e.g., forest, grassland, and ecosystem services) at the county level in the United States. We estimated green water resources from effective rain (ER) using three different methods: Smith, U.S. Department of Agriculture—Soil Conservation Service (USDA-SCS), and the NHD plus V2 dataset. The analysis illustrates that, if green water meets all crop water demands, the fraction of green water resources available to all other users varies significantly across regions, from the Northern Plains (0.71) to the Southeast (0.98). At the county level, this fraction varies from 0.23 to 1.0. Green water resources estimated using the three different ER methods present diverse spatiotemporal distribution patterns across regions, which could affect green water availability estimates. The water availability index for green water (WAI_R) was measured taking into account crop water demand and green water resources aggregated at the county level. Beyond these parameters, WAI_R also depends on the precipitation pattern, crop type and spatially differentiated regions. In addition, seasonal analysis indicated that WAI_R is sensitive to the temporal boundary of the analysis.

Keywords: green water availability; effective rain; crop water demand; water resources

---

1. Introduction

Fresh water is widely considered the most essential natural resource for human life and the ecosystem [1], yet the sustainable use of water resources is an increasing challenge [2–4]. Because water underpins agriculture, energy production, and municipalities, water overexploitation is becoming a threat to food security, energy production, and socioeconomic development in many parts of the world [4–6]. To be water secure, it is critical to manage natural water resources properly and to keep water consumption at a sustainable level [3–5].

The United States (U.S.) has relatively abundant freshwater resources, although there is significant regional variability [7]. Gerten et al. [8] estimated that blue water resources in the U.S., that is, fresh water from surface streams, reservoirs and groundwater, amount to about 1700–2000 m^3 per capita per year. However, county-level runoff (flow per unit area) [9] and per-capita blue water resources (calculated by dividing annual runoff volume [9] by population in each county [10]) range from 0.2 to 3040 mm/year and from 2.3 to 7,846,654 m^3/cap/year, respectively. In terms of water scarcity, Moore et al. [11] suggest that 81.9% of areas in the U.S. are in the low water scarcity category. However, 4.4% and 13.7% of areas in the U.S. are moderately and highly water scarce, respectively. In the summer, especially, hot spots (areas with high water scarcity) increase in the western regions [11]. Mekonnen et al. [4] estimated that about 130 million people, or 42% of the U.S. population, are facing moderate to severe water scarcity, mostly in western and southern states. Earlier studies also found that these regions would be particularly vulnerable to potential shifts in rainfall patterns [7,12]. To improve water management programs nationwide, it is important to examine the tensions between water demand and water resource supply in the energy and agriculture sectors.

A large body of literature has evaluated the impacts of water withdrawals [7,13] or water consumption [11] on water resource sustainability, under both current and future climate conditions [12,14,15]. However, the scope of these studies has traditionally been limited to blue water resources, even though it is primarily green water that sustains the terrestrial ecosystem [8]. Green water represents the precipitation on land that does not run off or recharge to groundwater [16]. In other words, it includes precipitation that temporarily stays on top of vegetation and precipitation stored in soil, which eventually will return to the atmosphere via evapotranspiration (ET) [16,17]. Liu et al. [18] suggested that previous studies skipped green water mainly due to different measurements of blue (flow) and green (storage) water resources. Another important factor is that blue water has a much higher opportunity cost than green water [19]. Unlike blue water, green water resources are spatially immobile, so they are only naturally available on land for plants, except when this water is lost to a blue water pool (e.g., contributed to aquifers via deep percolation). In contrast, blue water is important for many economic sectors and its consumption may affect many downstream users. Agricultural water use was often a focus of previous blue water scarcity studies, because irrigation consumption for agriculture composes about 80% to 90% of total blue water consumption [5,11]. Despite the crucial role of irrigation in agriculture, blue water accounts for only 16% of all water consumed by crops globally; the remaining 84% comes from green water [6]. Green water is the primary water resource used to meet the water demand for crops, forest, grassland, and ecosystem requirements; irrigation of crops becomes necessary when there is a soil moisture deficit.

Given the importance of green water in biomass production, previous studies have quantified both green and blue water footprints embedded in the production of various crops [20–23] and biomass feedstocks [24–26]. Recently, Argonne National Laboratory developed an online water footprint tool, Water Analysis Tool for Energy Resources (WATER) (http://water.es.anl.gov), to model water footprints of biofuels produced from various feedstocks via a range of conversion pathways in the U.S. at the county level. Nonetheless, water footprints of crop products alone are not sufficient for regional water scarcity assessment, because water scarcity is a function of relative water supply and water demand [7]. In addition, researchers [27–30] have made efforts to develop blue and green water scarcity footprints and examine how land-use change may affect surface runoff and green water flow. Still, the focus of the water scarcity footprint approach is quantifying water use impacts rather than water scarcity. Although many studies have assessed surface water or groundwater scarcity, previous work, with a few exceptions [19,31,32], rarely considered both crop water demand and green water resources in the same study. In fact, recent studies have repeatedly identified green water as a key challenge that needs to be addressed in future water scarcity assessments [18,33]. Núñez et al. [19], Rodrigues et al. [31] and Veettil et al. [32] have estimated green water scarcity for small watersheds, using the blue-green water footprint concept, but few have attempted to evaluate green water availability within the conterminous U.S. Because green water is critical to agriculture and terrestrial ecosystems, a spatially explicit quantification of green water availability in the U.S. is still needed for agricultural and bioenergy planning.

The main objective of this study was to estimate the amount of green water resources available for agricultural and bioenergy production, and to assess how crop and bioenergy feedstock production may affect green water resources available for other uses (e.g., forest, grassland, ecological needs) in the U.S. After a review of major existing water availability indices, we employed a modified green water availability index to assess how crop water demand may affect green water resources available to other users at local and regional scales, as well as implications for future water resource planning. Because effective rain (ER) (i.e., the portion of total precipitation that does not contribute to surface runoff or deep percolation) [34,35] is frequently used as a proxy for available green water resources [36], we also compared ER estimated from three different methods to evaluate variability and uncertainty in green water resource assessments.

2. Review of Current Water Availability Assessment Metrics

Among the proposed means of quantifying the water resources available for sustaining production at the regional level, the water availability index (WAI) is one of the key metrics that enable analysts to address regional water demand and water supply issues. Selecting a suitable WAI for a particular study could be confusing because a wide range of "water availability", "water stress", or "water scarcity" indices exist [37–40]. The three terms have been used frequently and interchangeably in the literature either to label a metric or to describe water resource problems [39]. According to the ISO (International Standardization Organization) 14046 standard [41], "water availability" describes whether humans and ecosystems have sufficient water resources to meet their needs, whereas "water scarcity" refers to volumetric abundance without considering water quality and environmental water requirements [41]. Boulay et al. [42] suggest that "water scarcity" and "water stress" have the same meaning. Still, other definitions for these terms exist. For instance, Schyns et al. [43] considered that "Water availability" refers to water supply only, whereas "water scarcity" considers both water supply and demand.

In general, existing water scarcity indices can be categorized into four major groups (Table 1): (1) indices measuring per-capita water availability [8,44], (2) indices based on the ratio of water withdrawal or consumption of water resource [45,46], (3) composite indices including socioeconomic factors (e.g., lack of infrastructure) [47], and (4) indices based on variation in ET . Studies by Balcerski [48] and Falkenmark and Lindh [49] in the 1970s were among the first to compare freshwater resources with human withdrawals. In the late 1980s, Falkenmark proposed the water stress indicator (WSI) [44], which measures water scarcity based on per-capita surface water resources. One of the limitations of the WSI index is that it assumes fixed, universal water demand. For instance, all areas with a per-capita water resource less than 1700 m^3 per year would be considered water scarce. To address that problem, various metrics based on use-to-availability ratios (Table 1), which incorporate spatial varying water demand from multiple economic sectors, have been proposed during the second wave of water resource assessments [40]. In general, a country or region is considered water scarce if annual withdrawals are higher than 20% of annual freshwater supply and severely water scare if this ratio exceeds 40% [40,50,51]. Falkenmark [6] and Kummu et al. [5] suggest that per-capita water availability metrics are helpful to identify "water shortage" driven by population growth, whereas use-to-resource-based indices measure "water scarcity" due to high demand relative to availability. In recent years, the use-to-resource-based indices have been modified to account for environmental water requirement (EWR) explicitly (e.g., aquatic habitat preservation) [4,52]. However, determination of an appropriate EWR remains challenging because the amount of water needed to sustain freshwater ecosystems is highly variable, depending on the region's flow season, and a consistent method to estimate and verify the EWR is lacking [53,54]. More extensive reviews of major existing blue and green WAIs can be found in references [37–39,43].

Table 1. Summary comparison of existing water availability or stress or scarcity studies.

Category	Blue Water Index			Green Water Index		
	Key Components	References	Pros and Cons	Key Components	References	Pros and Cons
Water crowding	Predefined thresholds of per-capita share of total annual runoff; human water demand. Data on annual runoff and population is needed.	Falkenmark [44]	Pros: provides an easy to use threshold to assess water scarcity status.Cons: focuses on basic human water demands only; ignores regional differences in per-capita water demand.	Demand of human diet vs. food production; data on consumptive water use of crops and meat production is needed.	Rockström et al. [55]; Gerten et al. [8]; Kummu et al. [56]	Pros: provides an easy-to-use and country-specific threshold to assess water scarcity status.Cons: focuses on basic human water demands only.
Use-to-resource ratio	Water withdrawal or consumption to streamflow/surface runoff or groundwater storage. Data on annual withdrawal and runoff is needed.	Vorosmarty et al. [45]; Pfister et al. [46]; Sun et al. [14]; Tidwell et al. [57]; Brauman et al. [58]	Pros: uses multi-sectoral water use and supply data to generate critical ratios for each region.Cons: does not consider EWR.	Crop ET/effective rain; green water footprint (WF)/green water resource. Data on crop ET and effective rain is needed.	Núñez et al. [19]; Rodrigues et al [31]; Veettil and Mishra [32]	Pros: clear and easy-to-use definitions for both demand and supply variables.Cons: does not address water scarcity at the field level; does not consider EWR.
	Environmental water requirement (EWR) is also included. Monthly streamflow data is often needed for EWR calculations.	Hoekstra [16]; Smakhtin [52]; Vanham et al. [33]	Pros: is eco-centric, explicitly considers EWR.Cons: it is often difficult to determine appropriate EWR for individual regions.	Data on ET, the area of land reserved for natural vegetation, and ET that cannot be made productive	Hoekstra et al. [16]	Pros: Explicitly considers EWR.Cons: it is hard to estimate land area that should be reserved for natural vegetation and to determine the portion of ET that is unproductive
Composite index	In addition to physical water demand/supply, considers social and economic factors (e.g., infrastructure). Data inputs vary by indicators, including hydrology, accessibility (e.g., distance, time) and economic and policy capability.	Sullivan et al. [47]; Chaves and Alipaz [59]	Pros: comprehensive, including social factors.Cons: requires extensive data input; may not be straightforward to interpret.			
Variation in ET	Actual ET (AET), reference ET, deficit in soil moisture supply. Data inputs include long-term climate data (e.g., precipitation, temperature) and crop ET.	Palmer [60]; Woli et al. [61]; Devineni et al. [62]	Pros: can identifies abnormal changes in crop ET.Cons: focuses on drought detection, rather than regional water scarcity.	Transpiration (T)/AET; AET/potential ET (PET). Data on T, AET, and PET are needed.	Meyer et al. [63]; Rockström et al. [55]; Wada et al. [64,65]	Pros: can examine changes in site-specific green water flow.Cons: does not provide a critical ratio to describe average local water scarcity status.

Although the blue-green paradigm is relatively new [17], the idea of evaluating soil moisture availability is not new. In fact, a large body of agricultural drought indicators exist in the literature, which often considers the impact of soil water supply on crop growth by comparing actual crop ET with a reference ET [43]. Nonetheless, the bulk of drought indicators focus on measuring plant water deficit and therefore say more about irrigation needs than green water availability [43]. Although some WAI explicitly considering green water have been proposed in the literature [43], none of them is widely adopted or currently operational due to the difficulty of obtaining required data, or other limitations [43,66]. Rockström et al. [55], Gerten et al. [8], and Kummu et al. [56] extended the Falkenmark Water Stress Indicator [45] to compare combined per-capita green-blue water resources with the amount of fresh water need to sustain a standard diet or balanced diet in each country. However, these indexes focus on basic human demands and thus fail to consider water demands from economic developments (e.g., bioenergy production). The Green Water Scarcity Index (GWSI) by Núñez et al. [19] was calculated as the ratio of green water footprint (GWF) to effective rain. GWF refers to the volume of green water consumed during the biomass (e.g., crops and woody biomass) production process [16]. Núñez et al. [19] applied the GWSI to cropland only. The GWSI by Hoekstra et al. [16] also compares GWF to available green water resources, but the definition of available green water resource is different from the Núñez et al. [19] definition. Specifically, Hoekstra et al. [16] defined available green water resources in a given catchment as total ET from all land area in that catchment, excluding the environmental ET requirements (i.e., ET from land area reserved for natural vegetation) and the portion of ET that is unproductive (i.e., ET from land area that cannot be productive) [16]. Although this definition of available green water resources is more comprehensive than the Núñez et al. [19] definition, it is not straightforward to determine how much land must be reserved as natural land and when the green water flow cannot be productive [17,44]. For this reason, the GWSI by Hoekstra [16] has not been operational.

The green water WAI mentioned above are all area-based and related to land use patterns; therefore they do not address green water scarcity at a particular site. Within a given land unit, there is usually no competition over green water resources, unless land-use change is considered. Falkenmark et al. [6] describe site-specific green water scarcity as a problem related to lower-than-potential plant-accessible water in the root zone. Rockström et al. [55] suggested that only transpiration by plants is a productive green water use, so they use a "transpiration efficiency" metric (calculated as the ratio of transpiration to evaporation) to assess green water use efficiency. Area-based scarcity indexes and site-specific metrics are not comparable, but they can be complementary to each other. For instance, areas with high green water scarcity and low transpiration efficiency may achieve better yields by improving their soil water management strategies [43]. Schyns et al. [43] presented a more comprehensive review of WAI focused on green water.

One of the limitations of current GWF based GWSI is that GWF measures actual green water consumption, which can be lower than crop water demand if green water resources are limited. Because GWF is calculated as the minimum of effective rain (green water resources) and crop water demand [36,66], a low annual GWF does not necessarily mean low green water resource demand, because effective rain may simply be limited during the crop growing stage. In other words, temporal aggregation of a green water footprint (e.g., annual basis) may not be representative, because green water demand may occur within in a short time period of the year. Consequently, areas with high crop water demand but low growing season effective rain may receive a low green water scarcity score that does not reflect the actual scarcity of the green water resource. To address this issue, Rodrigues et al. [31] proposed the concept of "potential green water footprint", which was estimated as the sum of "maximum transpiration" and "soil water evaporation", rather than actual consumption. Potential green water footprint is equivalent to crop water demand or crop water consumption when the soil moisture supply is unlimited. However, they did not explain the rationale for using the median (50th percentile) of daily soil water content at the beginning of a simulation period as the available green water resource.

Given that studies of green water scarcity are limited, there is a need for more systematic assessments of green water availability and use, as well as the continuing development of suitable green water scarcity indexes. In addition to metric-based assessment, agro-hydrological models that can systematically account for soil-plant-water interactions may provide a more robust assessment of green water resources and scarcity, since both natural (e.g., climate and soil) and human management factors (e.g., tillage, irrigation) will affect blue and green water flow. For instance, Mekonnen et al. [67], Faramarzi et al. [68], and Wada et al. [69] have utilized sophisticated hydrological models to assess crop water footprint and water scarcity. However, this level of investigation is beyond the scope of this study.

3. Method for Green Water Availability Assessment

A modified green water availability index (WAI_R) (Figure 1), which is an extension of the existing GWSI [16,19,31], was employed in this study. WAI_R is a metric that measures the fraction of green water resources, after the water demand of specified sectors (e.g., agriculture) is met by green water, available to all other remaining green water users (e.g., timber, pasture, ecosystem services). Like other area-based GWSI, WAI_R quantifies green water balance aggregated at a regional level (e.g., county). It does not consider green water availability at the field level. To estimate green water demand, we use total plant water demand rather than GWF. A companion green water WAI that uses GWF rather than crop water demand is also presented below (Equations (5) and (6)). For agricultural production, the green water resources used apply to all crops (rain-fed or irrigated). We assume that crop water demand will be met with green water resources first, and irrigation will be supplied only if there is a deficit in rainwater supply. In this sense, the application of irrigation water may affect yield but does not affect the portion of green water resources that would be available to other green water users. The WAI_R index proposed here is specifically designed to estimate the impacts of plant water demand on regional water resources (Figure 1). For available green water resources, Núñez et al. [19] use green water resources from existing cropland only; other studies consider all green water resources in a region, regardless of the land-use type [31,32]. In this study, we assume green water resources from all pervious land (e.g., cropland, pasture) are ultimately available for plant use; impervious land area (e.g., urban) and open water surfaces were excluded from green water resource calculations. We follow the suggestion of Liu et al. [18] to use annual green water resources, regardless of whether or not they are used by crops or other plants. Although some studies prefer to use growing season green water resources [19], off-season green water resources may be stored in soil or lost to deep percolation, depending on local soil and climate conditions. For instance, the portion of green water resources stored as soil moisture during the winter when the crop is dormant, which is also called carry-over soil moisture, can be used to meet the consumptive water needs of crops [70].

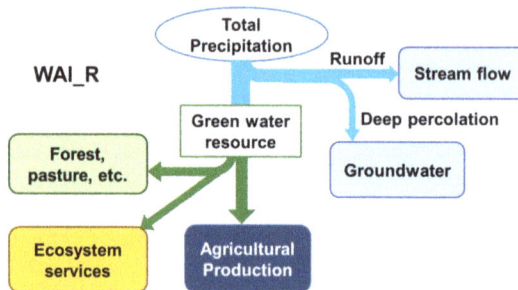

Figure 1. Conceptual diagram for the water availability assessment. Total precipitation is divided into green water resources (effective rain) and blue water resources (runoff, deep percolation to aquifers). This approach quantifies the water resource balance aggregated at the regional level (e.g., county) without considering water availability related to specific fields within each region.

For a given county j, the fraction of green water resources needed to meet the demand from a certain sector i ($WDI\_R_{i,j}$) is defined as the ratio of plant water demand from that sector to the total green water resources in county j (Equation (1)). Green water resources in a given county are defined as the volume of ER from all pervious land area in that county (Equation (1)). Pervious land area in county j ($A_{pervious,j}$) is the total surface area in the county minus the total open water surface area, which includes streams, ponds, lakes, swamps and costal water area, and impervious (urban) area in that county (Equation (2)).

$$WDI\_R_{i,j} = \frac{plant\ water\ demand_{i,j}}{green\ water\ resource_j} = \frac{plant\ water\ demand_{i,j}}{ER_j \times A_{pervious,j}} \qquad (1)$$

$$A_{pervious,j} = A_{total,j} - A_{water,j} - A_{impervious,j} \qquad (2)$$

where: $WDI\_R_{i,j}$ = the fraction of plant water demand of sector i in county j; ER_j = annual effective rainfall depth (m/year) in county j; $A_{total,j}$ = total surface area (m^2) of county j; $A_{water,j}$ = open water surface area (e.g., river, ponds) (in m^2) of county j; $A_{impervious,j}$ = impervious surface in urban area of county j (m^2).

Once WDI_R is defined, WAI_R is simply calculated as the difference between 1 and WDI_R. Specifically, WAI_R is a general metric that can be applied to multiple sectors. Let S be a set of sectors, where sector i belongs to S, or $i \in S$. Let $WAI\_R_{non\ i,j}$ (Equation (3)) be the fraction of green water available for remaining sectors in S after meeting the needs of sector i, and let $WAI\_R_{non\ S,j}$ (Equation (4)) be the fraction of green water resources available for remaining users after meeting the needs of all sectors in S. Then

$$WAI\_R_{non\ i,j} = 1 - WDI\_R_{i,j} \qquad (3)$$

$$WAI\_R_{non\ S,j} = 1 - WDI\_R_{S,j} = 1 - \sum_{i \in S} WDI\_R_{i,j} \qquad (4)$$

where $WAI\_R_{non\ i,j}$ = the fraction of green water available to the remaining sectors in S after meeting the needs of sector i; $WAI\_R_{non\ S,j}$ = the fraction of green water available after meeting the needs of all sectors in S; $WDI\_R_{S,j}$ = the fraction of green water resource needed to meet plant water needs of all sectors in S in county j.

The value of $WAI\_R_{non\ i,j}$ or $WAI\_R_{non\ S,j}$ range from 0 to 1. A value of 1 means that 100% of the green water resources are available to sectors other than the specific sector(s). Take the agriculture sector as an example, a value of 1 means there is no agricultural production in a given region; a value of 0 means there are no remaining green water resources after meeting the demand from specified economic activities. When plant water demand exceeds supply, additional water resources (e.g., irrigation water) may be required to make up the green water deficit to sustain the growth. However, a detailed discussion on blue water consumption is outside the scope of this analysis. Although some studies (e.g., Quinteiro et al. [71]) have started to consider the dynamics between green and blue water in water scarcity footprint analysis, this study focuses on estimating green water availability.

In addition to WAI_R, we also calculated green water availability based on GWF for comparison. Similar to WAI_R, the fraction of green water resources consumed by a certain sector ($WDI\_R\_F_{i,j}$) is defined as the ratio of the GWF of sector i (in m^3) to total green water resources (in m^3) in county j (Equation (5)). Once WDI_R_F has been defined, the GWF-based green water availability index (WAI_R_F) can be defined as the difference between 1 and WDI_R_F (Equation (6)), as follows:

$$WDI\_R\_F_{i,j} = WDI\_R\_F_{i,j} = \frac{GWF_{i,j}}{ER_j \times A_{pervious,j}} \qquad (5)$$

$$WAI\_R\_F_{non\ i,j} = 1 - WDI\_R\_F_{non\ i,j} \qquad (6)$$

Similar to $WAI\_R_{non\ i,j}$, the value of $WAI\_R\_F_{non\ i,j}$ also ranges from 0 to 1; a value of 0 means all green water resources consumed by sector i, and a value of 1 means the sector does not consume green water in region j.

3.1. Application to Agricultural Crop Production

The improved green water availability index (WAI_R) was applied to the production of three major crops (corn, soybeans and wheat) that represent the agriculture sector in the U.S. at the county level. We quantified the fraction of green water resources needed if the crop water demands of three crops in county j are met by green water ($WDI\_R_{ag,j}$) (Equation (7)), and the fraction of green water resources in county j that is available to remaining green water users (e.g., other crops, grassland, forest and ecosystem services) ($WAI\_R_{non\_ag,j}$) (Equation (8)). The water demands of crop production can be calculated from crop evapotranspiration (ET_c) and harvested acreages (Equation (7)):

$$WDI\_R_{ag,j} = \sum_c \frac{ET_{c,j} \times A_{harvest,c,j}}{ER_j \times A_{pervious,j}} \tag{7}$$

$$WAI\_R_{non\_ag,j} = 1 - WDI\_R_{ag,j} \tag{8}$$

where: $ET_{c,j}$ = annual crop evapotranspiration depth (m/year) of crop c (corn, soybean, and wheat) in county j; and $A_{harvest,c,j}$ = area (m²) of crop c harvested for all purposes in county j.

For comparison, we also applied the WAI_R_F metric to these three major crops. The fraction of green water resources consumed by the three crops ($WDI\_R\_F_{ag,j}$) in county j is defined as the ratio of total crop green water consumption over green water resource in county j (Equation (9)). Green water availability for sectors other than these three crops ($WAI\_R\_F_{non\_ag,j}$) is therefore the difference between 1 and $WDI\_R\_F_{ag,j}$ (Equation (10)):

$$WDI\_R\_F_{ag,j} = \sum_c \frac{GWF_{c,j} \times A_{harvest,c,j}}{ER_j \times A_{pervious,j}} \tag{9}$$

$$WAI\_R\_F_{non\_ag,j} = 1 - WDI\_R\_F_{ag,j} \tag{10}$$

where $GWF_{c,j}$ = annual crop GWF in depth (m/year) of crop c in county j.

3.2. Crop Water Requirement and Green Water Footprint

Consumptive water use for individual crops (i.e., corn, soybeans, and winter and spring wheats) was quantified by estimating crop ET (ET_c). For each crop, we estimated ET_c as the product of reference ET (ET_0) and crop coefficients (K_c) on a monthly basis at each county and summed to find annual crop ET [66]. Crop GWF was calculated as the minimum of crop water requirement (ET_c) and green water resources (estimated from effective rain) on a monthly basis in each county and summed to find annual GWF. Monthly ET_0 was computed using the American Society of Civil Engineers' (ASCE's) standardized Penman-Monteith method [72]. Similar to previous studies [20,25], ET outside the crop growing season was not counted as crop water use in this analysis.

The growing period of winter wheat spans two consecutive years, but the calculation method is the same with corn and soybeans. This is because we used 30-year (1971–2000) mean monthly climate data, so whether a month is in year 1 or year 2 does not affect crop ET calculation. For instance, if winter wheat spans from October in year 1 to March in year 2, we simply calculated annual wheat ET by summing January–March ET and October–December ET.

3.3. Green Water Resource Estimation

Green water resources can be estimated from ER using several existing methods, including field monitoring, empirical equations, and soil water balance models [73]. A detailed review of ER estimation methods can be found in Dastane [34]. Many water footprint studies have utilized empirical equations to estimate ER [36,74,75]. Given the importance of ER in green water resource assessment, we employed three alternative ER estimation methods in this study to estimate 30-year (1971–2000) mean ER depth (mm/month) for each county in the conterminous U.S. at monthly intervals. Two are empirical methods, including the U.S. Department of Agriculture—Soil Conservation Service (USDA-SCS) (also known as Technical Release (TR)-21) method [70] and the Smith method [76]. The latter is a simplified version the USDA-SCS method implemented in the CROPWAT model [76]. The third method is derived from a water balance dataset (National Hydrography Dataset (NHD) Plus V2) [77].

3.3.1. ER Based on the USDA-SCS Method

The USDA-SCS method [70] was developed with water balance calculations using 50 years of precipitation records at 22 locations throughout the U.S. The climate stations were selected to cover all climatic conditions across the 48 states in the continental U.S. Each of the stations has rainfall records of at least 25 years during the growing season of major crops. USDA scientists calculated daily soil water balance and related it to crop ET, precipitation, and soil water factors. Precipitation that is not lost to deep percolation or surface runoff is considered ER. The resulting equation for estimating effective rainfall is:

$$ ER = SF \times \left(0.70917 \times P^{0.82416} - 0.11556 \right) \left(10^{0.02426 \times ET_c} \right) \tag{11} $$

And the soil factor (SF),

$$ SF = (0.531747 + 0.295164 \times D - 0.057697 \times D^2 + 0.003804 \times D^3 \tag{12} $$

where P is 30-year average monthly precipitation. ET_c is average monthly crop evapotranspiration (inches). D is the "useable soil water storage" (inches), which is usually calculated as 40% to 60% of the available soil water capacity [35], depending on local irrigation practices. The management allowable soil water depletion for the three crops ranges from 50% to 65% [78]. In this study, we used 60% [75] because we assume farmers will use soil water first before applying irrigation water. However, using 50% or 60% does not make a noticeable difference in ER estimations; county level annual ER would decrease by 4.3% (SD = 2.46) if 50% is used. The soil water capacity layer was extracted from the Digital General Soil Map of the U.S. or STATSGO2 soil dataset [79].

3.3.2. ER Based on the Smith Method

The Smith method (Equation (13)), which is a simplification of the "USDA-SCS" method, assumes an average ET of 8 inches ($\approx$203.2 mm) per month and a "useable" soil water storage of 3 inches ($\approx$76.2 mm) [36]. The Smith method is more frequently used in the literature than the original USDA-SCS method [36,66,74,80], probably due to its simplicity and the wide application of the CROPWAT [78] model:

$$ ER = \begin{cases} \frac{P \times (125 - 0.2 \times P)}{125}, & for\ P \leq 250\ mm/month \\ 125 + 0.1 \times P, & for\ P > 250\ mm/month \end{cases} \tag{13} $$

3.3.3. ER Based on the NHDPlus V2 Data

The NHDPlus V2 dataset [77] provides simulated runoff at the catchment level, which is based on a soil water balance (WB) model developed by Wolock and MaCabe at USGS [81,82]. For this method, ER is the difference between precipitation (P) and runoff (RO) on a monthly basis (Equation (14)):

$$ ER = P - RO \tag{14} $$

where RO is model-simulated 30-year (1971–2000) average monthly runoff (mm/month), aggregated from original catchment level data using an area-weighting method. The weighting factors for a county that crossed the boundaries of multiple catchments were calculated separately based on the area of the county that fell inside each catchment and then aggregated to re-form the county-level mean runoff. The WB model uses monthly temperature and precipitation data to determine the proportions of monthly precipitation that are rain and snow [81,82]. Rainfall and melted snow contribute to runoff, which is calculated as the sum of direct runoff and surplus runoff, where direct runoff is computed from overland runoff. When soil moisture storage exceeds soil water capacity, the excess soil water contributes to runoff as surplus runoff [81,82]. Actual ET is equal to potential ET if rainfall and snow-melt exceed the potential ET. Soil moisture storage can be removed to support ET, but the fraction of moisture storage that can be removed decreases linearly with decreasing soil moisture [81,82]. This simplified scheme does not consider crop-specific ET. The WB model does not include a groundwater component, so deep percolation and base flow are not directly modeled.

3.3.4. Advantages and Limitations of the Three ER Estimation Methods

Among the three ER methods, the Smith method is the most convenient because it only requires monthly precipitation data; however, this method does not incorporate variations in local soil properties. The USDA-SCS method is conceptually more comprehensive because it includes a soil factor, but it still fails to account for soil water intake rates and rainfall intensity because of insufficient data and the complexity of these two factors [70]. In addition, although the USDA claimed that the 22 stations were selected to cover all climatic conditions in the U.S. [70], the USDA did not publish the data used for model development so the spatial and temporal pattern of the climate data is unclear. If climate and soil patterns of a given county are significantly different from those of the 22 stations, the USDA-SCS method may not work well. In addition, the experiments were published in 1970 and the 50 years of data reflect the period from the 1910s to the 1960s. Therefore, recent changes in soil infiltration rates caused by management practices (e.g., tillage) and rainfall intensity may require an update to the regression model published decades ago. In general, the USDA-SCS method is more applicable to regions with well-drained soil and low-intensity rainfall [34,73]. Unlike the two empirical approaches, the NHDPlus V2 dataset was derived from a WB model [81,82]. Although the model explicitly accounts for soil type and dynamics in water balance, it ignores the impact of plants and land management on runoff. In addition, the NHDPlus V2 dataset does not provide monthly changes in snow water storage, which means in areas with heavy snowfall it is difficult to differentiate runoff sourced from rainfall or snowmelt.

3.4. Study Area and Data Sources

Green water availability was analyzed at the county and regional level in the 48 continental states in the United States. We divided the 48 states into 10 major agricultural production regions (Figure 2) based on the boundaries of USDA farm production regions [83]. States within each region share similar farm production characteristics (e.g., crop types). Annual precipitation ranges from 100 to 3000 mm/year. Spatially, precipitation generally decreases from the southeastern U.S. to the western U.S., except in the northwestern costal area (Figure S1). Plantings of the major commodity crops (corn, soybeans, and wheat) are mostly concentrated in the Midwest (Figure 3) where soil is fertile and flat terrain is suitable for farming.

County-level corn, soybean, and wheat harvested acreages and yields in 2008 were collected from the USDA National Agricultural Statistics Service (NASS) (Table 2). The 30-year (1971–2000) mean monthly precipitation data at the county level were aggregated from the gridded Parameter-elevation Relationships on Independent Slopes Model (PRISM) dataset [84]. Ideally, the climate data period should cover the crop year (2008). However, the NHDPlus V2 dataset was based on 1971–2000 climate data and the potential ET (PET) data from the WATER model is only available for 1971–2000. To make sure that all methods use the same climate data, we used the 1971–2000 climate data for both ET and ER calculations. In fact, there is no significant change in precipitation patterns between the 1971–2000 and 1981–2010 PRISM datasets. For instance, county-level mean annual precipitation would decrease by 9.16 mm only (SD = 31.1) if 1981–2010 data is used. For the Smith and USDA-SCS methods, differences in annual ER calculated using the 1971–2000 versus 1981–2010 precipitation data are less than 10% for all but 24 counties.

Monthly potential ET and Crop coefficient (Kc) were provided by the WATER model [25,26,66] at the county and agricultural production region level, respectively (Table 2). Impervious land area and open water surface area (e.g., streams, lakes, swamps) for each county were extracted from the National Land Cover Database (NLCD) 2011 dataset [85] and the Cartographic Boundary Shapefiles [86], respectively.

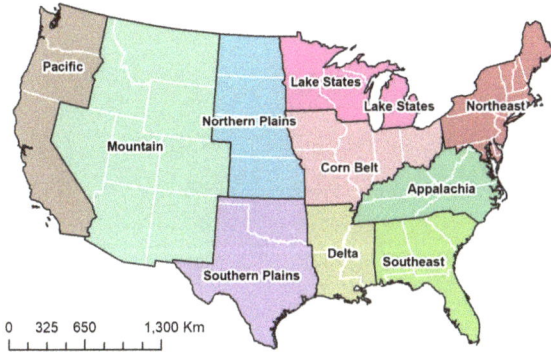

Figure 2. Ten agricultural production regions used in this study.

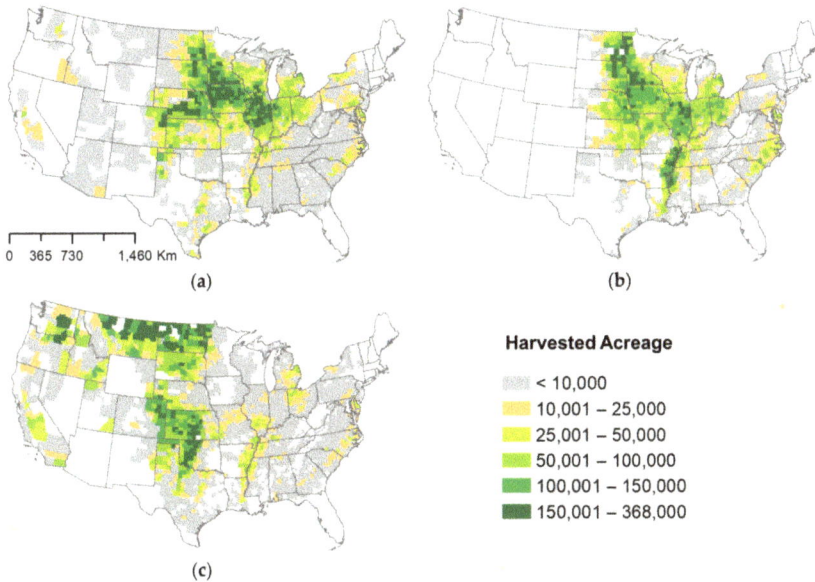

Figure 3. County-level harvested acreages for (**a**) corn, (**b**) soybean, and (**c**) wheat in the conterminous United States in 2008. Data is from U.S. Department of Agriculture (USDA) National Agricultural Statistics Service (NASS) reports [87].

Table 2. Input data for green water and crop water demand modeling.

Item	Timespan	Spatial Resolution	Temporal Resolution	Data Source
Corn, soybean and wheat acreages	2008	County	Annual	USDA NASS [87]
Precipitation	1971–2000	800 m	Monthly	PRISM [84]
Potential ET	1971–2000	County	Monthly	WATER [25,26,66]
Crop coefficient (Kc)	–	Farm production region	Monthly	WATER [25,26,66]
Impervious (urban) area	2011	30 m	–	NLCD 2011 [85]
Land and water surface area	2015	County	–	Cartographic Boundary Shapefiles [86]
Runoff	1971–2000	1 km	Monthly	NHDPlus V2 [77]
Soil water capacity	2016	1:250,000 (vector)	–	STATSGO2 [79]

4. Results and Discussion

4.1. Comparison of Green Water Resources Estimated by Three Methods

Geospatially, all three green water resource estimations—based on the Smith method (ER_Smith), the USDA-SCS method (ER_USDA), and the NHDPlus (ER_NHD) method—presented a decreasing trend from the southeast region to the western states, except in the Pacific Northwest (Figure 4). This pattern largely follows the distribution pattern of annual precipitation (Figure S1). Among the three ER methods, ER_Smith (Figure 4a) and ER_USDA (Figure 4c–f) tend to have the highest and lowest values, and ER_NHD (Figure 4b) falls in the middle (Figure 4). In addition, discrepancies in annual ER estimations are more evident in the eastern U.S. than regions in the central and western U.S. Because the USDA-SCS method is crop specific, ER_USDA based on ETc of three major crops (i.e., corn, soybean, and wheat [including winter and spring wheat]) was also generated. The resulting four ER_USDA maps are quite similar (Figure 4c–f). This suggests that the USDA-SCS method for ER is consistent regardless of crop type.

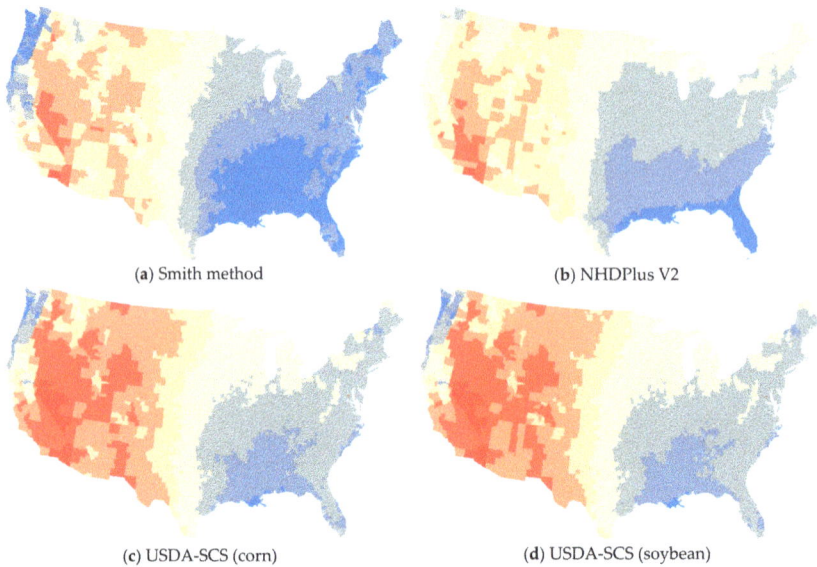

(**a**) Smith method

(**b**) NHDPlus V2

(**c**) USDA-SCS (corn)

(**d**) USDA-SCS (soybean)

Figure 4. *Cont.*

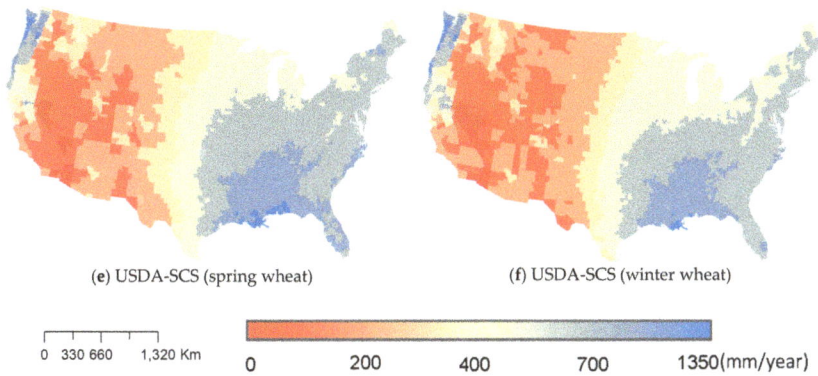

(e) USDA-SCS (spring wheat)   (f) USDA-SCS (winter wheat)

0 330 660   1,320 Km

0   200   400   700   1350(mm/year)

Figure 4. 30-year (1971–2000) average annual green water resources (mm/year) estimated using (**a**) the Smith method, (**b**) the NHDPlus V2 dataset, and the USDA-Soil Conservation Service (SCS) methods based on crop evapotranspiration of (**c**) corn, (**d**) soybeans, (**e**) spring wheat, and (**f**) winter wheat.

In addition to spatial variation, the three green water resource estimates also presented diverse temporal patterns (Figure 5). For each agricultural production region, we plotted average monthly precipitation, ET_c of corn, and three green water resource estimations, all weighted by county-level corn harvested acres (Figure 5). ET_c of corn is presented as an example to illustrate that, depending on the region, peaks of green water supply and crop water demand could vary significantly. We plotted corn ET_c only because it is the most widely planted commodity crop in the U.S. Still, using soybeans or wheat ET_c generated very close results (not presented). Across the 10 regions, ER_Smith consistently produces higher results than ER_USDA. This is largely because the Smith method assumes an average monthly ET_c of 200 mm throughout the year [37], which is much higher than the growing-season average monthly corn ET_c (average = 117 mm/month, SD = 39.46) as determined by the Penman-Monteith equation that was used for ER_USDA calculation. For most regions, even peak corn ET_c is less than 200 mm/month (Figure 5). On a monthly basis, the differences among the three estimations are generally smaller during the crop-growing seasons and higher during the non-growing seasons (Figure 5). Furthermore, months with peak corn ET_c and ER_USDA values generally match each other, except in the Pacific and Southern Plains regions, where precipitation during crop-growing season is limited. The Smith method, on the other hand, correlates more closely with monthly precipitation. Although ER_Smith and ER_USDA generally follow monthly precipitation distributions, ER_NHD shows a more dynamic temporal pattern (Figure 5). This is because the ER_NHD also factors in changes in monthly snowmelt and soil water content, using a water balance model. However, monthly ER_NHD in the spring and fall needs to be interpreted cautiously for some regions. For instance, in areas with snowpack, ER_NHD may underestimate green water resources in the spring because runoff includes input from snowmelt, but monthly precipitation data does not track changes in snowpack. In the fall, ER_NHD diverges from precipitation in several regions (Figure 5) because the soil may have been saturated; thus additional precipitation input will be classified as runoff rather than green water resources.

Differences in temporal patterns suggest that agriculture and bioenergy production may use green water more effectively if land use composition matches the temporal green water resource distribution pattern better. Ideally, peaks of crop water demand would match with those of green water resources to minimize reliance on irrigation water, especially in regions with high blue water scarcity. For instance, effective rain in the Pacific region is relatively more abundant in the winter but very limited in the summer. In this case, growing winter wheat may use green water more effectively than growing corn. In addition, in counties with multiple crops and land uses, adjusting the fractions of land uses (e.g., cropland, forest, grassland) and crops with varying growing seasons or improving soil water management practices could be options. For example, the Delta region, while green water resources are limited in July and August, there are excessive green water resources in the spring and winter (Figure 5). Land management strategies like cover crops may be used to conserve more soil moisture and reduce the negative impacts of green water variability [88].

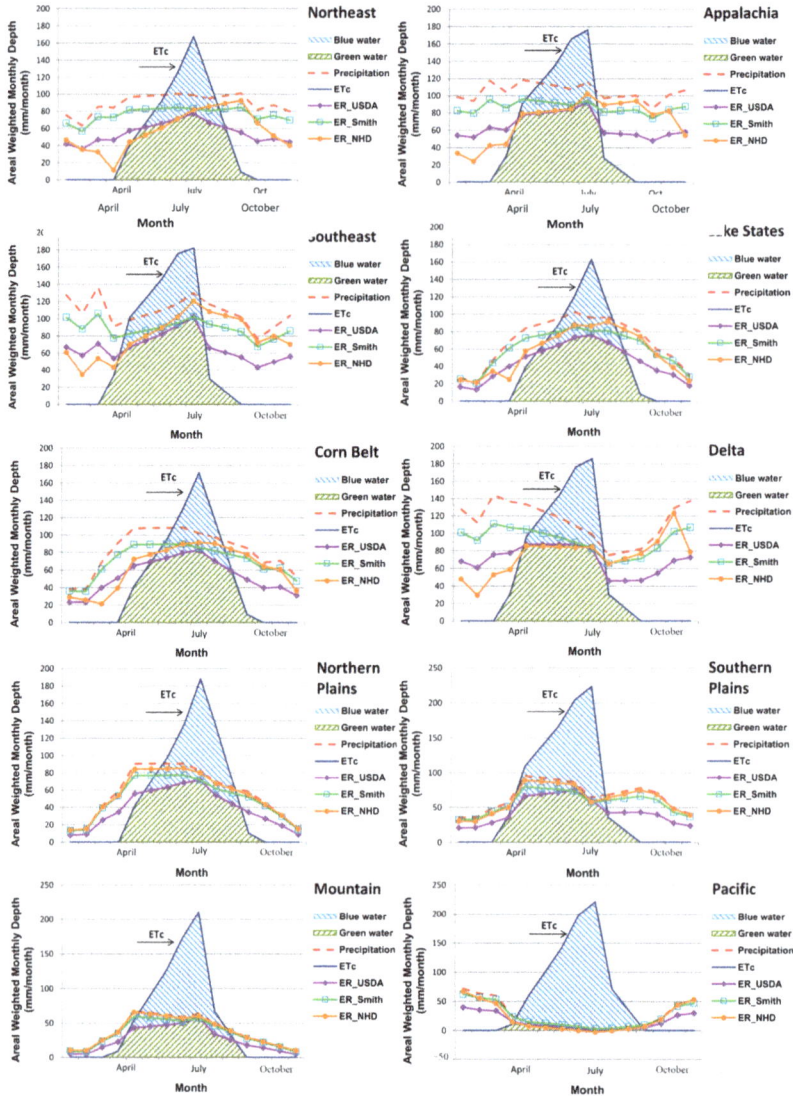

Figure 5. Comparison of monthly precipitation, corn evapotranspiration (ETc of corn), and green water resources (calculated from effective rain) based on three methods (Smith, USDA-SCS, and NHDPlus V2) for 10 major agricultural regions. Green water represents corn green water consumption and was calculated as the minimum of monthly green water resources and corn ETc. In a given month, if green water resource is lower than corn water demand (ETc), blue water represents the amount of irrigation water needed to make up the difference.

Few studies to date have verified alternative green water resource methods for a large study area, partly because a lack of field measurements at scale. Conceptually, the USDA-SCS and NHDPlus methods are more comprehensive, while the Smith method requires less input data. Wu et al. [66] found that the water footprints calculated based on the Smith method reasonably resemble peak monthly corn water use in the growing stage. A comparison study in the United Kingdom (U.K.) found that the Smith method turns out to be more accurate under U.K. conditions [36]. This is most likely because the USDA-SCS method was developed to fit U.S. conditions, while the Smith method was largely simplified for the ease of computation. For the NHDPlus method, interpreting results for spring in certain regions may be difficult. In short, considerable uncertainties remain in green water resource estimates. For county-level green water availability analysis, WAI_R was primarily calculated based on the Smith method, so the method for estimating crop water demand and green water footprint from this study would be consistent with our previous studies [25,26,66].

4.2. Regional and County-Level Green Water Availability

Using the WAI_R and *WAI_R_F* metrics, a regional and county-level green water availability analysis was applied to three major commodity crops (corn, soybean, and wheat) in the conterminous United States. The two metrics quantified the impacts of crop water demand and crop GWF on green water availability to all other remaining economic activities (e.g., other crops, grassland, and forest) and ecosystem services. These fractions reflect green water balance aggregated at the county or regional level, disregarding green water availability at the field scale. In this sense, the area-based analysis demonstrated how current land use composition and potential land use change (e.g., expansion of rain-fed cropland) may affect green water availability.

$WAI\_R_{non\_ag}$ and $WAI\_R\_F_{non\_ag}$ values suggest that crop production overall uses less than 30% of annual green water resources at the agricultural production region level, but substantial spatial variation exists at the county level (Table 3 and Figure 6). For the 10 agricultural production regions, $WAI\_R_{non\_ag}$ and $WAI\_R\_F_{non\_ag}$ ranged from 0.71 to 0.98 and from 0.82 to 0.99, respectively (Table 3). At the county level, $WAI\_R_{non\_ag}$ (Figure 6a) and $WAI\_R\_F_{non\_ag}$ (Figure 6b) ranged from 0.23 to 1.0 and from 0.56 to 1.0, respectively. Among the 2694 counties with major crop production in 2008, there are about five counties with high ($WAI\_R_{non\_ag} < 0.3$) and 106 counties with medium ($0.3 < WAI\_R_{non\_ag} < 0.5$) tensions between crop water demand and green water resource (Table 4). When measured by $WAI\_R\_F_{non\_ag}$, there are about 155 counties with moderately low green water availability ($0.5 < WAI\_R\_F_{non\_ag} < 0.7$) (Table 5). Overall, counties facing moderate or higher green water resource tensions are mostly located in Iowa, Illinois, Minnesota, Nebraska, and South Dakota (Figure 6). The $WAI\_R\_F_{non\_ag}$ values are significantly higher than $WAI\_R_{non\_ag}$ in several regions because $WAI\_R\_F_{non\_ag}$ is calculated based on GWF rather than crop water demand, since GWF can be significantly lower than crop water demand if precipitation is limited during crop-growing season. In this case, $WAI\_R_{non\_ag}$ and $WAI\_R\_F_{non\_ag}$ can be complementary to each other. $WAI\_R_{non\_ag}$ could reflect the tension between crop water demand and green water resources better when green water consumption is limited by supply of green water resources, however, $WAI\_R\_F_{non\_ag}$ provides the actual amount of green water resources available to other sectors.

County-level green water available to each economic sector in a region depends on both green water resources and crop water use in that county. Although annual precipitation is relatively abundant in the Corn Belt, high crop acreages drive down green water availability substantially in this region. For reference, we also present regional crop water demand (CWD), expressed as intensity or the annual volume of rainwater needed per volume of crop produced (in dry short tons, d.s.t, which is equivalent to 0.907 metric ton of dry biomass) (Figure 7). Results clearly indicate that CWDs for all three major crops in the Midwest are among the lowest in the U.S. (Figure 7), which means there is higher water use efficiency in this region, and is consistent with previous studies [22,89]. CWDs in the Northern Plains, which is the second largest crop production region in the U.S., are moderately higher —41%, 22%, and 17% for corn, soybean, and wheat, respectively—than in the Corn Belt. CWDs in the Southern

Plains for soybean and wheat are much greater than in the Midwest. These differences in regional CWDs can be attributed to spatial variability in soil and climate conditions, as well as agricultural managing practices. Other variables like crop varieties also affect crop water demand and consumption.

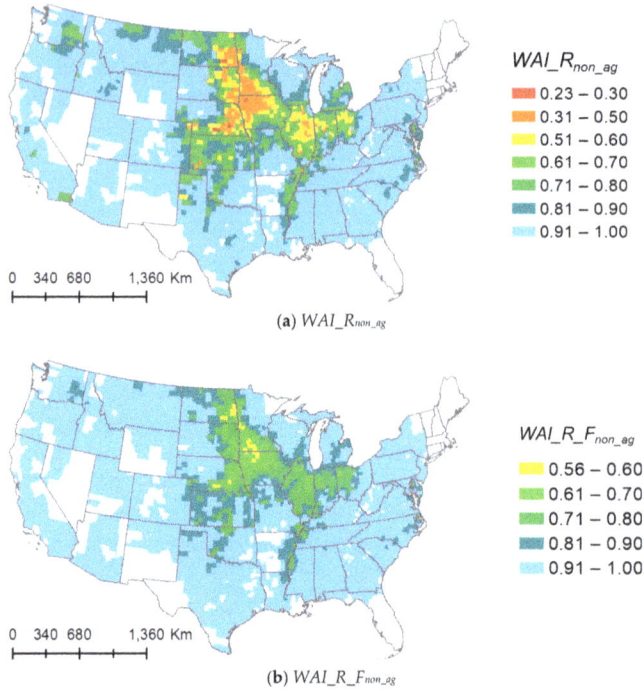

(a) *WAI_R~non_ag~*

(b) *WAI_R_F~non_ag~*

Figure 6. County-level (a) mean *WAI_R~non_ag~* and (b) mean *WAI_R_F~non_ag~*. Panel (a) shows the fraction of green water resources available for non-agriculture uses if green water meets the crop water demand of total corn, soybeans, and wheat production in 2008. Panel (b) shows, when green water consumption of three major crops is accounted for, the fraction of green water resources remaining for non-agriculture uses.

Table 3. Regional and national mean *WAI_R~non_ag~* and *WAI_R_F~non_ag~* with ranges of county-level values (dimensionless fractions). *WAI_R~non_ag~* is based on crop water demand of three crops (corn, soybeans, and wheat) and green water resources, and *WAI_R_F~non_ag~* is based on green water footprints of the three crops and green water resources.

Region	Mean $WAI\_R_{non\_ag}$	Range of $WAI\_R_{non\_ag}$	Mean $WAI\_R\_F_{non\_ag}$	Range of $WAI\_R\_F_{non\_ag}$
Northeast	0.96	0.68–1.0	0.97	0.79–1.0
Appalachia	0.96	0.62–1.0	0.97	0.70–1.0
Southeast	0.98	0.82–1.0	0.99	0.88–1.0
Lake States	0.83	0.33–1.0	0.88	0.57–1.0
Corn Belt	0.73	0.30–1.0	0.82	0.57–1.0
Delta	0.94	0.67–1.0	0.95	0.74–1.0
Northern Plains	0.71	0.23–1.0	0.83	0.56–1.0
Southern Plains	0.94	0.56–1.0	0.97	0.73–1.0
Mountain	0.96	0.50–1.0	0.98	0.76–1.0
Pacific	0.95	0.66–1.0	0.98	0.83–1.0
National	0.88	0.23–1.0	0.92	0.56–1.0

Table 4. Number of counties that fall within each $WAI\_R_{non\_ag}$ value (dimensionless fractions) band.

$WAI\_R_{non\_ag}$ Range (Unitless)	Number of Counties	Top Four States with Most Counties
0.23–0.3	5	Iowa, Nebraska, North Dakota
0.31–0.5	106	Iowa, Minnesota, Nebraska, South Dakota
0.51–0.6	138	Illinois, Iowa, Nebraska, South Dakota
0.61–0.7	184	Illinois, Indiana, Kansas, Iowa
0.71–0.8	244	Kansas, Indiana, Illinois, Montana
0.81–0.9	320	Kansas, Montana, Wisconsin, Indiana
0.91–1.0	1697	Texas, Georgia, Kentucky, Virginia

Table 5. Number of counties that fall within each $WAI\_R\_F_{non\_ag}$ value (dimensionless fractions) band.

$WAI\_R\_F_{non\_ag}$ Range (Unitless)	Number of Counties	Top Four States with Most Counties
0.51–0.6	17	Minnesota, Illinois, Nebraska, North Dakota
0.61–0.7	138	Iowa, Minnesota, Nebraska, Illinois
0.71–0.8	270	Illinois, Indiana, Ohio, Iowa
0.81–0.9	365	Kansas, Indiana, Montana, Illinois
0.91–1.0	1903	Texas, Georgia, Kentucky, Virginia

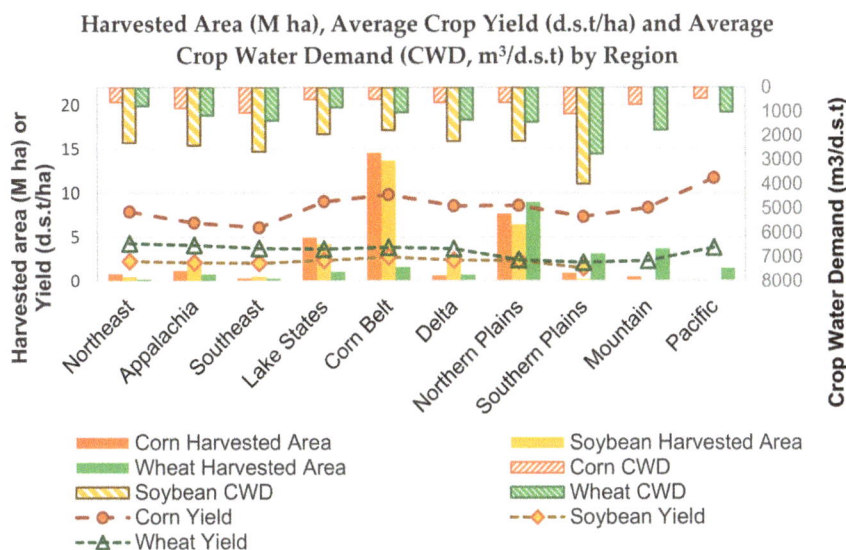

Figure 7. Regional crop production summary. Solid columns, dashed lines, and hatched columns show harvested crop area, average crop yield, and average crop water demand by crop by region, respectively. Harvested crop area and yields are based on 2008 county-level data reported by the USDA NASS.

4.3. Green Water Availability by Crop Type

Compared to green water resource distribution (Figure 4), variations in local $WAI\_R_{non\_ag}$ and $WAI\_R\_F_{non\_ag}$ strongly correlate to and are thus impacted by the spatial distribution of harvested crop acres (Figures 3 and 6). County-level $WAI\_R_{non\_ag}$ is often dominated by the water demands of different crops. This pattern is clearly demonstrated by crop-specific $WAI\_R_{non\_ag}$ (Figure 8) values, which measure the fraction of green water resources available for other uses after meeting crop water demand of a specific crop (e.g., corn). In the Midwest, corn and soybean acreages contribute the most to the volume of crop water demand (Figure 8a,b). Although wheat plays an important role in certain areas (e.g., Montana), total wheat acres are much less than corn and soybean acres in most areas.

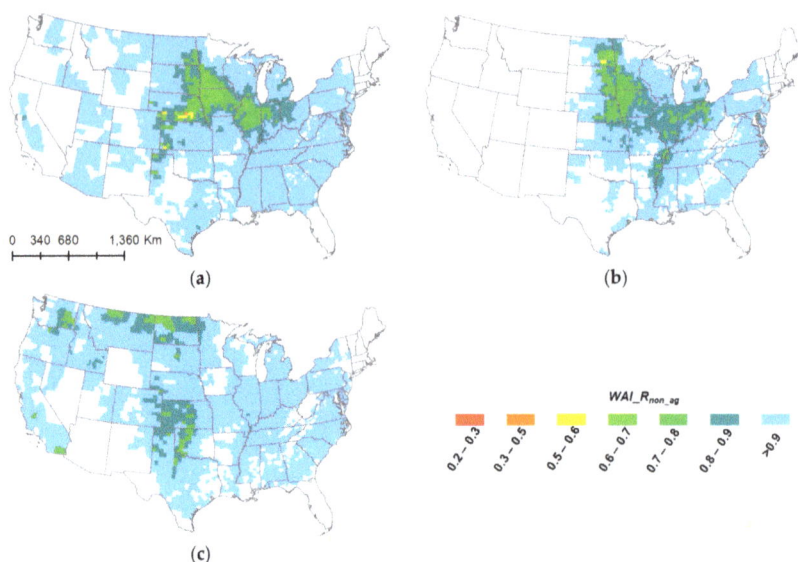

Figure 8. County-level crop specific $WAI\_R_{non\_ag}$ values based on harvested acres of (**a**) corn; (**b**) soybeans; and (**c**) wheat (including spring and winter wheats). Harvested acres include total production for all purposes (food, feed, and fuel) in 2008 in the U.S. Panels show green water resources available for other crops and sectors after meeting the crop water demand of corn, soybean, and wheat production.

The spatial pattern of crop-specific $WAI\_R_{non\_ag}$ does not mean crops in the Midwest use more water on a per-unit biomass basis. In fact, county-level CWD (m³/d.s.t) of the three crops (Figure 9) clearly indicate that counties in the Corn Belt are more water efficient than other areas. In addition, it seems that some counties located in northwestern states (Washington, Portland, Idaho, Oregon) are also water efficient in terms of corn and wheat production. Relatively low $WAI\_R_{non\_ag}$ values in these counties suggest it would be possible to increase crop acreages in these counties from a green water resource perspective, but land-use changes may cause other problems (e.g., deforestation).

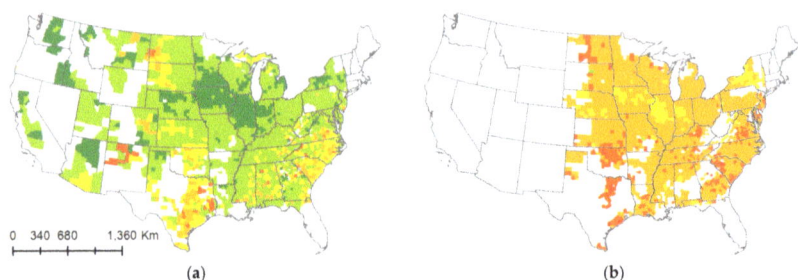

Figure 9. *Cont.*

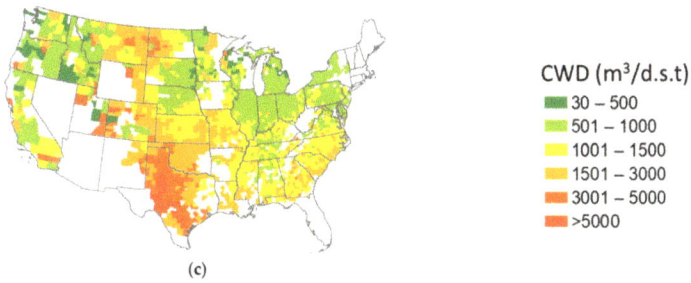

(c)

Figure 9. County-level per-unit crop water demand (CWD) (m³ per dry short ton (d.s.t)) calculated as the volume of water needed per d.s.t (0.907 metric ton of dry biomass) crop produced for (**a**) corn; (**b**) soybeans; and (**c**) wheat.

4.4. Annual Versus Growing Season Green Water Availability

Depending on local soil and climate conditions, off-season green water resources may lost to the atmosphere or contribute to blue water storage via deep percolation. In this case, it is helpful to estimate the green water resources available during the crop-growing season only if green water meets all crop water demand. Given that corn is the most widely produced commodity crop in the U.S., we also present the regional WAI_R based on growing season green water resources and of corn ($WAI\_R_{non\_corn}$) (Figure 10) as an example to illustrate how the temporal boundary of green water resources may affect the estimation of green water availability. Specifically, the $WAI\_R_{non\_corn}$ metric describes that, if all corn water demand met by green water, the fraction of green water resources are available to all other remaining green water users, aggregated at the county level. Results indicated that, when the green water resource is limited to the growing season only, total corn water demand alone would drive down green water availability substantially. A low $WAI\_R_{non\_corn}$ value means less green water available for other crops and plants in the region. Because of extremely low growing season precipitation, $WAI\_R_{non\_corn}$ in the Pacific could fall to 0.9 (Figure 10, Smith method) from an annual based WAI_R of 0.98 (Figure 8a). Growing-season $WAI\_R_{non\_corn}$ for the Corn Belt, Lake States, and Northern Plains would decrease 0.12, 0.06 and 0.07, respectively, compared with the annual-based index (Figure 8a).

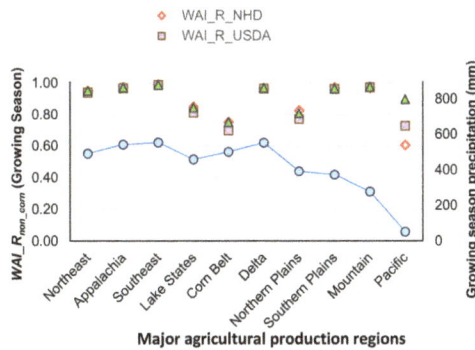

Figure 10. Regional growing-season rainwater available to other plants and uses after meeting corn water demand ($WAI\_R_{non\_corn}$) based on corn production. WAI_R_NHD, WAI_R_USDA, and WAI_R_Smith refer to $WAI\_R_{non\_corn}$ based on growing-season ER estimated using NHDPlus V2 data, the USDA-SCS method, and the Smith method, respectively.

Results in annual versus growing-season-based WAI_R values suggest that green water availability assessments can be sensitive to the temporal boundary of WAI_R analysis. In addition, it is important to note that the growing-season WAI_R could be conservative in certain soil conditions, because crops may utilize some of the non-growing season green water resource stored in the soil. Although field monitoring through sensor technology has been developed to guide precision agriculture programs and practices, it would be helpful if a national consistent soil water dynamic database could be developed for the U.S. in future studies.

4.5. Implications of Regional Water Resource Management for Bioenergy Production

Several regions produce the three major crops for feed, food, fiber, and fuel. To evaluate the impact of bioenergy feedstock production on green water availability, demand from the production of food, feed, and fiber is excluded. The resulting *WAI_R* metric—*WAI_R$_{non\_bioenergy}$*—describes the green water resources available for other uses if the water demand of biofuel feedstock production is met by green water resources only. For all but 149 counties (mostly located in Iowa, Nebraska, Minnesota, and Illinois), more than 90% of rainwater is still available to non-bioenergy productions (Figure 11). A majority of the 149 counties with *WAI_R$_{non\_bioenergy}$* values of 0.8–0.9, are concentrated in Iowa and Nebraska. If 24% of corn stover and 30% of wheat straw were also harvested as cellulosic biofuel feedstock [26,68] in 2008, holding total harvested crop acres and climate conditions constant, then regional mean *WAI_R$_{non\_bioenergy}$* would decrease slightly (0.03–0.05) for agricultural production regions in the Midwest, but *WAI_R$_{non\_bioenergy}$* would still be higher than 0.8 for all counties in these regions. These results suggest attributes of cellulosic feedstock to regional green water availability are small under the 2008 scenario.

From the perspective of water resource management, the production of the three major crops is most water efficient in the Corn Belt and Lake States because of their low CWD (Figure 7) and relatively abundant green water resources (Figure 4). Low green water availability for non-agriculture sectors in the Corn Belt is driven by food and feed production, with minimal contributions from biofuel feedstock production (Figure 11). However, when the production of food, fuel, feed, and fibers and the green water use by forestry and ecosystems are all accounted for, the aggregate impact on green water resource availability could become substantial. Therefore, it is critical to consider green water demands from multiple sectors when planning regional development.

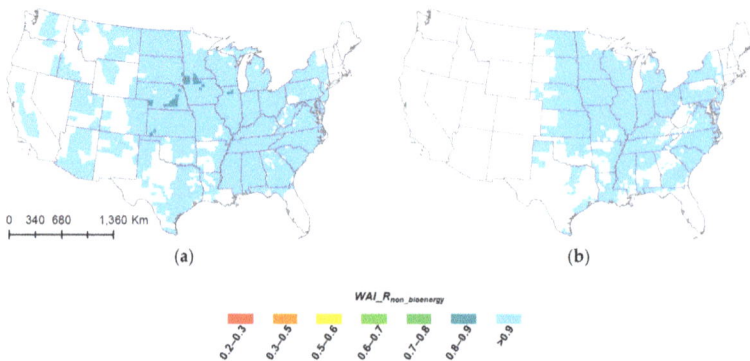

Figure 11. County-level crop-specific *WAI_R$_{non\_bioenergy}$* values based on acres of (**a**) corn grain and (**b**) soybeans harvested as biofuel feedstock only (i.e., acres harvested for food and feed purposes are excluded). For corn and soybeans, respectively, 30% and 12% of the total harvests in 2008 were biofuel feedstock. *WAI_R$_{non\_bioenergy}$* represents the fraction of green water resources available for other uses (e.g., food and feed), after meeting the crop water demand of bioenergy feedstock harvested from corn grain and soybeans.

4.6. Limitations and Future Work

The green water availability assessment approach presented in this study is applicable for large-scale assessments at varying spatial scales (e.g., watershed, state, and national). However, uncertainties exist, mainly associated with crop water demand calculation and green water resource estimations, due to incomplete data sources and assumptions made in estimation methods. County-level crop coefficients (Kc) for corn, soybean, and wheat are required in ET calculations, but are incomplete for all major crops. Therefore, we adopted regionalized crop coefficients for the 10 agricultural resource regions, but this practice may overlook variations in Kc within regions. In addition, monthly potential ET (ET$_o$) was computed using the ASCE's standardized Penman-Monteith method only. Although this is the dominant method used in the United States [26], other ET estimation methods exist and the uncertainties related to ET estimation by using different methods could be quite significant, especially in areas with high rainfall [90]. Finally, effective rain plays a central role in determining green water resources, and WAI_R based on varying temporal boundaries (annual vs. growing season) can lead to very different results for certain regions.

In this study, green water resources were quantified from effective rain using three alternative methods. The choice of effective rain methodology affects the spatial and temporal distribution patterns and the intensity of green water resources, the variation of which increases with increasing geospatial resolution. For future study, analyzing green water availability and competition on a monthly basis would be helpful for certain crops. However, determining monthly green water resources is more complicated than simply calculating monthly effective rain and crop water use, because green water resources can be stored in soil as carryover moisture. In this sense, a monthly analysis would require an integrated metric or framework that simultaneously considers the interactions among crop water use, soil moisture dynamics, and irrigation applications.

5. Conclusions

Although green water is vital for agricultural production and the terrestrial ecosystem, previous water resource assessments often focus on blue water. In this study, we present a first estimation of county-based green water availability, by applying a modified water availability index (WAI_R) to major crop (e.g., corn, soybeans, and wheat) production in the United States. The WAI_R metric was employed to quantify the fraction of green water resources available to non-agriculture sectors (e.g., grassland, forest, ecosystem services), assuming all crop water demands are met by green water. The metric quantifies green water balances aggregated at the county and farm production region level, disregarding green water availability at a field or land-use level. For comparison, a *WAI* (*WAI_R_F*) based on green water footprint was also presented, which quantifies the fraction of green water resources actually available to other users after accounting for the green water consumption of three crops.

Results highlight the spatial heterogeneity and temporal complexity of green water resources, as well as the heavy reliance of agriculture production on green water resources in the United States. In 2008, regional level mean *WAI_R$_{non\_ag}$* and *WAI_R_F$_{non\_ag}$* ranged from 0.71 to 0.98 and from 0.82 to 0.99, respectively. At the county level, however, fresh water was significantly constrained in five counties with a *WAI_R$_{non\_ag}$* value of lower than 0.3, which translates to a 70% demand for green water resources by the three crops. When measured by *WAI_R_F$_{non\_ag}$*, there are 17 counties with a values lower than 0.6, which means the three crops consumed more than 40% of annual green water resources in those counties. Geospatially, *WAI_R$_{non\_ag}$* and *WAI_R_F$_{non\_ag}$* are relatively higher in the Midwest, because of large crop acreages that are responsible for producing about 82% of the crops for the nation. However, crop production is also water efficient in this region, because water use per dry ton of biomass produced in the region is among the lowest across the 10 farm production regions. Seasonal analysis further revealed a substantial variability of crop water deficits during the growing season across regions, especially in Pacific west, which limits sustainable production of rain-fed crops. For areas with limited seasonal green water resources, adjusting land-use types to match seasons of high plant water demand with peaks of green water supply could be an option to use green water more effectively. In addition, improving soil water management strategies (e.g., adoption of cover crops) to increase carry-over soil moisture may also help with mitigating the negative impacts of seasonal variations in green water supply.

The development and scale-up of any land-based biomass production needs to be analyzed from the perspectives of resource availability and sustainability. Although surface (irrigation) water constraints have been stressed repeatedly, green water was usually taken for granted. Findings based on 2008 data suggest that bioenergy feedstock production demands a relatively small fraction of green water resources, but future large-scale biofuel feedstock production may substantially change land use composition (e.g., increase in woody crops for bioenergy), and therefore plant water consumption, in major agricultural regions [26]. To reduce competing use of surface and ground water (blue water consumption), rain-fed energy crops or biomass production is preferred in the bioenergy development in the U.S. In this case, availability of green water resources for large-scale feedstock production should be carefully evaluated. For agriculture and bioenergy production, the analysis presented can help decision makers consider geographical variations in green water availability when planning land-based biomass production sites, types, and production scales. However, analysis of green water availability alone is not sufficient for sustainable water management. Future studies should consider integrating green-blue water availability with land use at large scale, because land use patterns are likely to impact both blue and green flows.

Supplementary Materials: Figure S1: County-level 30-year (1971–2000) average annual precipitation for the conterminous United States.

Acknowledgments: This work was made possible by funding from the U.S. Department of Energy, Office of Energy Efficiency and Renewable Energy, Bioenergy Technologies Office (BETO), under contract #DE-AC02-06CH11357. We would like to thank Kristen Johnson of BETO for her valuable input and suggestions throughout the study. The authors are grateful to the three anonymous reviewers for their constructive suggestions.

Author Contributions: May Wu initiated project and directed analysis; Hui Xu performed the analysis; Hui Xu and May Wu analyzed the data; Hui Xu wrote the paper.

Conflicts of Interest: The authors declare no conflict of interest and the founding sponsors had no role in the design of the study; in the collection, analyses, or interpretation of data; in the writing of the manuscript; or in the decision to publish the results.

References

1. Vörösmarty, C.J.; McIntyre, P.B.; Gessner, M.O.; Dudgeon, D.; Prusevich, A.; Green, P.; Glidden, S.; Bunn, S.E.; Sullivan, C.A.; Liermann, C.R.; et al. Global threats to human water security and river biodiversity. *Nature* **2010**, *467*, 555–561. [CrossRef] [PubMed]

2. Rockström, J.; Steffen, W.; Noone, K.; Persson, A.; Chapin, F.S.; Lambin, E.F.; Lenton, T.M.; Scheffer, M.; Folke, C.; Schellnhuber, H.J.; et al. A safe operating space for humanity. *Nature* **2009**, *461*, 472–475. [CrossRef] [PubMed]

3. Gerten, D.; Hoff, H.; Rockström, J.; Jägermeyr, J.; Kummu, M.; Pastor, A.V. Towards a revised planetary boundary for consumptive freshwater use: Role of environmental flow requirements. *Curr. Opin. Environ. Sustain.* **2013**, *5*, 551–558. [CrossRef]

4. Mekonnen, M.M.; Hoekstra, A.Y. Four billion people facing severe water scarcity. *Sci. Adv.* **2016**, *2*, e1500323. [CrossRef] [PubMed]

5. Kummu, M.; Guillaume, J.H.A.; de Moel, H.; Eisner, S.; Flörke, M.; Porkka, M.; Siebert, S.; Veldkamp, T.I.E.; Ward, P.J. The world's road to water scarcity: Shortage and stress in the 20th century and pathways towards sustainability. *Sci. Rep.* **2016**, *6*, 38495. [CrossRef] [PubMed]

6. Falkenmark, M. Growing water scarcity in agriculture: Future challenge to global water security. *Philos. Trans. R. Soc. A Math. Phys. Eng. Sci.* **2013**, *371*, 20120410. [CrossRef] [PubMed]

7. Averyt, K.; Meldrum, J.; Caldwell, P.; Sun, G.; McNulty, S.; Huber-Lee, A.; Madden, N. Sectoral contributions to surface water stress in the coterminous United States. *Environ. Res. Lett.* **2013**, *8*, 35046. [CrossRef]

8. Gerten, D.; Heinke, J.; Hoff, H.; Biemans, H.; Fader, M.; Waha, K. Global Water Availability and Requirements for Future Food Production. *J. Hydrometeorol.* **2011**, *12*, 885–899. [CrossRef]

9. USGS (U.S. Geological Survey). WaterWatch. Available online: https://waterwatch.usgs.gov/new/index.php?id=ww_past (accessed on 16 July 2017).

10. US Census Bureau Census. 2010. Available online: http://quickfacts.census.gov/qfd/states/13/13135.html (accessed on 16 May 2015).

11. Moore, B.C.; Coleman, A.M.; Wigmosta, M.S.; Skaggs, R.L.; Venteris, E.R. A high spatiotemporal assessment of consumptive water use and water scarcity in the conterminous United States. *Water Resour. Manag.* **2015**, *29*, 5185–5200. [CrossRef]

12. Roy, S.B.; Chen, L.; Girvetz, E.H.; Maurer, E.P.; Mills, W.B.; Grieb, T.M. Projecting water withdrawal and supply for future decades in the U.S. under climate change scenarios. *Environ. Sci. Technol.* **2012**, *46*, 2545–2556. [CrossRef] [PubMed]

13. Tidwell, V.C.; Moreland, B.D.; Zemlick, K.M.; Roberts, B.L.; Passell, H.D.; Jensen, D.; Forsgren, C.; Sehlke, G.; Cook, M.A.; King, C.W.; et al. Mapping water availability, projected use and cost in the western United States. *Environ. Res. Lett.* **2014**, *9*, 64009. [CrossRef]

14. Sun, G.; Mcnulty, S.G.; Myers, J.A.M.; Cohen, E.C. Impacts of Climate Change, Population Growth, Land Use Change, and Groundwater Availability on Water Supply and Demand across the Conterminous U.S. *Water Supply* **2008**, *6*, 1–30.

15. Caldwell, P.V.; Sun, G.; McNulty, S.G.; Cohen, E.C.; Moore Myers, J.A. Impacts of impervious cover, water withdrawals, and climate change on river flows in the conterminous US. *Hydrol. Earth Syst. Sci.* **2012**, *16*, 2839–2857. [CrossRef]

16. Hoekstra, A.Y.; Chapagain, A.K.; Aldaya, M.M.; Mekonnen, M.M. *The Water Footprint Assessment Manual*; Earthscan: London, UK, 2011. ISBN 9781849712798.

17. Falkenmark, M.; Rockström, J. The New Blue and Green Water Paradigm: Breaking New Ground for Water Resources Planning and Management. *J. Water Resour. Plan. Manag.* **2006**, *132*, 129–132. [CrossRef]

18. Liu, J.; Yang, H.; Gosling, S.N.; Kummu, M.; Flörke, M.; Pfister, S.; Hanasaki, N.; Wada, Y.; Zhang, X.; Zheng, C.; et al. Water scarcity assessments in the past, present, and future. *Earth's Future* **2017**, *5*, 545–559. [CrossRef]

19. Núñez, M.; Pfister, S.; Antón, A.; Muñoz, P.; Hellweg, S.; Koehler, A.; Rieradevall, J. Assessing the Environmental Impact of Water Consumption by Energy Crops Grown in Spain. *J. Ind. Ecol.* **2013**, *17*, 90–102. [CrossRef]

20. Liu, J.; Zehnder, A.J.B.; Yang, H. Global consumptive water use for crop production: The importance of green water and virtual water. *Water Resour. Res.* **2009**, *45*. [CrossRef]

21. Mekonnen, M.M.; Hoekstra, A.Y. The green, blue and grey water footprint of crops and derived crop products. *Hydrol. Earth Syst. Sci.* **2011**, *8*, 1577–1600. [CrossRef]

22. White, M.; Gambone, M.; Yen, H.; Arnold, J.; Harmel, D.; Santhi, C.; Haney, R. Regional Blue and Green Water Balances and Use by Selected Crops in the U.S. *J. Am. Water Resour. Assoc.* **2015**, *51*, 1626–1642. [CrossRef]

23. Senay, G.B.; Friedrichs, M.; Singh, R.K.; Velpuri, N.M. Evaluating Landsat 8 evapotranspiration for water use mapping in the Colorado River Basin. *Remote Sens. Environ.* **2016**, *185*, 171–185. [CrossRef]

24. Gerbens-Leenes, P.W.; van Lienden, A.R.; Hoekstra, A.Y.; van der Meer, T.H. Biofuel scenarios in a water perspective: The global blue and green water footprint of road transport in 2030. *Glob. Environ. Chang.* **2012**, *22*, 764–775. [CrossRef]

25. Chiu, Y.W.; Wu, M. Assessing county-level water footprints of different cellulosic-biofuel feedstock pathways. *Environ. Sci. Technol.* **2012**, *46*, 9155–9162. [CrossRef] [PubMed]

26. Wu, M.; Ha, M. *Water Consumption Footprint of Producing Agriculture and Forestry Feedstocks, Chapter 8, 2016 Billion-Ton Report, Volume 2: Environmental Sustainability Effects of Select Scenarios from Volume 1*; Department of Energy Office of Energy Efficiency & Renewable Energy: Washington, DC, USA, 2017.

27. Núñez, M.; Pfister, S.; Roux, P.; Antón, A. Estimating water consumption of potential natural vegetation on global dry lands: Building an LCA framework for green water flows. *Environ. Sci. Technol.* **2013**, *47*, 12258–12265. [CrossRef] [PubMed]

28. Quinteiro, P.; Dias, A.C.; Silva, M.; Ridoutt, B.G.; Arroja, L. A contribution to the environmental impact assessment of green water flows. *J. Clean. Prod.* **2015**, *93*, 318–329. [CrossRef]

29. Lathuillière, M.J.; Bulle, C.; Johnson, M.S. Land Use in LCA: Including Regionally Altered Precipitation to Quantify Ecosystem Damage. *Environ. Sci. Technol.* **2016**, *50*, 11769–11778. [CrossRef] [PubMed]

30. Boulay, A.-M.; Hoekstra, A.Y.; Vionnet, S. Complementarities of Water-Focused Life Cycle Assessment and Water Footprint Assessment. *Environ. Sci. Technol.* **2013**, *47*, 11926–11927. [CrossRef] [PubMed]

31. Rodrigues, D.B.B.; Gupta, H.V.; Mendiondo, E.M. A blue/green water-based accounting framework for assessment of water security. *Water Resour. Res.* **2014**, *50*, 7187–7205. [CrossRef]

32. Veettil, A.V.; Mishra, A.K. Water security assessment using blue and green water footprint concepts. *J. Hydrol.* **2016**, *542*, 589–602. [CrossRef]

33. Vanham, D.; Hoekstra, A.Y.; Wada, Y.; Bouraoui, F.; de Roo, A.; Mekonnen, M.M.; van de Bund, W.J.; Batelaan, O.; Pavelic, P.; Bastiaanssen, W.G.M.; et al. Physical water scarcity metrics for monitoring progress towards SDG target 6.4: An evaluation of indicator 6.4.2 "Level of water stress.". *Sci. Total Environ.* **2018**, *613*, 218–232. [CrossRef] [PubMed]

34. Dastane, N.G. Effective rainfall in irrigated agriculture. In *Irrigation and Drainage Paper No. 25*; Food and Agriculture Organization of the United Nations: Rome, Italy, 1974.

35. USDA Soil Conservation Serivce. *Irrigation Water Requirements-Chapter 2, Part 623 of the National Engineering Handbook*; Natural Resources Conservation Service: Washington, DC, USA, 1993.

36. Hess, T. Estimating Green Water Footprints in a Temperate Environment. *Water* **2010**, *2*, 351–362. [CrossRef]

37. Brown, A.; Matlock, M.D. A Review of Water Scarcity Indices and Methodologies. *Sustain. Consort.* **2011**, *19*, White Paper #106. Available online: https://www.sustainabilityconsortium.org/downloads/a-review-of-water-scarcity-indices-and-methodologies/ (accessed on 15 October 2016).

38. Pedro-Monzonís, M.; Solera, A.; Ferrer, J.; Estrela, T.; Paredes-Arquiola, J. A review of water scarcity and drought indexes in water resources planning and management. *J. Hydrol.* **2015**, *527*, 482–493. [CrossRef]

39. Xu, H.; Wu, M. *Water Availability Indices—A Literature Review, ANL/ESD-17/5*; Argonne National Laboratory Techical Report; Argonne National Laboratory: Lemont, IL, USA, 2017.

40. Damkjaer, S.; Taylor, R. The measurement of water scarcity: Defining a meaningful indicator. *Ambio* **2017**, *46*, 513–531. [CrossRef] [PubMed]

41. International Organization for Standardization (ISO). *ISO 14046:2014 (E) Environmental Management. Water Footprint—Principles, Requirements and Guidelines*; International Organization for Standardization: Geneva, Switzerland, 2014.

42. Boulay, A.M.; Bare, J.; Benini, L.; Berger, M.; Lathuillière, M.J.; Manzardo, A.; Margni, M.; Motoshita, M.; Núñez, M.; Pastor, A.V.; et al. The WULCA consensus characterization model for water scarcity footprints: assessing impacts of water consumption based on available water remaining (AWARE). *Int. J. Life Cycle Assess.* **2018**, *23*, 368–378. [CrossRef]

43. Schyns, J.F.; Hoekstra, A.Y.; Booij, M.J. Review and classification of indicators of green water availability and scarcity. *Hydrol. Earth Syst. Sci.* **2015**, *19*, 4581–4608. [CrossRef]

44. Falkenmark, M. The massive water scarcity now threatening Africa—Why isnt it being addressed? *Ambio* **1989**, *18*, 112–118.

45. Vörösmarty, C.J.; Douglas, E.M.; Green, P.A.; Revenga, C. Geospatial Indicators of Emerging Water Stress: An Application to Africa. *AMBIO J. Hum. Environ.* **2005**, *34*, 230–236. [CrossRef]

46. Pfister, S.; Koehler, A.; Hellweg, S. Assessing the environmental impacts of freshwater consumption in LCA. *Environ. Sci. Technol.* **2009**, *43*, 4098–4104. [CrossRef] [PubMed]

47. Sullivan, C.A.; Meigh, J.R.; Giacomello, A.M.; Fediw, T.; Lawrence, P.; Samad, M.; Mlote, S.; Hutton, C.; Allan, J.A.; Schulze, R.E.; et al. The water poverty index: Development and application at the community scale. *Nat. Resour. Forum* **2003**, *27*, 189–199. [CrossRef]

48. Balcerski, W. Javaslat a vízi létesítmények osztályozásának új alapelveire/A proposal toward new principles underpinning the classification of water conditions. *Vízgazdálkodás: A vízügyi dolgozók lapja (Water Manag.)* **1964**, *4*, 134–136. (In Hungarian)

49. Falkenmark, M.; Lindh, G. How can we cope with the water resources situation by the year 2015? *Ambio* **1974**, *3*, 114–122.

50. Raskin, P.; Gleick, P.; Kirshen, P.; Pontius, G.; Strzepek, K. *Comprehensive Assessment of the Freshwater Resources of the World*; Stockholm Environmental Institute: Sweden, Stockholm, 1997.

51. Alcamo, J.; Döll, P.; Henrichs, T.; Kaspar, F.; Lehner, B.; Rösch, T.; Siebert, S. Development and testing of the WaterGAP 2 global model of water use and availability. *Hydrol. Sci. J.* **2003**, *48*, 317–337. [CrossRef]

52. Smakhtin, V.; Revanga, C.; Dol, P. *Taking into Account Environmental Water Requirements in Global-Scale Water Resources Assessments*; International Water Management Institute (IWMI): Colombo, Sri Lanka, 2005. ISBN 9290905425.

53. Arthington, A.H.; Bunn, S.E.; Poff, N.L.; Naiman, R.J. The challenge of providing environmental flow rules to sustain river ecosystems. *Ecol. Appl.* **2006**, *16*, 1311–1318. [CrossRef]

54. Pastor, A.V.; Ludwig, F.; Biemans, H.; Hoff, H.; Kabat, P. Accounting for environmental flow requirements in global water assessments. *Hydrol. Earth Syst. Sci.* **2014**, *18*, 5041–5059. [CrossRef]

55. Rockström, J.; Falkenmark, M.; Karlberg, L.; Hoff, H.; Rost, S.; Gerten, D. Future water availability for global food production: The potential of green water for increasing resilience to global change. *Water Resour. Res.* **2009**, *45*, W00A12. [CrossRef]

56. Kummu, M.; Gerten, D.; Heinke, J.; Konzmann, M.; Varis, O. Climate-driven interannual variability of water scarcity in food production potential: A global analysis. *Hydrol. Earth Syst. Sci.* **2014**, *18*, 447–461. [CrossRef]

57. Tidwell, V.C.; Kobos, P.H.; Malczynski, L.A.; Klise, G.; Castillo, C.R. Exploring the Water-Thermoelectric Power Nexus. *J. Water Resour. Plan. Manag.* **2012**, *138*, 491–501. [CrossRef]

58. Brauman, K.A.; Richter, B.D.; Postel, S.; Malsy, M.; Flörke, M. Water depletion: An improved metric for incorporating seasonal and dry-year water scarcity into water risk assessments. *Elem. Sci. Anthr.* **2016**, *4*, 83. [CrossRef]

59. Chaves, H.M.L.; Alipaz, S. An integrated indicator based on basin hydrology, environment, life, and policy: The watershed sustainability index. *Water Resour. Manag.* **2007**, *21*, 883–895. [CrossRef]

60. Palmer, W.C. Keeping Track of Crop Moisture Conditions, Nationwide: The New Crop Moisture Index. *Weatherwise* **1968**, *21*, 156–161. [CrossRef]

61. Woli, P.; Jones, J.W.; Ingram, K.T.; Fraisse, C.W. Agricultural reference index for drought (ARID). *Agron. J.* **2012**, *104*, 287–300. [CrossRef]

62. Devineni, N.; Lall, U.; Etienne, E.; Shi, D.; Xi, C. America's water risk: Current demand and climate variability. *Geophys. Res. Lett.* **2015**, *42*, 2285–2293. [CrossRef]

63. Meyer, S.J.; Hubbard, K.G.; Wilhite, D.A. A Crop-Specific Drought Index for Corn: I. Model Development and Validation. *Agron. J.* **1993**, *85*, 388. [CrossRef]

64. Wada, Y. *Human and Climate Impacts on Global Water Resources*; Utrecht University: Utrecht, The Netherlands, 2013.

65. Quinteiro, P.; Ridoutt, B.G.; Arroja, L.; Dias, A.C. Identification of methodological challenges remaining in the assessment of a water scarcity footprint: A review. *Int. J. Life Cycle Assess.* **2017**. [CrossRef]

66. Wu, M.; Chiu, Y.; Demissie, Y. Quantifying the regional water footprint of biofuel production by incorporating hydrologic modeling. *Water Resour. Res.* **2012**, *48*, 1–11. [CrossRef]

67. Mekonnen, M.M.; Hoekstra, A.Y. A global and high-resolution assessment of the green, blue and grey water footprint of wheat. *Hydrol. Earth Syst. Sci.* **2010**, *14*, 1259–1276. [CrossRef]

68. Faramarzi, M.; Abbaspour, K.C.; Schulin, R.; Yang, H. Modelling blue and green water resources availability in Iran. *Hydrol. Process.* **2009**, *23*, 486–501. [CrossRef]

69. Wada, Y.; Van Beek, L.P.H.; Viviroli, D.; Drr, H.H.; Weingartner, R.; Bierkens, M.F.P. Global monthly water stress: 2. Water demand and severity of water stress. *Water Resour. Res.* **2011**, *47*. [CrossRef]

70. USDA Soil Conservation Service. *Irrigation Water Requirements. Technical Release No.21*; Natural Resources Conservation Service: Washington, DC, USA, 1970.

71. Quinteiro, P.; Sandra, R.; Rey, P.V.; Arroja, L.; Dias, A.C. Addressing the green water scarcity footprint of eucalypt production in Portugal. In Proceedings of the 7th International Congress of Energy and Environment Engineering and Management, Universidade de, Las Palmas, Las Palmas, Spain, 17–19 July 2017.

72. ASCE-EWRI (Environmental & Water Resources Institute). *The ASCE Standardized Reference Evapotranspiration Equation. Report of the Task Committee on Standardization of Reference Evapotranspiration*; ASCE-EWRI: Reston, VA, USA, 2005.

73. Patwardhan, A.S.; Nieber, J.L.; Johns, E.L. Effective Rainfall Estimation Methods. *J. Irrig. Drain. Eng.* **1990**, *116*, 182–193. [CrossRef]

74. Chapagain, A.K.; Hoekstra, A.Y. The blue, green and grey water footprint of rice from production and consumption perspectives. *Ecol. Econ.* **2011**, *70*, 749–758. [CrossRef]

75. Obreza, T.A.; Pitts, D.J. Effective Rainfall in Poorly Drained Microirrigated Citrus Orchards. *Soil Sci. Soc. Am. J.* **2002**, *66*, 212. [CrossRef]

76. Smith, M. CROPWAT: A computer program for irrigation planning and management. In *Irrigation and Drainage Paper 46*; Food and Agriculture Organization of the United Nations: Rome, Italy, 1992.

77. McKay, L.; Bondelid, T.; Dewald, T.; Johnston, J.; Moore, R.; Rea, A. *NHDPlus Version 2: User Guide*; United States Environmental Protection Agency: Washington, DC, USA, 2012.

78. Ley, T.W.; Stevens, R.G.; Topielec, R.R.; Neibling, W.H. *Soil Water Monitoring and Measurement*; PNW0475; Washington State University: Washington, DC, USA, 1994.

79. Soil Survey Staff, Natural Resources Conservation Service, U.S. D. of A. Web Soil Survey. Available online: http://websoilsurvey.nrcs.usda.gov/ (accessed on 2 October 2016).

80. Pfister, S.; Bayer, P.; Koehler, A.; Hellweg, S. Environmental impacts of water use in global crop production: Hotspots and trade-offs with land use. *Environ. Sci. Technol.* **2011**, *45*, 5761–5768. [CrossRef] [PubMed]

81. Wolock, D.M.; McCabe, G.J. Explaining spatial variability in mean annual runoff in the conterminous United States. *Clim. Res.* **1999**, *11*, 149–159. [CrossRef]

82. McCabe, G.J.; Wolock, D.M. Independent effects of temperature and precipitation on modeled runoff in the conterminous United States. *Water Resour. Res.* **2011**, *47*. [CrossRef]

83. USDA NRCS USDA Farm Production Regions. Available online: https://www.ers.usda.gov/webdocs/publications/42298/32489_aib-760_002.pdf?v=42487 (accessed on 1 May 2017).

84. Daly, C.; Halbleib, M.; Smith, J.I.; Gibson, W.P.; Doggett, M.K.; Taylor, G.H.; Curtis, J.; Pasteris, P.P. Physiographically sensitive mapping of climatological temperature and precipitation across the conterminous United States. *Int. J. Climatol.* **2008**, *28*, 2031–2064. [CrossRef]

85. Homer, C.; Dewitz, J.; Yang, L.; Jin, S.; Danielson, P.; Xian, G.; Coulston, J.; Herold, N.; Wickham, J.; Megown, K. Completion of the 2011 national land cover database for the conterminous United States—Representing a decade of land cover change information. *Photogramm. Eng. Remote Sens.* **2015**, *81*, 346–354.

86. U.S Census Bureau Cartographic Boundary Shapefiles-Counties. Available online: https://www.census.gov/geo/maps-data/data/cbf/cbf_counties.html (accessed on 10 December 2016).

87. USDA NASS Quick Stats. https://quickstats.nass.usda.gov/ (accessed on 23 February 2017).

88. Basche, A.D.; Kaspar, T.C.; Archontoulis, S.V.; Jaynes, D.B.; Sauer, T.J.; Parkin, T.B.; Miguez, F.E. Soil water improvements with the long-term use of a winter rye cover crop. *Agric. Water Manag.* **2016**, *172*, 40–50. [CrossRef]

89. Mubako, S.T.; Lant, C.L. Agricultural Virtual Water Trade and Water Footprint of U.S. States. *Ann. Assoc. Am. Geogr.* **2013**, *103*, 385–396. [CrossRef]

90. Liu, W.; Yang, H.; Folberth, C.; Wang, X.; Luo, Q.; Schulin, R. Global investigation of impacts of PET methods on simulating crop-water relations for maize. *Agric. For. Meteorol.* **2016**, *221*, 164–175. [CrossRef]

MDPI

St. Alban-Anlage 66

4052 Basel

Switzerland

Tel. +41 61 683 77 34

Fax +41 61 302 89 18

www.mdpi.com

Water Editorial Office

E-mail: water@mdpi.com

www.mdpi.com/journal/water

www.ingramcontent.com/pod-product-compliance
Lightning Source LLC
Chambersburg PA
CBHW051850210326
41597CB00033B/5848